Advanced Nanoelectronics

Advanced Nanoelectronics

Post-Silicon Materials and Devices

Edited by Muhammad Mustafa Hussain

Editor

Prof. Muhammad Mustafa Hussain
KAUST
Integrated Nanotechnology Lab.
Bldg. 3 (Ibn Sina Bldg.)
Level 3, Rm 3274
Makkah Province
23955-690 Thuwal
Saudi Arabia

Cover image: © iStock.com/Jackie2k

Library of Congress Card No.:
applied for

British Library Cataloguing-in-Publication Data
A catalogue record for this book is available from the British Library.

Bibliographic information published by the Deutsche Nationalbibliothek
The Deutsche Nationalbibliothek lists this publication in the Deutsche Nationalbibliografie; detailed bibliographic data are available on the Internet at <http://dnb.d-nb.de>.

Print ISBN: 978-3-527-34358-4
ePDF ISBN: 978-3-527-81183-0
ePub ISBN: 978-3-527-81185-4
oBook ISBN: 978-3-527-81186-1

Cover Design Wiley
Typesetting SPi Global, Chennai, India
Printing and Binding C.O.S. Printers Pte Ltd, Singapore

Printed on acid-free paper

10 9 8 7 6 5 4 3 2 1

Contents

Preface

We live in the age of information where electronics, especially transistors, play a critical enabling role. Nearly 60 years ago, when Jack Kilby built the first integrated circuit (IC) that was also the beginning of today's complementary metal-oxide-semiconductor (CMOS) technology whose arts and science of miniaturization has enabled Moore's Law to physically scale transistors for the past 40 years. The dominant material vehicle for transistor technology has been silicon. However, as we are approaching the scaling limit in silicon CMOS transistor technology, we need to find new ways to keep the momentum in CMOS electronics technology advances. Therefore, in this edited book, I have closely worked with leading authorities to assemble a comprehensive collection of chapters dedicated to emerging technologies, which will take the CMOS electronics technology forward. The objective of such technology is simple: faster, multifunctional, energy-efficient computing and infotainment applications.

This book has been organized in such a manner that it covers materials, physics, architecture, and integration aspects of future generation CMOS electronics technology. We have discussed about emerging materials and architectures such as alternate channel materials like germanium, gallium nitride, 1D nanowires/tubes, 2D graphene and other dichalcogenide materials, ferroelectrics; new physics such as spintronics, negative capacitance, and quantum computing; and, finally, 3D-IC technology.

In 2009 fall, when I transitioned from industry to academia, I realized that there is a certain gap in our knowledge and understanding, which needs a bridge for emerging concepts and technological advances in the area of CMOS electronics technology. As a follow-up, I introduced a course on Advancement in CMOS Electronics Technology focusing on graduate and senior-level undergraduate students. I invited leading authorities to offer lectures through online and in person. Interestingly, we observed an increased number of class presence; especially, postdoctoral fellows and research scientists also started participating in the class. We kept enriching the course every year as new concepts and technological advancements were taking place in parallel and at a fast pace. In 2012 fall, United States National Academy of Engineering Member Prof. James Plummer (former Dean of Stanford Engineering) visited us and I discussed with him about the content of the course. He was vastly impressed and encouraged me to think about a book. Therefore, this current book was implanted as a seed in my mind in 2012 and today it has materialized to provide a comprehensive understanding about the future generation CMOS electronics technology to especially senior-level undergraduate students, graduate students, and professionals interested in the relevant technology.

Although the book in its current shape is rich in content, it was not an easy task to meet the timelines and to work with such a busy cohort of leading authorities as contributors of individual chapters. I definitely acknowledge their contributions along with the students and postdocs who have worked together to make it a comprehensive insightful resource on emerging CMOS electronics technology. I thank the proofreading service and especially my postdoctoral fellow Dr. Nazek Elatab. Her contribution as integrator has been instrumental for timely completion of this project. I also thank all my mentors and colleagues because it has been a privilege for me to serve with them in this exciting area of electronics with boundless promise. Understandably, my family has contributed by sacrificing long hours that I spent on this book instead of with them! Finally, I thank Wiley-VCH for encouraging me to lead this project and to finally realize a dream I have had for many years.

I sincerely hope that the readership will find this book a resourceful knowledge base for future generation CMOS electronics technology, which we are calling Advanced Nanoelectronics: Post-silicon Materials and Devices.

1

The Future of CMOS: More Moore or a New Disruptive Technology?

Nazek El-Atab and Muhammad M. Hussain

King Abdullah University of Science and Technology, Integrated Nanotechnology Lab, Thuwal, 4700, Saudi Arabia

For more than four decades, Moore's law has been driving the semiconductor industry where the number of transistors per chip roughly doubles every 18–24 months at a constant cost. Transistors have been relentlessly evolving from the first Ge transistor invented at Bell Labs in 1947 to planar Si metal-oxide semiconductor field-effect transistor (MOSFET), then to strained SiGe source/drain (S/D) in the 90- and 65-nm technology nodes and high-κ/metal gate stack introduced at the 45- and 32-nm nodes, then to the current 3D transistors (Fin field-effect transistors (FinFETs)) introduced at the 22-nm node in 2011 (Figure 1.1). In extremely scaled transistors, the parasitic and contact resistances greatly deteriorate the drive current and degrade the circuit speed. Thus, miniaturization of devices so far has been possible due to changes in dielectric, S/D, and contacts materials/processes, and innovations in lithography processes, in addition to changes in the device architecture [1, 2].

The gate length of current transistors has been scaled down to 14 nm and below, with over 10^9 transistors in state-of-the-art microprocessors. Yet, the clock speed is limited to 3–4 GHz due to thermal constraints, and further scaling down the device dimensions is becoming extremely difficult due to lithography challenges. In addition, further scaling down the complementary metal-oxide semiconductor (CMOS) technology is leading to larger interconnect delay and higher power density [3]. The complexity of physical design is also increasing with higher density of devices. So, what is next?

A promising More-than-Moore technology is the 3D integrated circuits (ICs) which can improve the performance and reduce the intra-core wire length, and thereby enable high transfer bandwidth with reduced latencies and power consumption, while maintaining compact packing densities [4]. Alternative technologies that could be promising for new hardware accelerators include resistive computing, neuromorphic computing, and quantum computing.

Resistive computing could lead to non–von Neumann (VN) computing and enforce reconfigurable and data-centric paradigms due to its massive parallelism and low power consumption [5]. Moreover, humans can easily outperform current high-performance computers in tasks like auditory and pattern recognition

Advanced Nanoelectronics: Post-Silicon Materials and Devices,
First Edition. Edited by Muhammad Mustafa Hussain.

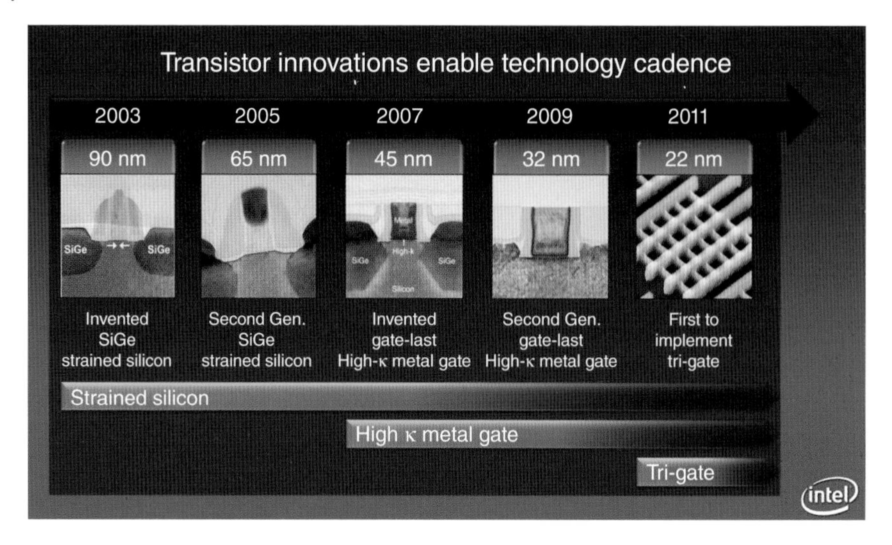

Figure 1.1 Intel innovation in process technology for the past decade. Source: www.intel.in.

and sensory motor control. Thus, neuromorphic computing can be promising for emulating such tasks due to its energy and space efficiency in artificial neural network applications [6]. Quantum computing can solve tasks that are impossible by classical computers, with potential applications in encryptions and cryptography, quantum search, and a number of specific computing applications [7].

In this chapter, four main technologies are discussed: FinFET, 3D IC, neuromorphic computing, and quantum computing. The state-of-the-art findings and current industrial state in these fields are presented; in addition, the challenges and limitations facing these technologies are discussed.

1.1 FinFET Technology

Over the past four decades, the continuous scaling of planar MOSFETs has provided an improved performance and higher transistor density. However, further scaling down planar transistors in the nanometer regime is very difficult to achieve due to the severe increase in the leakage current I_{off}. In fact, as the channel length in planar MOSFETs is reduced, the drain potential starts to affect the electrostatics in the channel and, consequently, the gate starts to lose control over the channel, which leads to increased leakage current between the drain and source. A higher gate-channel capacitance can relieve this problem using thinner and high-κ gate oxides; however, the thickness of the gate oxide is fundamentally restricted by the increased gate leakage and the gate-induced-drain leakage effect [8–10].

An alternative to planar MOSFETs is the multiple-gate FETs (MuGFETs) which demonstrate better electrostatics and better screening of the drain from the gate due to the additional gates covering the channel [11–14]. As a result, MuGFETs show better performance in terms of subthreshold slope, threshold voltage (V_t) roll-off, and drain-induced barrier lowering (DIBL). Another alternative to planar

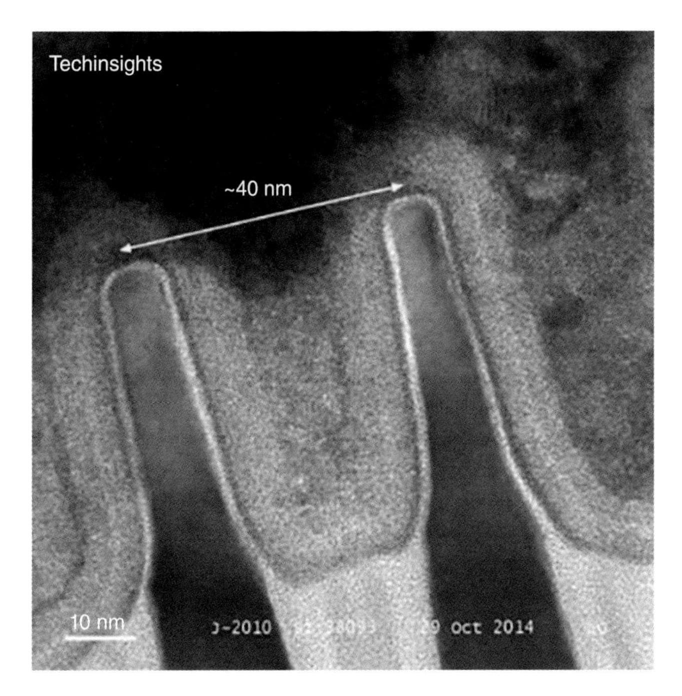

Figure 1.2 TEM image of Intel's 14-nm transistors with sub-40-nm fin pitch. Source: www.techinsights.com.

bulk MOSFETs is fully depleted silicon on insulator (FDSOI) MOSFETs, which reduce leakage between drain and source due to the removal of the substrate right below the channel [15]. The performance of the FDSOI MOSFETs is comparable with the double-gate field-effect transistors (DGFETs) in terms of SS, low junction capacitance, and high I_{on}/I_{off} ratio. Yet, the DGFETs have better scalability and can be manufactured on bulk Si wafers instead of silicon-on-insulator (SOI) wafers, which makes them more promising [16].

FinFETs or tri-gate FETs, which have three gates, have been found to be the most promising alternatives to MOSFETs due to their enhanced performance and simplicity of the fabrication process, which is compatible with and can be easily integrated into standard CMOS fabrication process (Figure 1.2) [17, 18]. In fact, an additional selective etch step is required in the FinFET fabrication process in order to create the third gate on top of the channel. FinFET devices have been explored carefully in the past decade. A large number of research articles that confirmed the improved short-channel behavior using different materials and processes have been published, as is shown in the following section. Next, the industrial state of FinFETs, their challenges, and limitations are discussed.

1.1.1 State-of-the-Art FinFETs

1.1.1.1 FinFET with Si Channel

In the semiconductor industry, silicon is the main channel material. The first FinFET technology (22-nm node) was produced by Intel in 2011. The second FinFET generation (14-nm node) published by Intel used strained Si channel [19].

The gate length was scaled from 26 to 20 nm in the second FinFET generation, which was possible due to new sub-fin doping and fin profile optimization. With a V_{DD} of 0.7 V, the saturation drive current is 1.04 mA μm^{-1} and the off current is 10 nA μm^{-1} for both nMOSFET (NMOS) and pMOSFET (PMOS). The SS is ~65 mV/decade, while the DIBL for N/PMOS is ~60/75 mV V^{-1}. High-density static random access memory (SRAM) having 0.0588 μm^2 cell size are also reported and fabricated using the 14-nm node. More recently, a research group from Samsung published a 7-nm CMOS FinFET using extreme ultraviolet (EUV) lithography instead of multiple-patterning lithography. This resulted in a reduction of the needed mask steps by more than 25%, in addition to providing smaller critical dimension variability and higher fidelity. The FinFET presented in this work consumes 45% less power and provides 20% faster speed than in the previous 10-nm technology. The reported SS is 65 and 70 mV/decade, and the DIBL is 30 and 45 mV V^{-1} for NMOS and PMOS, respectively. A 6T high-density and high-current SRAM memory has also been demonstrated using the 7-nm FinFET, and the results show a reduction in the bit line capacitance by 20% as a result of the reduction in the parasitic capacitance.

1.1.1.2 FinFET with High-Mobility Material Channel

The III–V materials gained growing attention for adoption as the channel material due to their promising characteristics such as high mobility, small effective mass, and, therefore, high injection velocity, in addition to near-ballistic performance. The first InGaSb pFET was demonstrated by Lu et al. [20], where a fin-dry etch technique was developed to obtain 15-nm narrow fins with vertical sidewalls. An equivalent oxide thickness (EOT) of 1.8 nm of Al_2O_3 was used as the gate oxide. The authors also demonstrated Si-compatible ohmic contacts that yielded an ultralow contact resistivity of 3.5×10^{-8} Ω cm^2. Devices with $L_g = 100$ nm and different fin widths (W_f) were demonstrated. The results show that with $W_f = 100$ nm, g_m of 122 μS μm^{-1} is achieved; while with $W_f = 30$ nm, g_m of 78 μS μm^{-1} is obtained.

Moreover, FinFETs with strained SiGe have lately attracted much interest due to their potential advantages such as higher mobility, built-in strain, and improved reliability with respect to conventional Si-based FETs. Very recently, a group of researchers at IBM demonstrated high-Ge-content strained SiGe FinFETs with replacement high-κ (HK)/metal gate (RMG). A long-channel subthreshold swing (SS) as low as ~68 mV/decade was reported [21]. This value is very competitive with other SiGe or Ge FinFETs with RMG process flow, where the reported SS values are in the range of 80–100 mV/decade [22]. In addition, a very high pFET hole mobility $\mu_{eff} = 235$ cm^2 V^{-1}s^{-1} was shown in a multi-fin device with average fin width of 4.6 nm and EOT of 7 Å which could be very promising for the sub-5-nm node FinFETs. Finally, in the same work, SiGe FinFETs with gate lengths $L_g = 25$ nm were fabricated using a gate-first flow. At a $V_{DD} = 0.5$ V, the devices showed DIBL = 40 mV, $SS_{lin}/SS_{sat} = 77/86$ mV/decade and $I_{on} = 430$ μA μm^{-1} at target high performance $I_{off} = 100$ nA μm^{-1}, which are among the largest reported values at such gate lengths (Figure 1.3).

In another work by Lei et al. [23], conducted in collaboration with Taiwan Semiconductor Manufacturing Co. (TSMC), the first GeSn FinFET device on

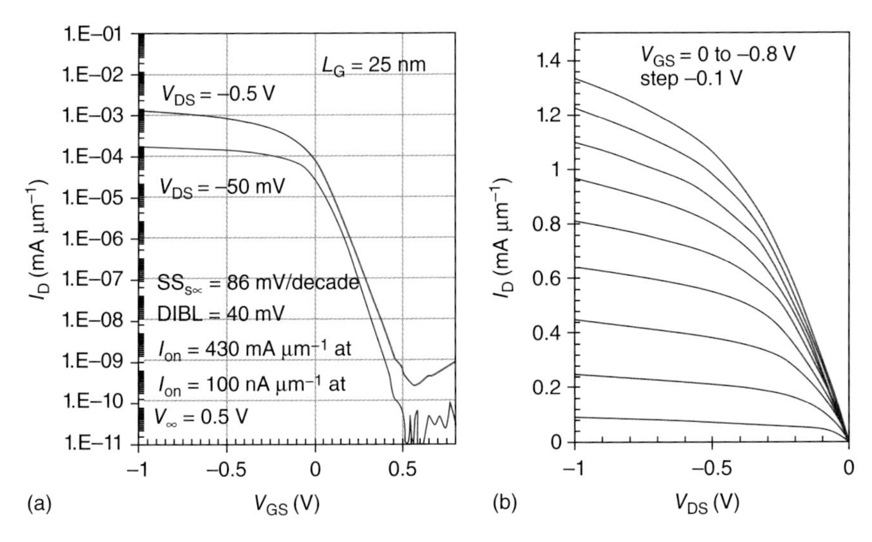

Figure 1.3 Transfer and output characteristics of high-Ge-content SiGe FinFETs with L_G 25 nm with gate first flow. Source: Hashemi et al. 2017 [21]. Reused with permision of IEEE.

a GeSnOI substrate was demonstrated with a channel length of 50 nm and W_{Fin} = 20 nm, and 4 nm HfO_2 was used as the gate oxide. The novel substrate was fabricated by the growth of high-quality GeSn by chemical vapor deposition (CVD) followed by a low-temperature process flow to get the GeSnOI. The GeSn pFET yielded the lowest SS of 79 mV/decade, the highest transconductance g_m of 807 µS µm^{-1}, and the highest hole mobility of 208 cm^2 V^{-1} s^{-1} (N_{inv} of 8×10^{12} cm^{-2}).

1.1.1.3 FinFET with TMD Channel

For sub-5-nm nodes, a body with sub-3-nm thickness is required to maintain good channel control. Most channel materials like the conventional Si or III–V face limitations in terms of mobility, quantum capacitance, or process at such ultrathin body (UTB) thickness. Advanced two-dimensional transition-metal dichalcogenide (TMD) is very promising in UTB thickness due to its sub-nanometer monolayer UTB thickness potential in addition to its good transport characteristics in nanometer thickness [24]. Chen et al. demonstrated the first 4-nm-thick TMD body FinFET with back gate control [25]. The main processes in the fabrication of the TMD FinFET is the compatibility of the CVD growth of TMD with CMOS processing, in addition to the reduction of the contact resistance by hydrogen plasma treatment of MoS_2. The V_t of this FinFET device can be adjusted dynamically by applying bias on the back gate. The front gate device showed an on/off current ratio over 10^5 with I_{on} of 200 µA µm^{-1} for V_{dd} = 1 V.

1.1.1.4 SOI versus Bulk FinFET

Bulk FinFETs are built on bulk-Si wafers, which are less expensive and have a lower defect density than do SOI wafers, while maintaining a better heat transfer rate to the Si substrate with respect to SOI FinFETs. The first Intel FinFET was a bulk FinFET. Lee [1] studied the 14-nm node FinFET technology and compared bulk and SOI FinFET in terms of scalability, heat dissipation, and parasitic capacitance. Lee showed that both 14-nm FinFETs with bulk and SOI substrates have the same $I-V$ characteristics when the same geometry and doping concentration are used. Therefore, both devices have similar scalability. Moreover, the fins in bulk FinFETs are easily depleted, which allows for the reduction of the S/D to fin body junction capacitance to values that are lower than in the case of SOI FinFETs. Finally, to increase the heat transfer rate in SOI FinFETs, the buried oxide should be made thinner than 20 nm, which could have a negative impact on the device performance such as an increase in the parasitic capacitance.

Finally, it is worth mentioning that there are many other factors that affect the performance and reliability of FinFET devices, such as the materials used for metal gate/gate oxide, the shape of the fins (trapezoidal versus rectangular), the spacing between the fins, the fin edge roughness, choice of FET structure (lateral, vertical), and so on, which are not discussed in this chapter.

1.1.2 Industrial State

In 2011, Intel was the first company to use the 22-nm bulk FinFETs in mass production of central processing units (CPUs), which is 18% and 37% faster at 1 and 0.7 V, respectively, than Intel's 32-nm transistors [26]. Intel reported at the International Electron Devices Meeting (IEDM) that these 3D tri-gate transistors have a saturation current that exceeds 2 mA μm^{-1}. Several companies then followed Intel and announced the production of 3D transistors such as Samsung, TSMC, and Global Foundries. In 2015, Samsung announced the first production of the 14-nm FinFET-based transistors for mobile applications followed by the first mass production of the 10-nm FinFET (10LPE) in October 2016. Samsung was able to show improvements in power (40% lower power consumption than their 14-nm FinFETs), performance (27% higher performance), and scalability of the 3D tri-gate transistors (30% higher area efficiency).

However, at the 10-nm node, only three companies were capable of manufacturing such transistors: Samsung, Intel, and TSMC (Global Foundries excluded). Moreover, the geometries of the transistors produced at the leading manufacturers are different. For instance, the 10-nm FinFETs produced at TSMC and Samsung are denser than Intel's 14-nm FinFETs; however, they are closer to Intel's 14-nm FinFETs than they are to the Intel's 10-nm (the metal pitch in the Samsung's 10-nm is just 1 nm shorter than Intel's 14-nm).

In addition, some foundries use a hybrid node while others execute full node shrinking, which results in different geometries. In hybrid node shrinking, a new structure for the transistor (or a smaller transistor) is used (front end of line (FEOL)) but employing a set of design rules established previously for connecting transistors together (back end of line (BEOL)). In full node shrinking, both FEOL and BEOL are shrinking. In fact, TSMC and Samsung used the hybrid nodes at 16/14 nm where they introduced the new FinFET structure, while Intel is the only company executing full node shrinking with every new technology. It is worth noting that hybridized nodes allow the foundries to tackle a single set of challenges since the whole design process is not fully scaled down at once.

During Intel's Technology and Manufacturing Day 2017, Intel announced the mass production of its 10-nm process which used self-aligned quad patterning (SAQP) for the first time. Intel's 10-nm technology showed 45% less power consumption and 25% better performance than their 14-nm transistors with a minimum gate pitch of 54 nm (versus 70 nm for Intel's 14 nm) and a metal pitch of 36 nm (versus 52 nm for Intel's 14 nm). Also, Intel's 10-nm density is 2.7× higher than the previous node (new density of 100.8 mega transistors mm^{-2}), with 25% taller (53-nm fin height) and more closely spaced fins (34-nm fin pitch).

Saumsung's 10 nm uses triple-patterning technology with a 68-nm contacted gate pitch, 51-nm metal pitch, dual-depth shallow trench isolation (STI) with a single dummy gate (ref Common Platform Alliance Paper which was presented in 2016), while TSMC's 10-nm used quad-patterning technology which allows a double increase in density compared to their 16-nm technology. TSMC claimed a poly pitch of 64 nm and a metal pitch of 42 nm with 35% less power consumption and 15% higher performance than their 16-nm technology.

In June 2017, Global Foundries announced the mass production of its 7-nm FinFET technology which offers 40% improvement in performance with volume production ramping in the second half of 2018. The initial production ramp of the 7-nm technology employs triple and quadruple patterning technology using a 193-nm excimer laser. Global Foundries will introduce EUV to its manufacturing process to accelerate the production ramp and improve the yield.

TSMC announced recently that its 7-nm FinFET will offer around 25% speed enhancement or a 35% power reduction over its 10-nm FinFETs, while Samsung announced the addition of the 8- and 6-nm process technologies to its current process roadmap with an aim of improving the cost competitiveness over its 10- and 7-nm technologies. It is also worth noting that Samsung's 7-nm will be its first technology to use EUV lithography.

1.1.3 Challenges and Limitations

The introduction of the FinFET technology has enabled the gate length scaling down to 7 nm with a 48-nm contacted poly pitch (CPP) due to improved device electrostatics [27]. The improved performance has been achieved through the "Fin Effect" boost (effective fin width/fin pitch) which increased the drive current for a certain capacitive load. However, the restrictions on the fin thickness are being rapidly approached, which would lead to a faster scaling in S/D sizes versus the contacted gate pitch. An increasing "Fin Effect" will thus result, which in combination with a plateau in the gate length would put pressure on the conduction path from contacts to S/D. In a work conducted by a group of researchers from Global Foundries and IBM, current contact resistivity of $\sim 2 \times 10^{-9}\ \Omega\,cm^2$ [28] will significantly deteriorate the performance of FinFETs below 40-nm CPP, while fully ohmic contacts with resistivity of $\sim 1 \times 10^{-10}\Omega\,cm^2$ [29] might push the CPP to below 30 nm. The work concluded that in order to further improve the performance and power consumption in future CMOS in the 30–40 nm CPP, industry will face pressure to use new device architectures or scaling choices [2].

Another challenge is that sub-5-nm nodes would need sub-3-nm body thickness for maintaining good channel control [25]. However, most of the channel materials such as Si, Ge, and other III–V materials face fabrication, mobility, and quantum capacitance challenges at such small body thicknesses [30]. In addition, a group of researchers from IBM have fabricated test structures to unambiguously observe quantum confinement effects. The structures included fins with 40-nm fin pitch, 20-nm L_g, and 4- to 30-nm W_{Fin}. The measurements showed performance/mobility degradation, increase in series resistance, increase in variability, DIBL, and in V_t of NMOS/PMOS as the W_{Fin} is reduced [31], which confirms the challenges to be faced when further scaling down the FinFET technology.

Wavy FinFET has been proposed by Fahad et al. [32] as a promising structure for the high-performance technology node. The wavy transistor integrates 2D UTBs with the fin structure which maximizes the chip area utilization resulting in higher density, higher gain, and back bias capability. The structure was simulated using the 2013 International Technology Roadmap for Semiconductors (ITRS) specifications for the 7-nm node with UTB thickness of 2.5 nm and fin thickness of 6.8 nm. The authors reported an improved SS and DIBL performance of the wavy channel with 109% higher non-normalized ON-state drive performance as opposed to conventional FinFETs.

1.2 3D Integrated Circuit Technology

3D integration technology can denote either 3D packaging or 3D IC, which can be defined in different ways. In general, in 3D packaging, the vertical stacks are achieved via traditional methods of interconnects such as wire bonding and flip chip [33, 34]. However, in 3D IC, interconnections between different stacking layers are formed via through-silicon-vias (TSVs) [35]. Die stacking can be achieved by connecting separately manufactured dies or wafers vertically through one of three integration schemes: die-to-die, die-to-wafer, and wafer-to-wafer. The contacts (mechanical and electrical) can be achieved using either microbumps or by wire bonding as used in system-in-package (SIP) and package-on-package (POP) devices. Even though SIP is sometimes referred to as a 3D stacking technology, it is better referred to as a 2.5D technology. Another approach is to integrate dies horizontally on a silicon substrate using interposers. The benefits of using interposers are several: (i) lower communication power consumption due to the short communication distance between dies, (ii) the possibility of stacking separately manufactured dies from heterogeneous technologies to get the best out of all technologies, and (iii) enhanced yield and cost of the system due to the ability of fabricating and testing the smaller dies separately before integrating them into a silicon substrate instead of fabricating very large dies with much lower yield. The most promising approach of 3D integration is the monolithic approach, where active layers are vertically grown on top of each other and interconnects are made through TSVs which provide the densest connectivity.

There are several topics related to 3D integration that have recently gained a lot of attention in research. In the following, the main research topics with corresponding state-of-the-art technology are presented, followed by the industrial state and the main challenges of this technology.

1.2.1 Research State

1.2.1.1 Thermal Management

The biggest obstacle to the commercialization of 3D IC is the thermal management problem. As a matter of fact, the very thin thickness of chips in the 3D IC (<50 μm) in addition to the very high density of devices results in an increase in the temperature of the dies which are not close to the heat sink, and thereby deteriorating the performance of the system. In the past few years, research addressing thermal problems in 3D IC has gained growing attention. Goplen et al. reported that TSVs can act as a vertical path for heat flow [36]; therefore, thermal TSVs in addition to signal TSVs can be used to vertically transfer the heat and thereby reduce the die temperatures [37, 38]. Another study done by Lee et al. [39] showed that the heat transfer is directly proportional to the size of the via islands. In addition, it was found that a large number of TSVs can lead to routing congestion in the 3D ICs; thus, in addition to being expensive to fabricate, an optimization algorithm is needed to find the needed number of TSVs and their locations in order to be able to reduce the temperature of the dies. Moreover, Furumi et al. [40] proposed new cooling architectures for 3D ICs based on thermal sidewalls, interchip plates, and a bottom plate (thermal SIB). The experimental results conducted using a 3D thermal solver show that the thermal SIB can reduce the temperature in a 3D IC by over 40% when compared with structures that used a conventional heat sink only.

1.2.1.2 Through-silicon-vias

Using TSVs in 3D ICs and 3D packaging is very promising since it allows higher integration density, higher clock rate, and lower power dissipation [41]. In addition, TSVs are used in the 2.5D through-silicon interposers which enable the integration of heterogeneous dies on a silicon substrate. However, the fabrication of TSVs can be challenging: the etch process of the high-aspect-ratio TSVs should lead to scallop-free Si [26] and the Cu-filled TSV should be void-free [42]. This is in addition to challenges related to Cu protrusion affecting the BEOL reliability [43], thinning of TSV wafer [44], revealing of the backside of the TSV, and the bonding process [45]. In general, TSV fabrication requires the following steps: patterning of the via, etching the via, depositing the dielectric liner, metallization, and, finally, chemical-mechanical planarization (CMP) for planarization [46].

Currently, scaling down the TSVs is driven by the need to lower the thermal-mechanical stress in addition to its effect on the BEOL performance. The depth of the TSVs is limited, constrained by the wafer thinning (usually fixed at 50 μm). For a higher aspect ratio TSV (beyond 10 : 1), using the physical vapor deposition (PVD) barrier and seed process might lead to non-conformal films. IMEC and Lam Research Corp developed a low-cost process for getting conformal deposition of a very thin barrier and seed layer in high-aspect-ratio TSVs. The process consists of depositing a highly conformal thin oxide liner using atomic layer deposition (ALD), followed by the ALD deposition of the WN barrier, electroless plating NiB seed, and, finally, filling the TSV with copper using electrochemical deposition (ECD) [47]. Tokyo Electron Limited also reported another method to deposit highly conformal barrier and seed layers using electroless plating of Cu on

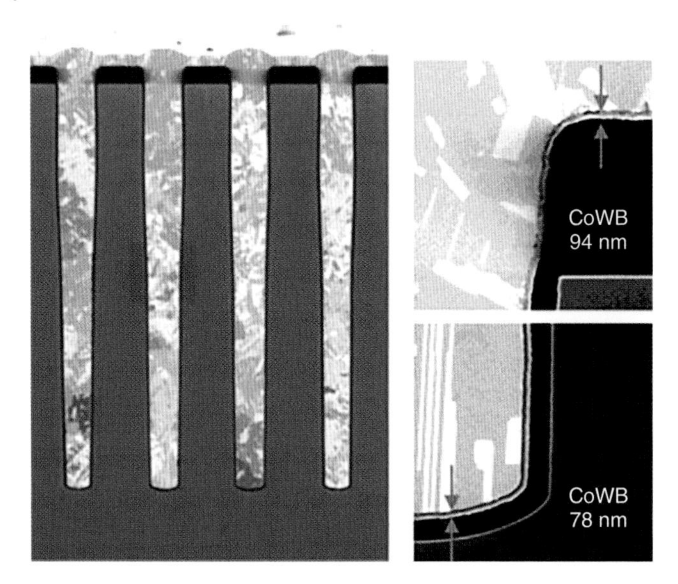

Figure 1.4 FIB-SEM after ECD-Cu filling on Eless-Cu/CoWB layers of 5 × 50 μm TSV. Source: Tanaka et al. 2015 [48]. Reused with permission of IEEE.

CoWB followed by Cu filling the TSV using ECD. Figure 1.4 shows the 5 × 50 μm TSV reported by Tokyo Electron Limited [48].

Another innovative metallization process was developed by Aveni (previously known as Alchimer). This metallization method is based on molecular engineering, where the film is grown molecule by molecule and can be applied in industry. First, a barrier layer is deposited by grafting and the NiB compound is used to make a Cu diffusion barrier which maintains the resistivity levels such that Cu can fill the high-aspect-ratio TSV using electrografting without the need for a cupper seed layer. The final fill process results in large, uniform, and high-purity grains of Cu, which could increase the yield due to eliminated voids, shorts, and opens [49].

1.2.1.3 Bonding in 3D IC

As already mentioned, the most important aspect of 3D IC is the ability to integrate heterogeneous dies fabricated at different foundries without performance degradation. The integration can be achieved either through wafer-on-wafer (WoW) bonding, chip-on-wafer (CoW) bonding, or chip-on-chip (CoC) bonding (Figure 1.5) [50]. WoW is the most preferred bonding due to its precise alignment [51]; more specifically, Cu metal-to-metal thermocompression bonding is the most favored among all bonding methods as it provides excellent electrical conductivity and mechanical strength after bonding [52]. During the thermocompression of Cu—Cu bonding, interdiffusion of Cu atom and grain growth across the bonding interface takes place. However, the main challenge to this process is to achieve it at low pressure and low temperature in order to avoid damaging the devices underneath or cause any reliability issues. But the Cu—Cu bonding process requires high temperature and pressure (or either of them) as native

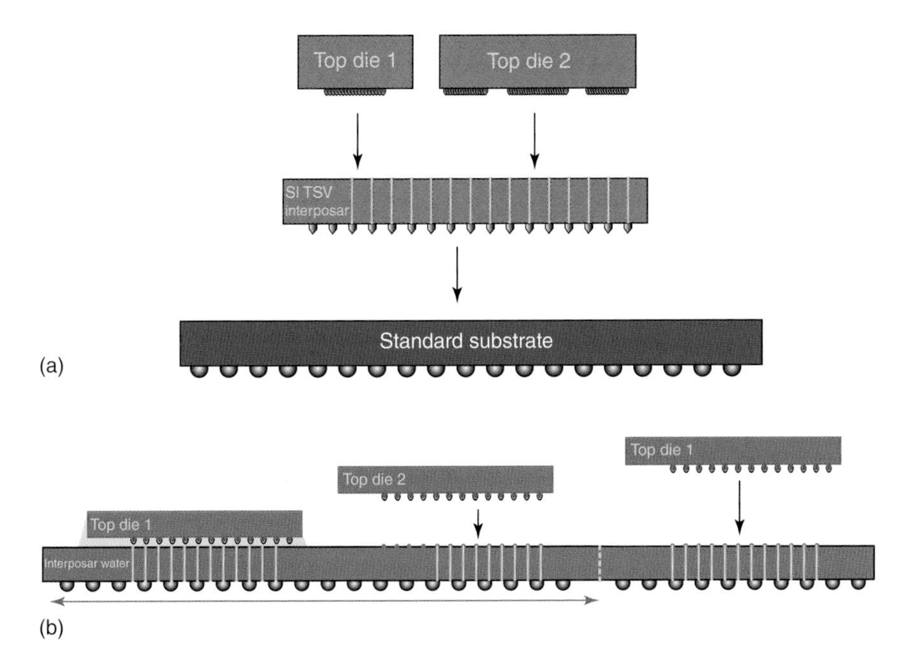

Figure 1.5 Assembly die stacking process flow: (a) CoS and (b) CoW. Source: https://amkor.com/.

oxide can be easily grown on the Cu surface, which inhibits the Cu interdiffusion and degrades the bonding quality [53]. Therefore, achieving high-quality Cu—Cu bonding at low temperature and pressure is needed in 3D IC.

Researchers have worked on several ways to avoid the surface oxidation of Cu and to remove the already grown native oxide. Shigetou et al. reported a bonding method based on surface-activated bonding (SAB). First, the native oxide is removed using argon bombardment at ultra-high vacuum (UHV), and then the bonding takes place at room temperature using a SAB flip-chip bonder [54]. However, the need for UHV increases the complexity of the process, and as a result becomes unattractive for manufacturing. Also, different chemistries have been proposed to do wet etching/cleaning of the native oxide such as hydrochloric acid [55], citric acid [56], sulfuric acid [57], and acetic acid [58]. Although some wet etching chemistries succeeded in removing the native oxide, immersing the wafer in such chemistries for a prolonged time might lead to etching the Cu and deteriorating the performance of the devices underneath. In another work conducted by Tan et al., a self-assembled monolayer (SAM) of alkane thiol, an organic monolayer, is deposited on the Cu surface to passivate it; and then the SAM is desorbed before the Cu—Cu bonding [59]. The use of the SAM passivation layer protected the Cu from growing native oxide, and the SAM removal process can be done at 250 °C [60]. However, all passivation based on SAM are not CMOS compatible.

In a work addressing this problem, a 3-nm Ti layer was used instead to passivate the Cu surface at 160 °C and 2.5 bar [61]. However, the Ti materials are challenging to be used in the damascene process; in addition, Ti can oxidize if exposed to air for more than two days, which is not favorable for the 3D IC process. In a continuation to this work, it has been reported that a 3-nm of manganin alloy passivation layer deposited at 150 °C and low pressure led to a strong Cu—Cu bonding of 5 kN force, in addition to being damascene compatible.

1.2.1.4 Test and Yield

Every additional manufacturing step introduces a risk for defects and complicates the testing of the system. Yield is based on test results, and the cost is based on the test, yield, and throughput. 3D high yield is challenging to achieve, which is why the wire bonding of "known good dies" in 3D packages first found its application in mobile devices.

Any 3D IC process would be considered feasible only if its manufacturability yield is high. A group from Xilinx Inc. reported the key challenges faced during fabricating a 28-nm 3D IC with chip-on-wafer-on-substrate process [62]. During the initial ramp stage, most of the observed failures were related to the assembly at the interposer level such as open microbumps, opens and shorts in the interposer metal line, and TSV opens. Another failure mode is the deterioration of the transistors during the assembly of the 3D IC. The group developed a failure analysis technique based on a closed loop feedback, which resulted in improved yields.

1.2.2 Industrial State

Samsung is already using the monolithic approach to die stacking in 3D flash memory and smart sensors. The first commercial prototype of 3D IC (microcontroller) dates back to 2004 when Tezzaron released it [63]. In 2006, Intel assessed 3D chip stacking in Pentium 4 [64]. In 2011, IBM announced the introduction of the 3D chip production process [65]. Also, in 2012, Tezzaron released a prototype for its multicore design, which includes 64 core 3D-MAPS (*MA*ssively *P*arallel processor with *S*tacked memory) (http://arch.ece.gatech.edu/research/3dmaps/3dmaps.html)[66]. In 2013, a 128-Gb 3D NAND chip was introduced by Samsung which has 2× transistor density, 50% lower power consumption, 2× data storage speed, and 10× better retention characteristics compared to the planar version.

In 2015, Intel also introduced the 3D XPoint memory with 10× higher capacity than DRAM and 1000× faster than NAND flash [67]. Moreover, NVIDIA and AMD manufactured a high bandwidth memory (HBM) using 3D stacked memories, which is already used in the AMD GPU based on the Fiji architecture since 2015 (https://images.nvidia.com/content/pdf/tesla/whitepaper/pascal-architecture-whitepaper.pdf). A high-performance RAM competing with HBM is the hybrid memory cube (HMC), which was introduced in 2011 by Micron and is based on DRAM stacked using TSVs (https://www.micron.com/products/hybrid-memory-cube). SanDisk and Toshiba announced in 2015 the production of the world's first 3D NAND with 48 layers and using BiCS (Bit-Cost Scalable) technology. The 3D NAND achieved 32 GB capacity with a storage of 3 bits per transistor. The latest version is called BiCS3, which will have 64 layers and will show a 40% larger capacity than the BiCS2, according to Toshiba.

Moreover, Micron reported the mass production of its 64-layer 3D NAND by the end of 2017, while Western Digital began mass production of its 64-layer

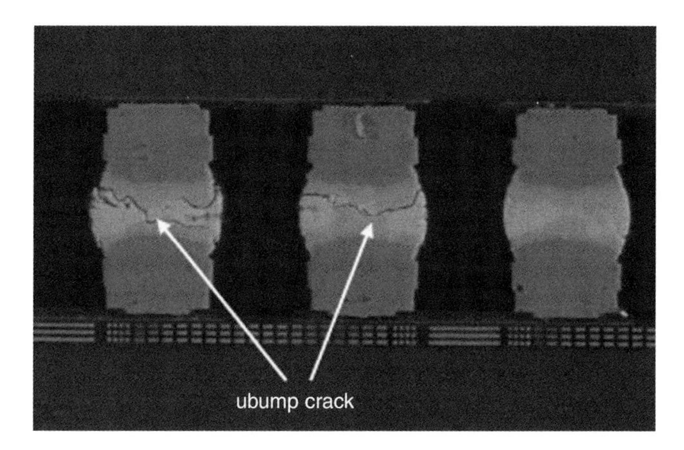

Figure 1.6 Picture of a microbump crack. Source: Yip et al. 2017 [68]. Reused with permission of IEEE.

3D NAND flash chips in 2017 (https://www.anandtech.com/show/10274/the-crucial-mx300-750gb-ssd-review-microns-3d-nand-arrives). Also, in early 2017, Intel announced the world's first commercial solid-state drive (SSD) based on 64-layer 3D NAND with a capacity of 512 GB (https://www.anandtech.com/show/11571/the-intel-ssd-545s-512gb-review-64layer-3d-tlc-nand-hits-retail).

1.2.3 Challenges and Limitations

Several challenges face the commercialization of 3D IC or 3D packaging. In a work done by Intel, it was found that yield estimates modeled using traditional methods can be pessimistic by as much as 50%. New analytical models have to be established to take into consideration other effects such as defect clustering and systematic defects introduced by equipment and handling issues during manufacturing. Moreover, it has been reported that the electrical performance of 3D IC with TSVs is affected due to the structure of the TSV with microbumps. TSV and microbump structures lead to local mechanical stress and strain due to the mismatch in the coefficient of thermal expansion (CTE) of Si, Cu TSV, and microbump (Figure 1.6) [68]. Moreover, crystal defects and stress can be induced in the Si chip as it is thinned down to less than a couple of tens of micrometers. Also, the gettering layers used to avoid the contamination of the metal and crystal defects can be removed from the Si chip as it is thinned down.

1.3 Neuromorphic Computing Technology

The flexibility of the VN architecture for "stored program" has led to enormous improvements in system performance for more than five years. However, since miniaturizing devices have slowed down in the past years, the energy and time used to transport data between memory and processor has become difficult, especially for data-centric applications such as real-time pattern recognition where state-of-the-art VN systems try hard to meet the performance of a human being. The human brain outperforms advanced processors on many tasks such as unstructured data classification due to its parallel architecture connecting low-power neurons and synapses which act as computing and adaptive memory elements, respectively. The human brain performance is actually inspiring for novel non–VN computing models needed in future computing systems.

Even though designing neural circuits using electronic components dates back to the implementation of retinas [69] and perceptrons [70], modern research about very-large-scale integration (VLSI) technology using the nonlinear current characteristics began in the mid-1980s through collaboration between Richard Feynman, Carver Mead, Max Delbrück, and John Hopfield [71]. In fact, Mead tried to imitate the gradual synaptic transmission in the retina using the analog properties of transistors rather than operating them as digital switches. Mead was able to demonstrate that neuromorphic circuits using analog transistors instead of digital ones can match the physical properties of the proteic channels in neurons [72], leading to the need for a much smaller number of transistors to emulate neural systems.

In the neural system, neurons are connected to many other neurons, and they pass electrical and chemical signals to each other via synapses. These connections are either strengthened or weakened through a process called spike-timing-dependent plasticity (STDP), which is biologically observed [73, 74]. STDP changes depending on the timing between spikes (action potentials) within the input neuron (presynaptic) and output neuron (postsynaptic). In long-term potentiation (LTP), causal spiking strengthens synapses; while in long-term depression (LTD), the synaptic strength is weakened by causal spiking [75]. The change in the weight of synapses, also called synaptic plasticity, explains how the brain learns and memorizes [76].

Neuromorphic computing technology is considered a promising candidate for implementing applications such as self-learning, recognition of patterns, gestures, and speech using energy-efficient/low-power spiking networks. However, the progress in this technology faces two main challenges: (i) the lack of a full understanding of how the brain works and (ii) the lack of agreement on which technology can achieve synaptic and neural circuits with the best balance between cost, performance, and power consumption. Currently, a great deal of research is being conducted on different technologies for neuromorphic computing including mathematical and machine learning algorithms, neuromorphic datasets, field programmable gate array (FPGA) codes, photonic neuromorphic signal processing, nonvolatile memory (NVM) solutions, and so on. In this chapter, NVM for neuromorphic computing is discussed. In addition, the current industry state of neuromorphic computing, its challenges, and limitations are discussed.

1.3.1 State-of-the-Art Nonvolatile Memory as a Synapse

Around 10^{11} neurons and 10^{14} synapses exist in the human brain. In order to be able to implement brain-like processing architectures without using large and expensive areas on the silicon wafer, highly scalable and low-power memory devices are needed.

Different NVM devices have different physical properties and switching behaviors, and thus can be used to emulate synapses in different ways. For instance, when synapses are connected or not, an on/off NVM response would be sufficient; and this can be achieved using conductive-bridging random access memory (CBRAM). In other cases, synaptic weights are needed; therefore, an NVM with adjustable conductance would be required and this can be achieved using phase change memory (PCM) or memristor/resistive-random access memory (RRAM). In the following, the different types of NVM used to emulate synapses are briefly explained with state-of-the-art examples from the literature.

1.3.1.1 Phase Change Memory

In PCM, the state of the memory, whether programmed/SET or erased/RESET, depends on the difference in electrical resistivity between the amorphous and crystalline phases of the "phase change materials" leading to low (RESET) and high conductance (SET), respectively [77, 78] (Figure 1.7a).

PCM is attractive for neuromorphic applications where "device history" is needed, since the SET state can be achieved gradually by applying repetitive pulses to crystallize the phase of the plug in the device, resulting in a high-resistance state [84]. However, the RESET process can be only done sharply, since it involves melt and quench. The STDP can be implemented using a two-PCM approach: when an input neuron spikes, it outputs a signal (read pulse) and enters the LTP mode for a period of time t_{LTP}. If the postsynaptic neuron spikes during this period, a SET pulse is then sent to the LTP synapse. If not, then the LTD synapse is programmed, as shown in Figure 1.8a,b.

Suri et al. demonstrated that by adding a thin HfO_2 layer to the $Ge_2Sb_2T_5$ (GST)–based PCM, their synaptic performance can be improved [85, 86]. The addition of the interface layer affects the nucleation and growth activation energies, and thereby the crystallization kinetics, resulting in an increased dynamic range. In a later work, the authors developed a circuit model including

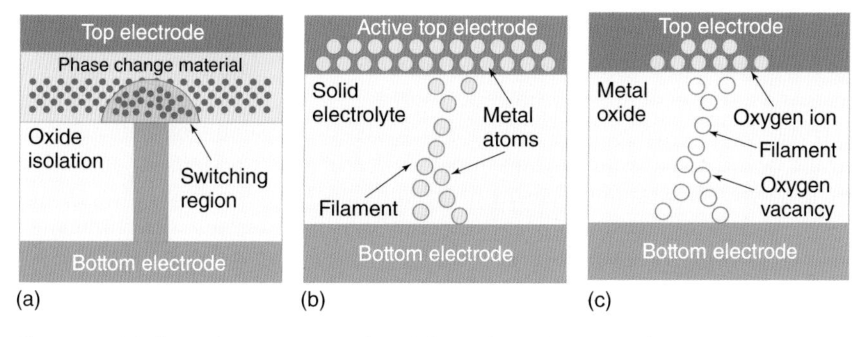

Figure 1.7 (a) Phase change memory (PCM) depends on the large difference in electrical resistivity between the amorphous (low-conductance) and crystalline (high-conductance) phases of so-called phase change materials [79, 80]. (b) Conductive-bridging RAM is based on the electrochemical formation of conductive metallic filaments through an insulating solid electrolyte or oxide [81]. (c) The conductive filaments in a filamentary RRAM are chains of defects through an otherwise insulating thin-film oxide [82]. Source: Reused with permission from [83].

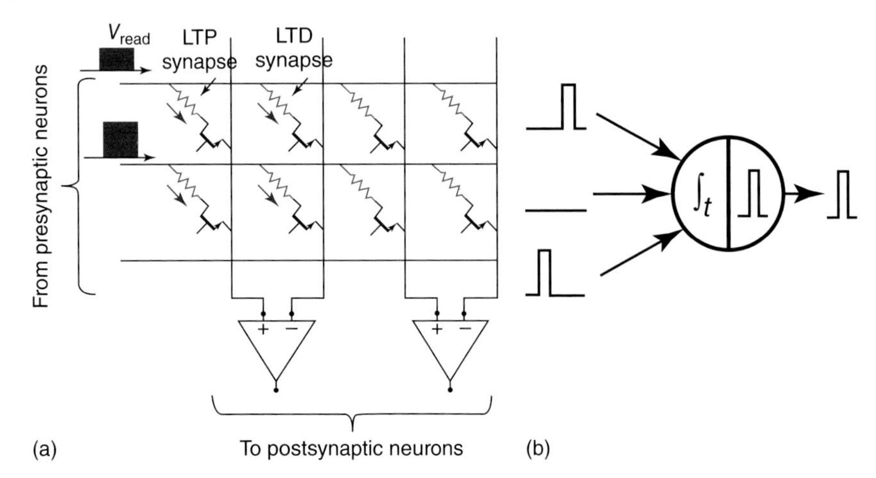

Figure 1.8 (a) Implementation of STDP with two-NVM-per-synapse scheme. Due to abrupt RESET in PCM devices, LTD and LTP are implemented with SET switching in different devices, with total weight of the synapse depending on the difference between these two conductances. (b) Key spiking characteristics of spiking neural network: downstream spikes depend on the time integration of continuous inputs, with synaptic weight change dependent on relative spike timing. Source: Reused with permission from [83].

the electrical and thermal characteristics of both top and bottom contacts with the PCM [79]. The authors showed that by enhancing the growth or nucleation rate, the maximum conductance can be reached in fewer pulses. They have also suggested that GST can offer more conductance states than GeTe, since GeTe (growth-dominated) saturated in conductance faster than the nucleation-dominated GST. Pattern learning and recognition was experimentally shown by Eryilmaz et al. using a 10×10 array of transistor-selected PCM cells [80]. They have also shown that longer training leads to lower initial resistance variation. Ambrogio et al. simulated larger networks of 28×28 pre- and 1 postneuron transistor-selected PCM cells (45-nm node) [87]. With two layers, the achieved MNIST (Modified National Institute of Standards and Technology) digit recognition probability was 33% with an error of 6%; while with three layers of networks, the recognition probability increased to 95.5% for 256 neurons with an error of 0.35%. The authors also demonstrated the capability of their network to learn new data in sequence and in parallel and forget previous data. Jackson et al. achieved STDP schemes using programming energies below 5 pJ in 10-nm pore PCM devices. Then, 100 leaky integrate-and-fire neurons were simulated. The authors showed the successful prediction of the next item in a sequence of four stimuli [88].

1.3.1.2 Conductive-Bridging RAM

Conductive-bridging random access memory (CBRAM) is based on the formation of conductive metallic filaments through an insulating solid electrolyte or oxide (Figure 1.7b) [77, 78]. CBRAM is promising for future NVM due to its characteristics such as extremely low power consumption (~nW), fast speed (~ns), and scalability to the nanometer range. However, the SET state in the CBRAM is achieved abruptly as the formed filaments are quite conductive, leading to high currents for neuromorphic devices. Integrate-and-fire neurons would need larger capacitors.

STDP synaptic performance was achieved by Ohno et al. using an Ag_2S atomic switch [89]. The amplitude and widths of the pulses are found to affect the short-term memory formation. Then, the authors experimentally demonstrated the learning and forgetting mechanisms of two patterns using a 7×7 array of organic synapses [90]. Yu et al. demonstrated STDP with 1.5×10^8 cycles of depression and potentiation without noticeable degradation [91] using a 100×100 nm CBRAM-based memristor device that is connected to integrate and fire neurons. Using CBRAM devices as binary synapses and applying the STDP learning rule, Suri et al. demonstrated the recognition and extraction of real-time visual and audio patterns in an unsupervised manner [92]. Nonassociative and associative types of learning are shown in single $Pt/Ge_{0.3}Se_{0.7}/SiO_2/Cu$ memristive device by Ziegler et al. [93]. Sillin et al. developed a numerical model imitating the synapse-like properties of single atomic switches [94].

1.3.1.3 Filamentary RRAM

Filamentary resistive random access memory (F-RRAM) is similar to CBRAM; however, the conductive filament in this device is due to a chain of defects in an oxide once triggered by electrical field and/or local temperature increases rather than by metallic atoms (Figure 1.7c) [81]. F-RRAM is attractive because it requires metal-oxides, most of which are already used in CMOS fabrications, such as HfO_X, TiO_X, WO_X, TaO_X, FeO_X, and AlO_X in addition to laminates of such films. Adaptive synaptic changes leading to gradual memory have been demonstrated in such materials. The structure of an F-RRAM is based on a metal-insulator-metal structure which is CMOS compatible and highly scalable, in addition to achieving very low energy consumption per synaptic operation (sub-pJ), fast switching (<10 ns) [95], extremely small size (<10 nm), very low currents (1 μA programming current), and multibit storage [96]. However, similar to the CBRAM, the SET function is abrupt to the rapid formation of the filament.

A group of researchers at University of Michigan led by Prof. Wei Lu recently demonstrated a prototype memristor network to experimentally process natural images using the sparse-coding algorithm. In this study, a 16×32 sub-array from the 32×32 WO_x–based memristor array was used, corresponding to a 2× over-complete dictionary with 16 inputs and 32 output neurons and dictionary elements. The dictionary elements were learned offline using a realistic memristor model and an algorithm based on the "winner-take-all" (WTA) approach and Oja's learning rule. After training, they successfully experimentally reconstructed grayscale images using the 16×32 memristor crossbar [97]. Choi et al. demonstrated a multilevel RESET switching with continuously increasing RESET voltages in a GdO_x-based F-RRAM but with rapid SET switching [98]. Yu et al. and Wu et al. showed gradual switching in SET operation with constantly increasing external currents, and in RESET with uninterruptedly increasing reset voltages in $TiN/HfO_x/AlO_x/Pt$ and $TiN/Ti/AlO_x/TiN$ RRAM devices, respectively [99, 100].

Yu et al. used an F-RRAM with multilayer oxide-based Pt/HfO_x/TiO_x/HfO_x/TiO_x/TiN to achieve hundreds of resistance states during the RESET [101, 102]. Sub-pJ energy per spike was obtained with 10-ns short pulses. Finally, the multilevel resistance modulation was modeled using a stochastic model and was applied to a visual system simulation; a two-layer neural network was simulated using 1024 neurons and 16 348 oxide-based synapses, achieving up to 10% tolerance to resistance variations. Piccolboni et al. recently demonstrated an HfO_2-based vertical resistive random access memory (VRRAM) technology, each exhibiting two distinct states [103]. A stack of VRRAM devices forms a single synapse, with one common select transistor, exhibiting gradual conductance behavior. Simulation was used to demonstrate real-time auditory and visual pattern recognition.

1.3.2 Research Programs and Industrial State of Neuromorphic Computing

With the availability and advances in deep submicron CMOS technology, developing brain-like structures on electronic substrates has recently received growing attention, and large research projects on brain-like systems have been launched internationally. Currently, the two largest programs in this field worldwide are the SyNAPSE program (Systems of Neuromorphic Adaptive Plastic Scalable Electronics) in the United States (started in 2009, (http://www.artificialbrains.com/darpa-synapse-progra)) and the European Commission flagship Human Brain Project (started in 2013 (http://www.humanbrainproject.eu)). Funded by the Defense Advanced Research Projects Agency (DARPA), the SyNAPSE program aims to emulate a mammalian brain in terms of power consumption, size, and function using an electronic neuromorphic machine. Then, robots with the intelligence of cats and mice would be built using such artificial brains. The neuromorphic microprocessor should be able to simulate the activity of 10 billion neurons and 100 trillion synapses using less than two liters of space and 1 kW of power (http://www.artificialbrains.com/darpa-synapse-progra). A project funded by DARPA's SyNAPSE initiative is the "Cognitive Computing via Synaptronics and Supercomputing" (C2S2) program, which is headed by IBM. A remarkable outcome of this project is the "True North chip," which is the largest chip fabricated at IBM and the second largest CMOS chip worldwide. This chip includes a 64 × 64 network of cores for digital applications, 256 millions of programmable synapses, and over 400 million bits of on-chip SRAM memory as storage space for neuron and synapse parameters. The 28-nm CMOS technology node with a die size of 4.3 cm^2 is used to fabricate the 5.4 billion transistors on the chip. The "True North chip" consumes 70 mW power (or 20 mW cm^{-2}), which is comparable to the cortex; however, the conventional CPU consumes at least 3 orders of magnitude higher power (50–100 W cm^{-2}) [104].

The Human Brain Project (HBP) is a European Commission (EC) flagship project with goals of increasing world awareness about the fields of neuroscience and brain-related medicine. This program has several subprojects, and one of them (called SP9) aims to develop a neuromorphic computing system using (i) physical brain-emulation models (with 200 000 neurons fabricated using 180-nm

CMOS technology), (ii) real-time numerical models (with 18-Advanced Reduced instruction set computer Machines (ARM) cores fabricated using the 130-nm CMOS technology, and (iii) software tools to design, run, and record the performance of the system [104]. The Blue Brain Project (launched in 2005 and led by EPFL and IBM) aims to understand the structure and functionality of the brain using simulations of the rodent and the brain. The simulations are conducted using an IBM supercomputer (Blue Gene, 10TB) with 8K CPUs to simulate artificial neural networks (http://bluebrain.epfl.ch/page-56882-en.html). Closely related to this project is the BrainScaleS (brain-inspired multiscale computation in neuromorphic hybrid systems), which is European funded. The BrainScaleS project uses Petaflop supercomputers to run numerical simulations to emulate and understand the brain-information processing. The hardware consists of the HICANN (High Input Count Analog Neural Network) chip, which has 112K synapses and 512 neuron circuits fabricated in a 180-nm CMOS technology (http://brainscales.kip.uni-heidelberg.de). Another impressive neuromorphic computing project is the SpiNNaker project [96], which consists of multiple core chips with multi-ARM interconnected through a specific communication technology. An 18-ARM9 CPU is included in each SpiNNaker package with a DRAM memory of 128 Mbyte, and each ARM core can real- time simulate 1000 neurons. The current full SpiNNaker board consists of 47 packages with goals of assembling 1200 boards with 90-kW power consumption.

1.4 Quantum Computing Technology

Yuri Manin and Richard Feynman independently reported that simulating physics using quantum computers would be more beneficial than using classical computers. Other than simulating physics, the question arose regarding whether quantum computers could outperform classical computers in solving other problems too. Paul Benioff and David Deutsch [105] later designed a layout for the quantum computer, while P. Shor and L. Grover developed the first algorithms that could run more efficiently on such quantum computers than on classical ones [106, 107].

In classical computers, the unit of information is the bit, which exists in two states: 0 and 1. The computations in such computers are a sequence of operations known as gates which are applied to bits. The computer's size and clock rate vary with the physical medium in which the bits are stored; however, the computational power of the computer is not affected by the bits' physical medium. Thus, two computers with the same storage capacity (bits) and set of operations (gates) are considered equivalent. In quantum computing, however, the unit of information is called "qubit" and the relevant operations are the "quantum gates." Unlike the bit, the qubit can exist in the state $|0\rangle$, $|1\rangle$ (labeled using Dirac's "bra-ket") or a superposition of the two states. Different approaches are used to design the physical medium of the qubit. Nevertheless, approaches based on semiconductors are gaining growing attention since they can be produced easily using lithography technology. The favored quantum degree of freedom in semiconductors is the spin since it does not interact with the environment. To be specific, silicon is an excellent candidate for spin qubits since it can be chemically purified, resulting in long-spin coherence time (in the seconds range) [108–110].

Different schemes have been proposed to implement qubits and quantum gates such as optics, ion traps, and nuclear spins in nuclear magnetic resonance devices. All of these schemes face several challenges and are still under development. Other researchers are focusing on developing advanced algorithms and mathematical models to run quantum computers. In this chapter, the qubits based on spins and superconducting materials are discussed.

1.4.1 Quantum Bit Requirement

The quantum bit implementation requires a system that can hold two states 0 and 1, and that can be initialized, acted on, and read [111]. Unlike the conventional electronics where the bits are transferred through wires from the processor to the memory, the qubits actually do not move; however, the control signals (logical gates) are brought close to the qubits to operate on and control them. Like digital electronics, an arbitrary logic can be implemented using a discrete set of logical operations [112]. At least two qubits are needed to be acted on at the same time by the set of operations, and the state of one qubit affects the state of the other. Therefore, computation requires qubits that can be coupled in a scalable manner and with high fidelity. Solid-state approaches are promising for the integration of a large number.

1.4.2 Research State

Quantum computers are able to solve problems related to chemistry, materials science, and mathematics that are beyond the capabilities of any supercomputer. The power of the quantum computers arises from the nature of the quantum bits that can exist in both states 0 and 1 at the same time, which is called the quantum superposition state. As a result, the computing power doubles with each additional qubit. Promising areas of research include superconducting circuits, electron spins in impurities, electron spins in semiconductor quantum dots, single photons [113], trapped ions [114], single defects or atoms in diamond [115, 116] and silicon [117], and so on, with single-qubit fidelities exceeding the threshold needed for fault-tolerant quantum computing.

Here, the two most promising systems which are the most similar to current solid-state circuits are discussed: superconducting circuits and electron spins in semiconductor quantum dots. It is worth mentioning that both of these qubits require cryogenic temperatures for operation, depend on analog control signals, and use radio frequency (RF) circuits to read the qubit state.

1.4.2.1 Spin-Based Qubits

Spin qubits are based on the intrinsic properties of semiconductors, such as electron spins trapped in the potential of chemical impurity or quantum dot. Spins are indeed protected from charge noise as a result of the weak spin-orbit coupling.

Loss and DiVincenzo focused on semiconductor quantum dots patterned using lithography. They reported the initialization of the ground state of the spin at low temperatures and high magnetic fields, the control of the spin using the electron spin resonance (ESR) toolbox, and the read based on the spin-to-charge conversion process [118]. The electrical control of the overlap in wave function results in an exchange coupling that can be tuned by the gate voltage. When combined with ESR, a controlled-not (CNOT) gate can be implemented, which is an essential logic operation in the implementation of a quantum computer. This was first demonstrated in GaAs quantum dots [119, 120]. Also, a high-fidelity two-qubit gate was recently demonstrated in a silicon device [121].

In addition, latest experiments have reported that the lifetime of the electron spin limits the high-fidelity readout of the qubits. Using a nanodevice, T. Watson et al. reported the longest lifetime of any electron spin qubit (30 seconds). The researchers engineered the electron wave function within phosphorous atom quantum dots such that the spin relaxation is minimized. Due to the longer lifetimes of the electron spin, the authors reported the readout of two sequential qubits with 99.8% fidelities, which are above the surface-code fault-tolerant threshold [122].

In another work, Veldhorst et al. reported the control over the spin states of the qubits by applying voltages with GHz frequencies. The authors used a phosphorous single-atom transistor with all epitaxial monolayer-doped gates (Figure 1.9a,b) and pulsed spectroscopy with selective transport via excited states which enabled the differentiation between the excited states of the single P atom. [121]

1.4.3 Superconducting Circuits for Quantum Information

Superconducting quantum circuits consist of a high number of atoms (usually aluminum) assembled with metallic wire/plate shapes and are based on the electric LC oscillator [121]. Two phenomena form the basis for the operation of the superconducting qubits: (i) superconductivity, which is the frictionless flow of electrical fluid through metals at low temperatures, and (ii) the Josephson effect, which provides nonlinearity to the circuit without causing dephasing or dissipation. The electron fluid motion around the circuit is denoted with the flux F reaching the inductor, which acts as the center-of-mass position in a mass-spring mechanical oscillator [123]. The Josephson tunnel junction converts the circuit into an artificial atom which can be selectively excited from the ground state to an excited state and used as a qubit. By changing the relative strengths of the three energies associated with the capacitance, inductance and tunnel element, different shapes of potential energies can be achieved. The performance of the qubits has drastically enhanced as the fabrication, measurements, and materials affecting coherence have been understood and enhanced. Moreover, other design variations have been introduced such as quantronium [124], fluxonium [125], and hybrid qubits [126], all of which are fabricated using the same materials but aim to enhance the performance by lowering their sensitivity to decoherence mechanisms in the environment.

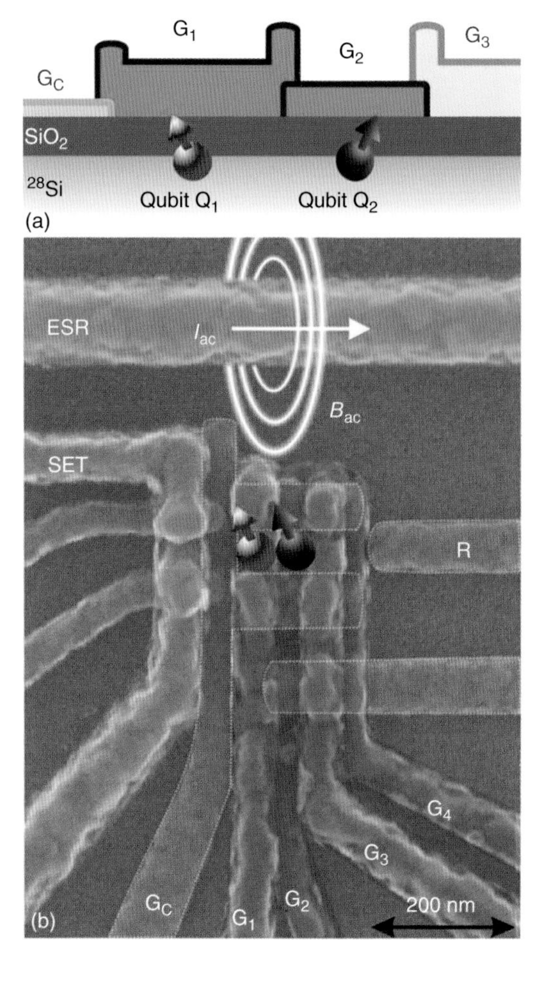

Figure 1.9 Silicon two-qubit logic device, incorporating SET readout and selective qubit control. Schematic (a) and scanning electron microscope colored image (b) of the device. The quantum dot structure (labels G_C and G_{1-4}) can be operated as a single or double quantum dot by appropriate biasing of gate electrodes G_1–G_4, where the dots $D_{1,2}$ are confined underneath gates $G_{1,2}$, respectively. The confinement gate G_C runs underneath G_1–G_3 and confines the quantum dot on all sides except on the reservoir (R) side. Qubit operation is achieved via an ac current I_{ac} through the ESR line, resulting in an ac magnetic field B_{ac}. Source: Devoret and Schoelkopf 2013 [123]. Reused with permission of Nature Publishing Group.

1.4.4 Industry State

In March 2017, IBM introduced two of its most powerful quantum computing processors for the IBM Q to help researchers and scientists solve problems that are not possible with today's most powerful computer. The two new IBM quantum processors include the following:

- A processor with 16 qubits which will enable solving of more complex experimentations than the previous 5-qubit processor.
- A processor with 17 qubits which is the first commercial prototype from IBM. This processor is the most powerful quantum processor invented by IBM to date: it is at least 2× more powerful than what is available to users on the IBM Cloud (https://phys.org/news/2017-05-ibm-powerful-universal-quantum-processors.html).

Also, in October 2017, Intel announced its new 17-qubit chip, which was delivered to QuTech in the Netherlands. It is worth mentioning that these quantum computers still cannot compete with current classical computers; however, the future is bright, especially with superconducting qubits (https://newsroom.intel.com/press-kits/quantum-computing/).

Moreover, IBM and Intel are not the only two companies working on building quantum computers. Google is also preparing a 50-qubit quantum computer, which is going to be used to solve a scientific previously unsolvable problem. Other companies working on creating quantum computers include Tigetti Computing, which is a startup in Berkeley, CA, and Microsoft Corp ((https://www.sciencealert.com/google-s-quantum-announcement-overshadowed-by-something-even-bigger) and (https://news.microsoft.com/features/new-microsoft-breakthroughs-general-purpose-quantum-computing-moves-closer-reality/)).

In addition, the European Commission is funding a €1 billion flagship project on quantum computing to launch in 2018 (https://ec.europa.eu/digital-single-market/en/news/european-commission-will-launch-eu1-billion-quantum-technologies-flagship).

1.4.5 Challenges and Limitations to Quantum Computing

Some of the challenges facing quantum computing technology are discussed in this section. First of all, there is the need for quantum error correction since the qubit states change in time in uncontrolled ways due to their interaction with the environment (aka decoherence). The error probability calculated by the quantum error correction algorithm must be below 1% (accuracy threshold for fault tolerance) [123]. This would require additional qubits for encoding and decoding. However, the number of qubits needed should be reduced; in fact, an estimated number of qubits needed to compute a molecule reaches millions. Therefore, this number should be brought down by several orders of magnitude.

Moreover, specific electronics should be built to produce the control signals, and to store and process the output signals. These electronics include analog-to-digital converters (ADCs), digital-to-analog converters (DACs), RF sources, amplifiers, multiplexer circuits, digital data processing units, and so on. The electronics need to be low cost (at $1.00 per qubit) and show high accuracy (exceeding the 1% accuracy threshold by 2 orders of magnitude). In addition, some electronics might require cryogenic temperatures to operate, which poses a tight power budget. Also, qubits receive control signals from outside; therefore, multiplexing strategies must be employed in the interconnect technology between the qubits and the control and output electronics [127].

References

1 Lee, J.H. (2016). Bulk FinFETs: design at 14 nm node and key characteristics. In: *Nano Devices and Circuit Techniques for Low-Energy Applications and Energy Harvesting*, 33–64. Netherlands: Springer.

2 Razavieh, A., Zeitzoff, P., Brown, D.E. et al. (2017). Scaling challenges of FinFET architecture below 40nm contacted gate pitch. In: 2017 75th Annual Device Research Conference (DRC), pp. 1–2.

3 Joyner, J.W., Venkatesan, R., Zarkesh-Ha, P. et al. (2001). Impact of three-dimensional architectures on interconnects in gigascale integration. *IEEE Transactions on Very Large Scale Integration (VLSI) Systems* 9 (6): 922–928.
4 Beyne, E. (2006). The rise of the 3rd dimension for system intergration. In: 2006 International IEEE Interconnect Technology Conference, pp. 1–5.
5 Pershin, Y.V. and Di Ventra, M. (2011). Solving mazes with memristors: a massively parallel approach. *Physical Review E* 84 (4): doi: 10.1103/physreve.84.046703.
6 Pickett, M.D., Medeiros-Ribeiro, G., and Williams, R.S. (2013). A scalable neuristor built with Mott memristors. *Nature materials* 12 (2): 114–117.
7 Pershin, Y.V. and Di Ventra, M. (2012). Neuromorphic, digital, and quantum computation with memory circuit elements. *Proceedings of the IEEE* 100 (6): 2071–2080.
8 Hu, C. (1996). Gate oxide scaling limits and projection. In: International Electron Devices Meeting, 1996, IEDM'96, pp. 319–322. IEEE.
9 Yeo, Y.C., King, T.J., and Hu, C. (2003). MOSFET gate leakage modeling and selection guide for alternative gate dielectrics based on leakage considerations. *IEEE Transactions on Electron Devices* 50 (4): 1027–1035.
10 Chen, J., Chan, T.Y., Chen, I.C. et al. (1987). Subbreakdown drain leakage current in MOSFET. *IEEE Electron Device Letters* 8 (11): 515–517.
11 Ferain, I., Colinge, C.A., and Colinge, J.P. (2011). Multigate transistors as the future of classical metal-oxide-semiconductor field-effect transistors. *Nature* 479 (7373): 310–316.
12 Wong, H.S., Chan, K.K., and Taur, Y. (1997). Self-aligned (top and bottom) double-gate MOSFET with a 25 nm thick silicon channel. In: International Electron Devices Meeting, 1997, IEDM'97, Technical Digest, pp. 427–430. IEEE.
13 Choi, Y.K., Lindert, N., Xuan, P. et al. (2001). Sub-20 nm CMOS FinFET technologies. In: International Electron Devices Meeting, 2001, IEDM'01. Technical Digest, pp. 421–424. IEEE.
14 Mitard, J., Witters, L., Loo, R. et al. (2014). 15nm-W FIN high-performance low-defectivity strained-germanium pFinFETs with low temperature STI-last process. In: 2014 Symposium on VLSI Technology (VLSI-Technology), Digest of Technical Papers, pp. 1–2. IEEE.
15 Choi, Y.K., Asano, K., Lindert, N. et al. (1999). Ultra-thin body SOI MOSFET for deep-sub-tenth micron era. In: International Electron Devices Meeting, 1999, IEDM'99, Technical Digest, pp. 919–921. IEEE.
16 Doris, B., Cheng, K., Khakifirooz, A. et al. (2013). Device design considerations for next generation CMOS technology: Planar FDSOI and FinFET. In: 2013 International Symposium on VLSI Technology, Systems, and Applications (VLSI-TSA), pp. 1–2. IEEE.
17 Auth, C. (2012). 22-nm fully-depleted tri-gate CMOS transistors. In: 2012 IEEE Custom Integrated Circuits Conference (CICC), pp. 1–6. IEEE.
18 Guillorn, M., Chang, J., Bryant, A. et al. (2008). FinFET performance advantage at 22nm: An AC perspective. In: 2008 Symposium on VLSI Technology, pp. 12–13. IEEE.

19 Natarajan, S., Agostinelli, M., Akbar, S. et al. (2014). A 14nm logic technology featuring 2 nd-generation FinFET, air-gapped interconnects, self-aligned double patterning and a 0.0588 μm 2 SRAM cell size. In: 2014 IEEE International Electron Devices Meeting (IEDM), pp. 3–7. IEEE.

20 Lu, W., Kim, J.K., Klem, J.F. et al. (2015). An InGaSb p-channel FinFET. In: 2015 IEEE International Electron Devices Meeting (IEDM), pp. 31–36. IEEE.

21 Hashemi, P., Ando, T., Balakrishnan, K. et al. (2017). High performance PMOS with strained high-Ge-content SiGe fins for advanced logic applications. In: 2017 International Symposium on VLSI Technology, Systems and Application (VLSI-TSA), pp. 1–2. IEEE.

22 Hashemi, P., Ando, T., Balakrishnan, K. et al. (2016). Replacement high-K/metal-gate High-Ge-content strained SiGe FinFETs with high hole mobility and excellent SS and reliability at aggressive EOT ~ 7Å and scaled dimensions down to sub-4nm fin widths. In: 2016 IEEE Symposium on VLSI Technology, pp. 1–2. IEEE.

23 Lei, D., Lee, K.H., Bao, S. et al. (2017). The first GeSn FinFET on a novel GeSnOI substrate achieving lowest S of 79 mV/decade and record high Gm, int of 807 μS/μm for GeSn P-FETs. In: 2017 Symposium on VLSI Technology, pp. T198–T199. IEEE.

24 Lee, Y.J., Luo, G.L., Hou, F.J. et al. (2016). Ge GAA FETs and TMD FinFETs for the applications beyond Si: a review. *IEEE Journal of the Electron Devices Society* 4 (5): 286–293.

25 Chen, M.C., Li, K.S., Li, L.J. et al. (2015). TMD FinFET with 4 nm thin body and back gate control for future low power technology. In: 2015 IEEE International Electron Devices Meeting (IEDM), pp. 32–2. IEEE.

26 Morikawa, Y., Murayama, T., Sakuishi, Y.N.T. et al. (2013). Total cost effective scallop free Si etching for 2.5 D & 3D TSV fabrication technologies in 300mm wafer. In: 2013 IEEE 63rd Electronic Components and Technology Conference (ECTC), pp. 605–607. IEEE.

27 Xie, R., Montanini, P., Akarvardar, K. et al. (2016). A 7nm FinFET technology featuring EUV patterning and dual strained high mobility channels. In: 2016 IEEE International Electron Devices Meeting (IEDM), pp. 2–7. IEEE.

28 Niimi, H., Liu, Z., Gluschenkov, O. et al. (2016). Sub- 10^{-9} Ω -cm^2 n-type contact resistivity for FinFET technology. *IEEE Electron Device Letters* 37 (11): 1371–1374.

29 Maassen, J., Jeong, C., Baraskar, A. et al. (2013). Full band calculations of the intrinsic lower limit of contact resistivity. *Applied Physics Letters* 102 (11): 111605.

30 Liu, W., Kang, J., Cao, W. et al. (2013). High-performance few-layer-MoS_2 field-effect-transistor with record low contact-resistance. In: 2013 IEEE International Electron Devices Meeting (IEDM), pp. 19–4. IEEE.

31 Chang, J.B., Guillorn, M., Solomon, P.M. et al. (2011). Scaling of SOI FinFETs down to fin width of 4 nm for the 10nm technology node. In: 2011 Symposium on VLSI Technology (VLSIT), pp. 12–13. IEEE.

32 Fahad, H.M., Hu, C., and Hussain, M.M. (2015). Simulation study of a 3-D device integrating FinFET and UTBFET. *IEEE Transactions on Electron Devices* 62 (1): 83–87.

33 Mahajan, R., Sankman, R., Patel, N. et al. (2016). Embedded multi-die interconnect bridge (EMIB)--a high density, high bandwidth packaging interconnect. In: 2016 IEEE 66th Electronic Components and Technology Conference (ECTC), pp. 557–565. IEEE.
34 Zhang, D. and Lu, J.J.Q. (2017). 3D integration technologies: an overview. In: *Materials for Advanced Packaging*, 1–26. Springer International Publishing.
35 Patti, R.S. (2006). Three-dimensional integrated circuits and the future of system-on-chip designs. *Proceedings of the IEEE* 94 (6): 1214–1224.
36 Goplen, B. and Sapatnekar, S. (2005). Thermal via placement in 3D ICs. In: Proceedings of the 2005 international symposium on Physical design, pp. 167–174. ACM.
37 Kandlikar, S.G. and Ganguly, A. (2017). Fundamentals of heat dissipation in 3D IC packaging. In: *3D Microelectronic Packaging*, 245–260. Springer International Publishing.
38 Cong, J., Wei, J., and Zhang, Y. (2004). A thermal-driven floorplanning algorithm for 3D ICs. In: IEEE/ACM International Conference on Computer Aided Design, 2004. ICCAD-2004, pp. 306–313. IEEE.
39 Lee, S., Lemczyk, T.F., and Yovanovich, M.M. (1992). Analysis of thermal vias in high density interconnect technology. In: Eighth Annual IEEE Semiconductor Thermal Measurement and Management Symposium, 1992. SEMI-THERM VIII., pp. 55–61. IEEE.
40 Furumi, K., Imai, M., and Kurokawa, A. (2017). Cooling architectures using thermal sidewalls, interchip plates, and bottom plate for 3D ICs. In: 2017 18th International Symposium on Quality Electronic Design (ISQED), pp. 283–288. IEEE.
41 Karnezos, M., Carson, F., Pendse, R., and ChipPAC, S.T.A.T.S. (2005). 3D packaging promises performance, reliability gains with small footprints and lower profiles. *Chip Scale Review* 1: 29.
42 Wolf, M.J., Dretschkow, T., Wunderle, B. et al. (2008). High aspect ratio TSV copper filling with different seed layers. In: 58th Electronic Components and Technology Conference, 2008. ECTC 2008. pp. 563–570. IEEE.
43 Che, F.X., Putra, W.N., Heryanto, A. et al. (2013). Study on Cu protrusion of through-silicon via. *IEEE Transactions on Components, Packaging and Manufacturing Technology* 3 (5): 732–739.
44 Che, F.X. (2014). Dynamic stress modeling on wafer thinning process and reliability analysis for TSV wafer. *IEEE Transactions on Components, Packaging and Manufacturing Technology* 4 (9): 1432–1440.
45 Huang, B.K., Lin, C.M., Huang, S.J. et al. (2013). Integration challenges of TSV backside via reveal process. In: 2013 IEEE 63rd Electronic Components and Technology Conference (ECTC), pp. 915–917. IEEE.
46 Redolfi, A., Velenis, D., Thangaraju, S. et al. (2011). Implementation of an industry compliant, $5 \times 50\mu m$, via-middle TSV technology on 300mm wafers. In: 2011 IEEE 61st Electronic Components and Technology Conference (ECTC), pp. 1384–1388. IEEE.
47 Van Huylenbroeck, S., Li, Y., Heylen, N. et al. (2015). Advanced metallization scheme for $3 \times 50\mu m$ via middle TSV and beyond. In: 2015 IEEE 65th

Electronic Components and Technology Conference (ECTC), pp. 66–72. IEEE.

48 Tanaka, T., Iwashita, M., Toshima, T. et al. (2015). Electro-less barrier/seed formation in high aspect ratio via. In: 2015 IEEE 65th Electronic Components and Technology Conference (ECTC), pp. 78–82. IEEE.

49 3D TSVs, aveni (2016). http://aveni.com/wet-deposition/3d-tsvs/ (accessed 31 May 2018).

50 Lee, S.H., Chen, K.N., and Lu, J.J.Q. (2011). Wafer-to-wafer alignment for three-dimensional integration: a review. *Journal of Microelectromechanical Systems* 20 (4): 885–898.

51 Lu, J.Q., McMahon, J.J., and Gutmann, R.J. (2012). Hybrid metal/polymer wafer bonding platform. In: *Handbook of Wafer Bonding*, 215–236. Wiley-VCH.

52 Cho, S. (2011). Technical challenges in TSV integration to Si. In: Sematech Symposium Korea, pp. 1–33.

53 Pangracious, V., Marrakchi, Z., and Mehrez, H. (2015). Three-dimensional integration: a more than moore technology. In: *Three-Dimensional Design Methodologies for Tree-based FPGA Architecture*, 13–41. Springer International Publishing.

54 Shigetou, A., Itoh, T., and Suga, T. (2006). Bumpless interconnect of Cu electrodes in millions-pins level. In: 56th Electronic Components and Technology Conference, 2006. Proceedings. pp. 4. IEEE.

55 Chen, K.N., Tan, C.S., Fan, A., and Reif, R. (2005). Copper bonded layers analysis and effects of copper surface conditions on bonding quality for three-dimensional integration. *Journal of Electronic Materials* 34 (12): 1464–1467.

56 Swinnen, B., Ruythooren, W., De Moor, P. et al. (2006). 3D integration by Cu-Cu thermo-compression bonding of extremely thinned bulk-Si die containing 10 μm pitch through-Si vias. In: International Electron Devices Meeting, 2006. IEDM'06. pp. 1–4. IEEE.

57 Huffman, A., Lannon, J., Lueck, M. et al. (2009). Fabrication and characterization of metal-to-metal interconnect structures for 3-D integration. *Journal of Instrumentation* 4 (03): P03006.

58 Fan, J., Lim, D.F., and Tan, C.S. (2013). Effects of surface treatment on the bonding quality of wafer-level Cu-to-Cu thermo-compression bonding for 3D integration. *Journal of Micromechanics and Microengineering* 23 (4): 045025.

59 Tan, C.S., Lim, D.F., Singh, S.G. et al. (2009). Cu–Cu diffusion bonding enhancement at low temperature by surface passivation using self-assembled monolayer of alkane-thiol. *Applied Physics Letters* 95 (19): 192108.

60 Lim, D.F., Wei, J., Leong, K.C., and Tan, C.S. (2013). Cu passivation for enhanced low temperature ($\leqslant 300\,°C$) bonding in 3D integration. *Microelectronic Engineering* 106: 144–148.

61 Panigrahi, A.K., Bonam, S., Ghosh, T. et al. (2016). Ultra-thin Ti passivation mediated breakthrough in high quality Cu-Cu bonding at low temperature and pressure. *Materials Letters* 169: 269–272.

62 Chaware, R., Hariharan, G., Lin, J. et al. (2015). Assembly challenges in developing 3D IC package with ultra high yield and high reliability. In: 2015 IEEE 65th Electronic Components and Technology Conference (ECTC), pp. 1447–1451. IEEE.
63 Tezzaron 3D-IC Microcontroller Prototype [Online]. (2016). http://www.tachyonsemi.com/OtherICs/3D-IC_8051_prototype.htm (accessed 11 February 2016).
64 Black, B., Annavaram, M., Brekelbaum, N. et al. (2006, December). Die stacking (3D) microarchitecture. In: *Proceedings of the 39th Annual IEEE/ACM International Symposium on Microarchitecture*, 469–479. IEEE Computer Society.
65 IBM Press Release [Online], in German. http://www-03.ibm.com/press/de/de/pressrelease/36129.wss (accessed 11 February 2016).
66 Kim, D.H., Athikulwongse, K., Healy, M. et al. (2012). 3D-MAPS: 3D massively parallel processor with stacked memory. In: 2012 IEEE International Solid-State Circuits Conference Digest of Technical Papers (ISSCC), pp. 188–190. IEEE.
67 Intel® Optane™ (2016). Supersonic memory revolution to take-off in 2016. http://www.intel.eu/content/www/eu/en/it-managers/non-volatile-memory-idf.html (accessed 11 February 2016).
68 Yip, L., Hariharan, G., Chaware, R. et al. (2017). Board level reliability optimization for 3D IC packages with extra large interposer. In: 2017 IEEE 67th Electronic Components and Technology Conference (ECTC), pp. 1269–1275. IEEE.
69 Fukushima, K., Yamaguchi, Y., Yasuda, M., and Nagata, S. (1970). An electronic model of the retina. *Proceedings of the IEEE* 58 (12): 1950–1951.
70 Rosenblatt, F. (1958). The perceptron: a probabilistic model for information storage and organization in the brain. *Psychological Review* 65 (6): 386.
71 Hey, T. (1999). Richard Feynman and Computation. *Contemporary Physics* 40 (4): 257–265.
72 Mead, C. and Ismail, M. (2012). *Analog VLSI Implementation of Neural Systems*, vol. 80. Springer Science & Business Media.
73 Markram, H., Lübke, J., Frotscher, M., and Sakmann, B. (1997). Regulation of synaptic efficacy by coincidence of postsynaptic APs and EPSPs. *Science* 275 (5297): 213–215.
74 Markram, H., Gerstner, W., and Sjöström, P.J. (2011). A history of spike-timing-dependent plasticity. *Frontiers in Synaptic Neuroscience* 3 (4): 1–24.
75 Bi, G.Q. and Poo, M.M. (1998). Synaptic modifications in cultured hippocampal neurons: dependence on spike timing, synaptic strength, and postsynaptic cell type. *Journal of Neuroscience* 18 (24): 10464–10472.
76 Morrison, A., Diesmann, M., and Gerstner, W. (2008). Phenomenological models of synaptic plasticity based on spike timing. *Biological Cybernetics* 98 (6): 459–478.
77 Raoux, S., Burr, G.W., Breitwisch, M.J. et al. (2008). Phase-change random access memory: a scalable technology. *IBM Journal of Research and Development* 52 (4.5): 465–479.

78 Burr, G.W., Brightsky, M.J., Sebastian, A. et al. (2016). Recent progress in phase-change memory technology. *IEEE Journal on Emerging and Selected Topics in Circuits and Systems* 6 (2): 146–162.

79 Suri, M., Bichler, O., Querlioz, D. et al. (2012). Physical aspects of low power synapses based on phase change memory devices. *Journal of Applied Physics* 112 (5): 054904.

80 Eryilmaz, S.B., Kuzum, D., Jeyasingh, R.G. et al. (2013). Experimental demonstration of array-level learning with phase change synaptic devices. In: 2013 IEEE International Electron Devices Meeting (IEDM), pp. 621–624. IEEE.

81 Valov, I., Waser, R., Jameson, J.R., and Kozicki, M.N. (2011). Electrochemical metallization memories—fundamentals, applications, prospects. *Nanotechnology* 22 (25): 254003.

82 Wong, H.S.P., Lee, H.Y., Yu, S. et al. (2012). Metal–oxide RRAM. *Proceedings of the IEEE* 100 (6): 1951–1970.

83 Burr, G.W., Shelby, R.M., Sebastian, A. et al. (2017). Neuromorphic computing using non-volatile memory. *Advances in Physics: X* 2 (1): 89–124.

84 Eryilmaz, S.B., Kuzum, D., Yu, S., and Wong, H.S.P. (2015). Device and system level design considerations for analog-non-volatile-memory based neuromorphic architectures. In: 2015 IEEE International Electron Devices Meeting (IEDM), pp. 64–67. IEEE.

85 Suri, M., Bichler, O., Hubert, Q. et al. (2012). Interface engineering of pcm for improved synaptic performance in neuromorphic systems. In: 2012 4th IEEE International Memory Workshop (IMW), pp. 1–4. IEEE.

86 Suri, M., Bichler, O., Hubert, Q. et al. (2013). Addition of HfO_2 interface layer for improved synaptic performance of phase change memory (PCM) devices. *Solid-State Electronics* 79: 227–232.

87 Ambrogio, S., Ciocchini, N., Laudato, M. et al. (2016). Unsupervised learning by spike timing dependent plasticity in phase change memory (PCM) synapses. *Frontiers in Neuroscience* 10 (56): 1–12.

88 Jackson, B.L., Rajendran, B., Corrado, G.S. et al. (2013). Nanoscale electronic synapses using phase change devices. *ACM Journal on Emerging Technologies in Computing Systems (JETC)* 9 (2): 12.

89 Ohno, T., Hasegawa, T., Nayak, A. et al. (2011). Sensory and short-term memory formations observed in a Ag_2S gap-type atomic switch. *Applied Physics Letters* 99 (20): 203108.

90 Ohno, T., Hasegawa, T., Tsuruoka, T. et al. (2011). Short-term plasticity and long-term potentiation mimicked in single inorganic synapses. *Nature Materials* 10 (8): 591–595.

91 Jo, S.H., Chang, T., Ebong, I. et al. (2010). Nanoscale memristor device as synapse in neuromorphic systems. *Nano Letters* 10 (4): 1297–1301.

92 Suri, M., Bichler, O., Querlioz, D. et al. (2012). CBRAM devices as binary synapses for low-power stochastic neuromorphic systems: auditory (cochlea) and visual (retina) cognitive processing applications. In: 2012 IEEE International Electron Devices Meeting (IEDM), pp. 10–13. IEEE.

93 Ziegler, M., Soni, R., Patelczyk, T. et al. (2012). An electronic version of Pavlov's dog. *Advanced Functional Materials* 22 (13): 2744–2749.

94 Sillin, H.O., Aguilera, R., Shieh, H.H. et al. (2013). A theoretical and experimental study of neuromorphic atomic switch networks for reservoir computing. *Nanotechnology* 24 (38): 384004.
95 Xu, Z., Mohanty, A., Chen, P.Y. et al. (2014). Parallel programming of resistive cross-point array for synaptic plasticity. *Procedia Computer Science* 41: 126–133.
96 Orchard, G., Lagorce, X., Posch, C. et al. (2015). Real-time event-driven spiking neural network object recognition on the spinnaker platform. In: 2015 IEEE International Symposium on Circuits and Systems (ISCAS), pp. 2413–2416. IEEE.
97 Sheridan, P.M., Cai, F., Du, C. et al. (2017). Sparse coding with memristor networks. *Nature Nanotechnology* 12: 784–789.
98 Choi, H., Jung, H., Lee, J. et al. (2009). An electrically modifiable synapse array of resistive switching memory. *Nanotechnology* 20 (34): 345201.
99 Yu, S., Wu, Y., Jeyasingh, R. et al. (2011). An electronic synapse device based on metal oxide resistive switching memory for neuromorphic computation. *IEEE Transactions on Electron Devices* 58 (8): 2729–2737.
100 Wu, Y., Yu, S., Wong, H.S.P. et al. (2012). AlO_x-based resistive switching device with gradual resistance modulation for neuromorphic device application. In: 2012 4th IEEE International Memory Workshop (IMW), pp. 1–4. IEEE.
101 Yu, S., Gao, B., Fang, Z. et al. (2013). A low energy oxide-based electronic synaptic device for neuromorphic visual systems with tolerance to device variation. *Advanced Materials* 25 (12): 1774–1779.
102 Yu, S., Gao, B., Fang, Z. et al. (2012). A neuromorphic visual system using RRAM synaptic devices with Sub-pJ energy and tolerance to variability: Experimental characterization and large-scale modeling. In: 2012 IEEE International Electron Devices Meeting (IEDM), pp. 10–14. IEEE.
103 Piccolboni, G., Molas, G., Portal, J.M. et al. (2015). Investigation of the potentialities of Vertical Resistive RAM (VRRAM) for neuromorphic applications. In: 2015 IEEE International Electron Devices Meeting (IEDM), pp. 447–450. IEEE.
104 Merolla, P.A., Arthur, J.V., Alvarez-Icaza, R. et al. (2014). A million spiking-neuron integrated circuit with a scalable communication network and interface. *Science* 345 (6197): 668–673.
105 Deutsch, D. (1985, July). Quantum theory, the Church-Turing principle and the universal quantum computer. *Proceedings of the Royal Society of London A: Mathematical, Physical and Engineering Sciences* 400 (1818): 97–117, The Royal Society.
106 Shor, P.W. (1999). Polynomial-time algorithms for prime factorization and discrete logarithms on a quantum computer. *SIAM Review* 41 (2): 303–332.
107 Grover, L.K. (1997). Quantum mechanics helps in searching for a needle in a haystack. *Physical Review Letters* 79 (2): 325.
108 Veldhorst, M. et al. (2014). An addressable quantum dot qubit with fault-tolerant fidelity. *Nature Nanotechnology* 9: 981–985.

109 Itoh, K.M. and Watanabe, H. (2014). Isotope engineering of silicon and diamond for quantum computing and sensing applications. *MRS Communications* 4 (4): 143–157.

110 Maune, B.M., Borselli, M.G., Huang, B. et al. (2012). Coherent singlet-triplet oscillations in a silicon-based double quantum dot. *Nature* 481 (7381): 344–347.

111 DiVincenzo, D.P. (2000). *The Physical Implementation of Quantum Computation*. Wiley-VCH. *arXiv preprint quant-ph/0002077*.

112 Nielsen, M.A. and Chuang, I.L. (2000). *Quantum computation and Quantum Information*. Cambridge: Cambridge University Press.

113 Kok, P., Munro, W.J., Nemoto, K. et al. (2007). Linear optical quantum computing with photonic qubits. *Reviews of Modern Physics* 79 (1): 135.

114 Brown, K.R., Wilson, A.C., Colombe, Y. et al. (2011). Single-qubit-gate error below 10– 4 in a trapped ion. *Physical Review A* 84 (3): 030303.

115 Waldherr, G., Wang, Y., Zaiser, S. et al. (2014). Quantum error correction in a solid-state hybrid spin register. *Nature* 506 (7487): 204–207.

116 Dolde, F., Bergholm, V., Wang, Y. et al. (2014). High-fidelity spin entanglement using optimal control. *Nature communications* 5: 3371.

117 Muhonen, J.T., Dehollain, J.P., Laucht, A. et al. (2014). Storing quantum information for 30 seconds in a nanoelectronic device. *Nature nanotechnology* 9 (12): 986–991.

118 Loss, D. and DiVincenzo, D.P. (1998). Quantum computation with quantum dots. *Physical Review A* 57 (1): 120.

119 Koppens, F.H.L., Buizert, C., Tielrooij, K.J. et al. (2006). Driven coherent oscillations of a single electron spin in a quantum dot. *Nature* 442 (7104): 766–771.

120 Petta, J.R., Johnson, A.C., Taylor, J.M. et al. (2005). Coherent manipulation of coupled electron spins in semiconductor quantum dots. *Science* 309 (5744): 2180–2184.

121 Veldhorst, M., Yang, C.H., Hwang, J.C.C. et al. (2015). A two-qubit logic gate in silicon. *Nature* 526 (7573): 410–414.

122 Watson, T.F., Weber, B., Hsueh, Y.L. et al. (2017). Atomically engineered electron spin lifetimes of 30 s in silicon. *Science Advances* 3 (3): e1602811.

123 Devoret, M.H. and Schoelkopf, R.J. (2013). Superconducting circuits for quantum information: an outlook. *Science* 339 (6124): 1169–1174.

124 Vion, D., Aassime, A., Cottet, A. et al. (2002). Manipulating the quantum state of an electrical circuit. *Science* 296 (5569): 886–889.

125 Manucharyan, V.E., Koch, J., Glazman, L.I., and Devoret, M.H. (2009). Fluxonium: single cooper-pair circuit free of charge offsets. *Science* 326 (5949): 113–116.

126 Steffen, M., Kumar, S., DiVincenzo, D.P. et al. (2010). High-coherence hybrid superconducting qubit. *Physical Review Letters* 105 (10): 100502.

127 Vandersypen, L. and van Leeuwenhoek, A. (2017). 1.4 Quantum computing-the next challenge in circuit and system design. In: 2017 IEEE International Solid-State Circuits Conference (ISSCC), pp. 24–29. IEEE.

2

Nanowire Field-Effect Transistors

Debarghya Sarkar[1], Ivan S. Esqueda[2], and Rehan Kapadia[1]

[1] *Ming Hsieh Department of Electrical Engineering, University of Southern California, Los Angeles, CA 90089, USA*
[2] *Information Sciences Institute, University of Southern California, Marina del Rey, CA 90292, USA*

2.1 General Scaling Laws Leading to Nanowire Architectures

2.1.1 Scaling of Planar Devices and Off-state Leakage Current

Over the past several decades of metal-oxide semiconductor field-effect transistor (MOSFET) existence [1] and development, the semiconductor device community has continually been working to improve device characteristics, while reducing device dimensions for increased density of device integration to enhance system functionalities.

In the planar MOSFET device (Figure 2.1a), as the channel length is reduced keeping any other design parameter constant, the depletion regions at the source/drain and channel junctions become comparable to the intended channel length, thereby resulting in further shortened "effective channels," as schematically depicted in Figure 2.1b. This leads to "short-channel" effects where the gate electrode starts losing control of the channel current switching, and is more dominated by the drain. Some of these effects include reduction of the junction barrier between source and channel (drain-induced barrier lowering (DIBL)) as drain voltage is increased, in turn leading to reduction in gate threshold voltage (Figure 2.1c) and increased leakage current [3–5].

In order to address these issues, Dennard et al. [2] introduced a scaling design rule (Table 2.1) that would enable scaling of devices while addressing the short-channel effects.

The Dennard scaling rule advised scaling in dimensions, where each linear dimension such as channel length, width, junction depth, and gate insulator thickness would be scaled by the same factor. Further, the voltage applied to the device would be reduced and the substrate doping concentration increased by the same amount. If κ is the scaling factor, this would increase integration density by κ^2, increase speed by κ, reduce power consumption per transistor by κ^2, while keeping the same electric field and power density in the channel.

Advanced Nanoelectronics: Post-Silicon Materials and Devices,
First Edition. Edited by Muhammad Mustafa Hussain.

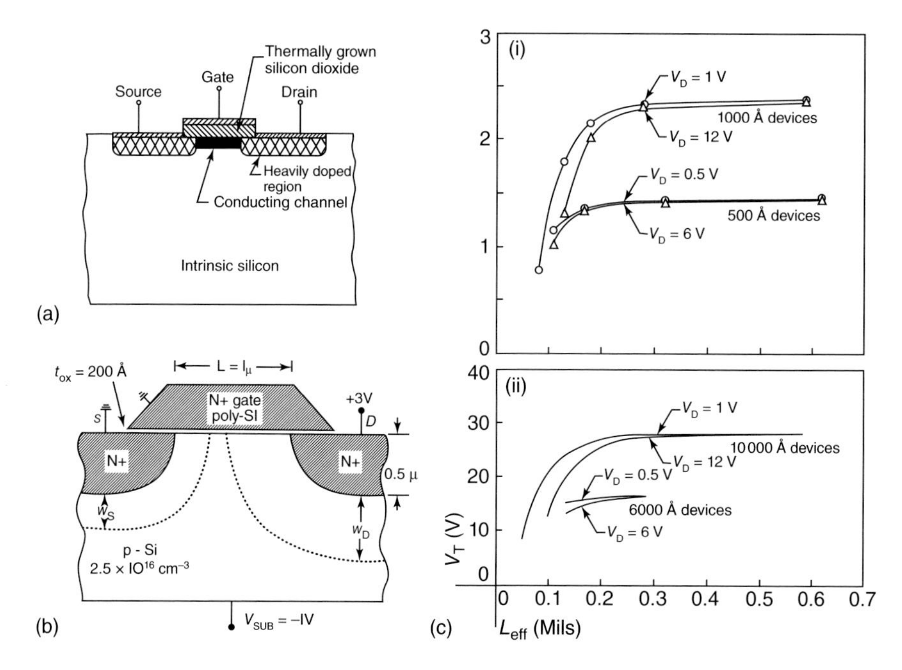

Figure 2.1 MOSFET structure and short-channel effects. (a) Schematic of MOSFET (or insulated-gate field-effect transistor (IGFET)) as it appeared in Hofstein and Heiman [1]. Source: Hofstein and Heiman 1963 [1]. Reproduced with permission of IEEE. (b) Schematic of effective channel length reduction with source/drain depletion widths being comparable to channel length. Source: Dennard et al. 1974 [2]. Reproduced with permission of IEEE. (c) Short-channel effect of gate threshold voltage reduction with reduction of nominal channel length. Higher drain voltage causes more reduction of threshold voltage (DIBL). Source: Critchlow et al. 1973 [3]. Reproduced with permission of IBM.

Table 2.1 Dennard scaling rules.

Scaling results for circuit performance	
Device or circuit parameter	**Scaling factor**
Device dimension t_{ox}, L, W	$1/\kappa$
Doping concentration N_a	κ
Voltage V	$1/\kappa$
Current I	$1/\kappa$
Capacitance $\varepsilon A/t$	$1/\kappa$
Delay time/circuit VC/I	$1/\kappa$
Power dissipation/circuit VI	$1/\kappa^2$
Power density VI/A	1

Source: Dennard et al. 1974 [2]. Reproduced with permission of IEEE.

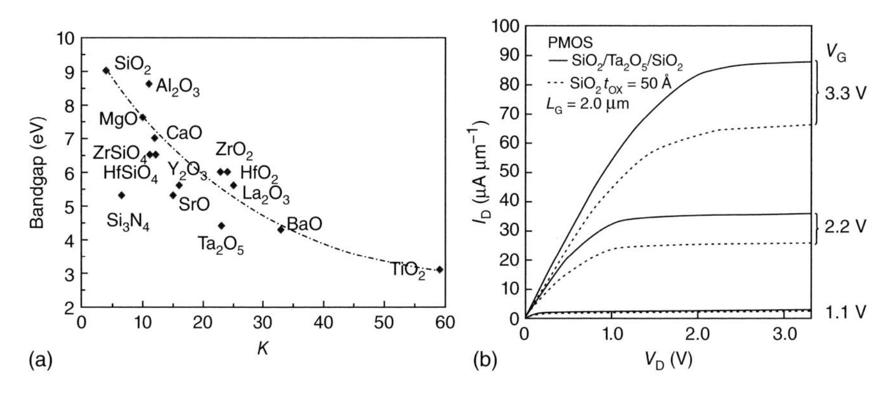

Figure 2.2 High-k dielectrics and effect on output characteristics. (a) Several common dielectrics in the bandgap and relative permittivity landscape. Source: John 2006 [6]. Reproduced with permission of IOP Publishing. (b) Increased drain current drive and saturation with incorporation of high-k Ta_2O_5. Source: Wilk et al. 2001 [7]. Reproduced with permission of AIP Publishing.

This scaling rule could be very well followed to produce new generations of devices with improved performance until the conventionally used material limits were reached, such as impossibility of scaling dielectric thickness without inducing breakdown, which would have otherwise been needed for the small channel length transistor.

These issues led to the discovery and engineering of other material properties, such as usage of high-k dielectrics as the gate insulator (several common examples are shown in Figure 2.2a). This enabled the dielectric to be thicker and yet have an "effective oxide thickness" of much less than that possible with SiO_2, reducing leakage and breakdown [6, 7]. Higher gate dielectric capacitance and thus higher drain currents could be obtained (Figure 2.2b) using high-k dielectrics compared to same thickness of SiO_2. Carrier mobility improvements were achieved by introducing strain and thus increasing current density [8, 9].

Nevertheless, at very short channel lengths, below 20 nm, it becomes increasingly difficult to engineer planar transistors that suitably address the short-channel effects while continuing the gate length reduction. Thus the state-of-the-art planar devices all have short-channel effects, and a redesign of the transistor is necessary for continued scaling.

2.1.2 FinFET and UTB Devices for Improved Electrostatics

It was realized that the issue of short-channel effects may be reduced by modifying the geometry of the device in a way that would enhance the electrostatic control of gate electric field over the source-drain electric field. For the same gate capacitance, the overall channel control is greater if the device's body is made thinner [10], and if the gate can physically wrap around the channel. Both of these concepts have been implemented, and are popularly known in the literature as ultrathin body (UTB) devices, and Fin field-effect transistors (FinFETs).

UTB devices (schematically shown in Figure 2.3a) of gate length less than 100 nm have traditionally been fabricated from a variety of starting substrates such as silicon-on-insulator (SOI) wafers, epitaxially transferred III–Vs, and

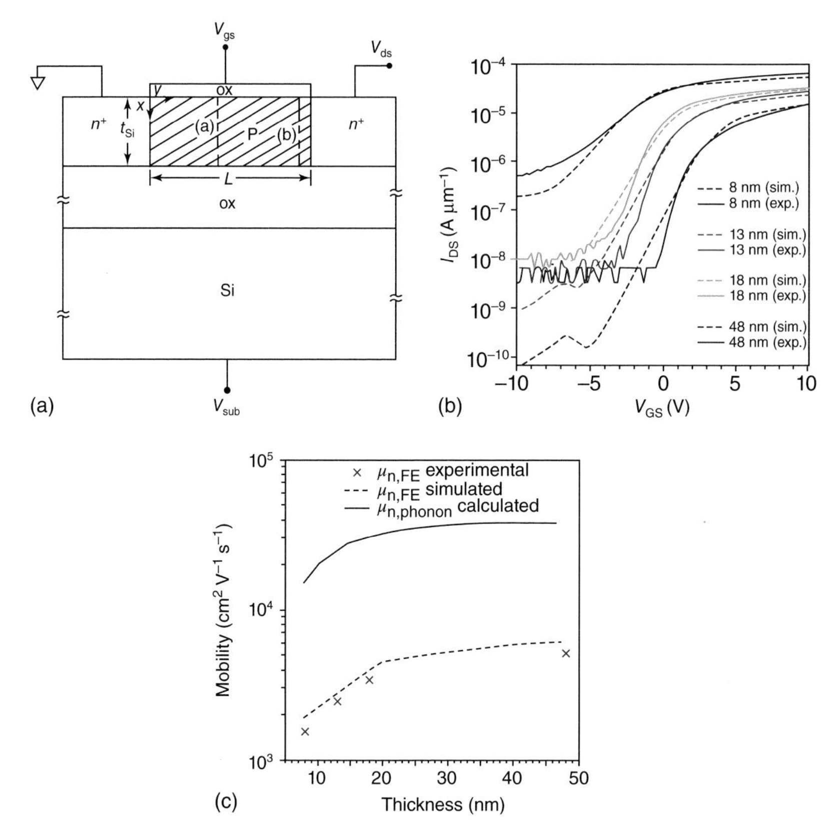

Figure 2.3 UTB devices. (a) Schematic of an SOI UTB MOSFET. Source: Young 1989 [10]. Reproduced with permission of IEEE. (b) Transfer characteristics of InAs UTB devices. ON–OFF ratio increases and subthreshold swing (SS) decreases with decrease in channel thickness. (c) Mobility decreases with increased proportion of surface roughness scattering with channel thickness reduction. Source: Ko et al. 2010 [11]. Figure 2.3b,c reproduced with permission of Springer Nature.

deposited and laterally crystallized ultrathin (~15–20 nm) Si [12] or SiGe [13] layers on patterned SOI wafers. The ultrathin layer acts as the channel, enabling significantly improved gate control. More recently, the same concept has been extended to UTB field-effect transistors (FETs) fabricated from epitaxially grown compound semiconductor layers (18- and 48-nm thick) transferred onto Si/SiO_2 wafers, termed XOI [11, 14–19], or "X-on-insulator." An improvement in ON–OFF ratio and subthreshold slope (Figure 2.3b) is seen with increased electrostatic control as the channel is made thinner.

It is worthwhile to note that because of the inherently thin nature of these channels, quantum effects start having a non-negligible effect on the charge dynamics, and hence on the transport characteristics [20]. Increased proportion of surface roughness also leads to decrease in mobility as the channel thickness is reduced (Figure 2.3c).

FinFETs have a vertical "fin" etched out of a silicon bulk (for bulk FinFET) or etched out of the top wafer-bonded silicon layer of an SOI wafer (for SOI FinFET, Figure 2.4a). This fin is surrounded on three sides by the gate dielectric followed by the gate metal, which results in the depletion layer from the gate field closing in from multiple directions, essentially leading to higher electrostatic control [21, 26, 27].

In a commercial FinFET (Figure 2.4d), the fin is made of relatively high aspect ratio, where the width of the fin is made thin to improve electrostatic control, whereas the height of the fin is kept high to have high current drive [24] (output and transfer characteristics are shown in Figure 2.4g,h). Although it has the gate on three sides, it has many similarities to a double-side gate device, but typically offers better scaling performance. The exact cross-sectional geometry of the fin also has a significant effect toward device performance. With the same base width, a trapezoidal fin would offer better electrostatic control, and thus reduced short-channel effects than a rectangular fin [28]. For even higher electrostatic control, further channel dimension engineering, such as reducing the fin height, and/or giving rise to a rib-waveguide structure of Si channel to design triple-gate [25] or Ω-gate device [29], are used.

2.1.3 Nanowires as the Ultimate Limit of Electrostatic Control

A channel thinned in both dimensions perpendicular to current flow, and a gate physically fully wrapping around the channel, would thus give the highest level of electrostatic control. This thin-channel gate-all-around (GAA) [22] architecture is essentially provided by nanowire field-effect transistors (NWFETs; Figure 2.4b,c,f). The limit of this architecture, leading to the best electrostatic control, would be for a cylindrical nanowire [23], with gate control improving as the wire diameter reduces.

To quantify the electrostatic field effect of the gate on the channel carrier concentration, a parameter λ or *natural length* is often used. It depends on device geometry and material parameters such as the thickness and relative permittivity of the gate dielectric and the channel. For the same effective gate length, a device geometry with a smaller natural length can be scaled further than a device with a larger natural length.

For comparison, for the same semiconductor material (Si), gate dielectric (SiO_2), and the same dimensions for channel (100 nm) and for gate dielectric (10 nm), the natural length of a planar single-gate MOSFET [30] is about 55 nm, that of a triple-gate FET [31] (Figure 2.4e) is 43 nm, and of a cylindrical nanowire GAA FET [32] (Figure 2.4c,f) is 31 nm, with the specific numbers arising from the details of the channel thickness and wire diameter. Figure 2.4i gives a similar comparative view of transconductance variation between planar, double-gate, and GAA architectures. For a more extensive discussion on the evolution of electrostatic control and its mathematical expressions in different FET device geometries, the reader is referred to other books on the subject such as [33–35].

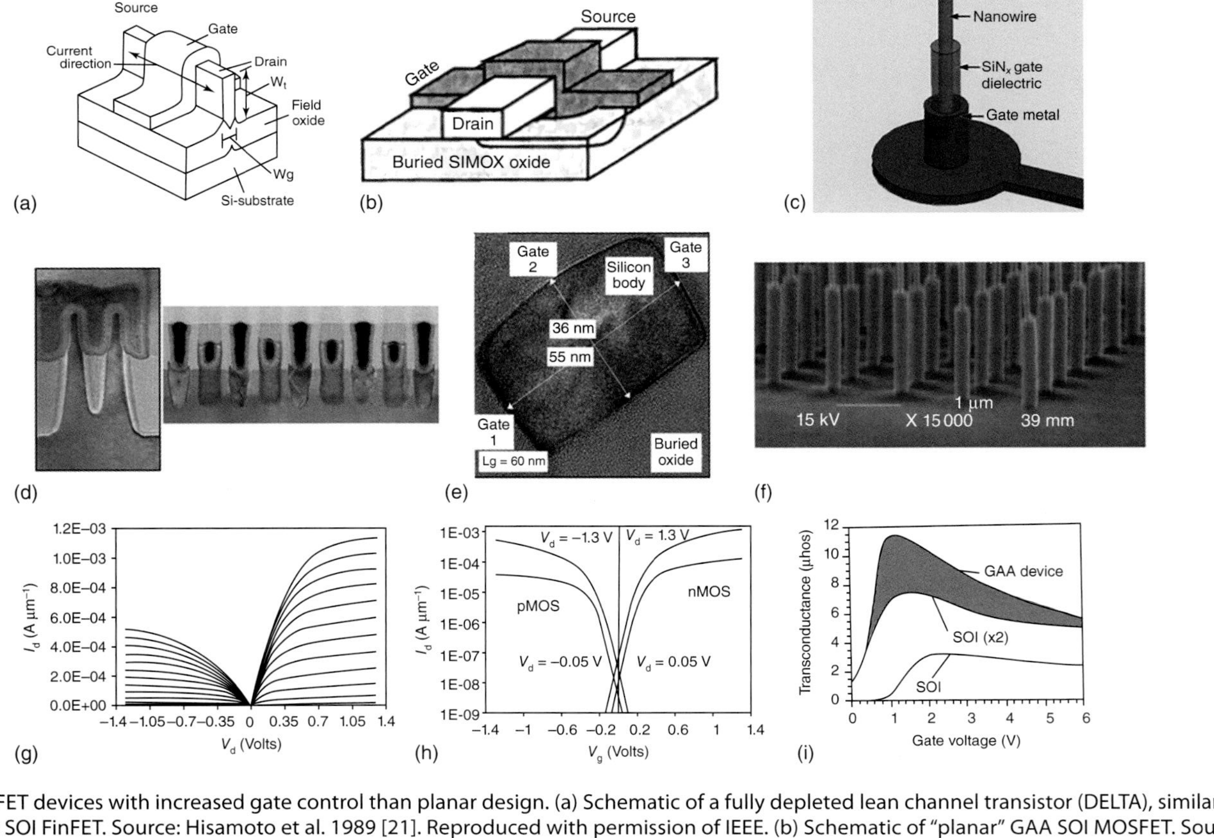

Figure 2.4 MOSFET devices with increased gate control than planar design. (a) Schematic of a fully depleted lean channel transistor (DELTA), similar to presently known SOI FinFET. Source: Hisamoto et al. 1989 [21]. Reproduced with permission of IEEE. (b) Schematic of "planar" GAA SOI MOSFET. Source: Colinge et al. 1990 [22]. Reproduced with permission of IEEE. (c) Schematic of vertical Si nanowire GAA MOSFET. Source: Bryllert et al. 2006 [23]. Reproduced with permission of IOP Publishing. (d) TEM micrograph of Intel 14 nm second generation FinFET. Source: Natarajan et al. 2014 [24]. Reproduced with permission of IEEE. (e) TEM cross-section of tri-gate Si MOS device. Source: Doyle 2003 [25]. Reproduced with permission of IEEE. (f) SEM micrograph showing nanowires with wrap gates. Source: Bryllert et al. 2006 [23]. Reproduced with permission of IOP Publishing. (g) Output and (h) Transfer characteristics of 60 nm tri-gate CMOS devices with DIBL = 48 mV V^{-1}, SS = 69.5 mV/decade, ON–OFF = ~1e4, ON current = ~500 mA mm^{-1}. Source: Doyle 2003 [25]. Reproduced with permission of IEEE. (i) Increasing transconductance (V_{ds} = 100 mV) from a single gate to double gate to GAA MOS device. Source: Colinge et al. 1990 [22]. Reproduced with permission of IEEE.

2.1.4 Quantum Effects

From the preceding discussion, it is quite clear that reduced cross-sectional dimensions lead to higher electrostatic control of the channel. As we continuously reduce the wire lateral dimension to an extent that it becomes comparable to the lateral spread of the electron wavefunction, quantum confinement effects start to play a significant role in determining the charge dynamics in the channel.

In a globally uniform potential as in the bulk of a semiconductor, the solution of Schrodinger's equation for the electron wavefunction gives a traveling wave in all three directions. We may note that at the surface of the semiconductor, the electron wavefunction sees a large step change in the potential (from semiconductor conduction band to the vacuum level), so that Schrodinger's equation gives an exponentially decaying bound-state solution in that direction. As we shrink down the semiconductor piece in two of the three dimensions, to an extent that the spatial difference between the vacuum level on both sides is comparable to the length scale that an electron in the middle of the channel would have a non-negligible probability of existence outside the channel, the electron wavefunction becomes bound in the two dimensions, while still being that of a traveling wave in the third dimension (as the electron still sees a uniform potential in that direction). The potential profile along any of the two directions of the channel along which it has been reduced is that of a finite potential well (although with a very high potential barrier). This gives rise to the formation of a finite number of allowed discrete energy states associated with the wavevector in the directions of confinement. The other direction still has a continuum of energy states (energy band) associated with it. This gives rise to the formation of energy sub-bands, where the energy levels are continuous, except for having step rises at discrete energy levels associated with the quantized electron in the two lateral directions [36]. The density of states in each sub-band falls off as the inverse square root of the energy, and having van-Hove singularities at the beginning of each sub-band (Figure 2.5).

The dimension at which the quantization effects become important depends on the particular material. Channels made of materials with smaller effective electron mass show quantization effects at larger dimensions than that made with materials of larger electron mass.

The projection in the freely propagating direction, of the electron wavefunction associated with the lowest sub-band, is a cosine curve centered in the middle of the channel, with exponentially falling off tails at the surface. Any electron in the first sub-band has this same cosine curve wavefunction along any of the lateral cross-sections (while having varying wavefunctions in the propagation direction). Thus, the probability density associated with any electron in the first sub-band is localized in the middle of the channel [37] (Figure 2.6).

If we imagine filling the channel with electrons gradually, starting with zero electrons, the first several electrons will start occupying states in the first sub-band, until it gets filled, then populating the second sub-band, and so on. Thus, at low electron concentrations, if the chemical potential lies such that primarily the first sub-band gets filled, then all these electrons have a high probability density in the middle of the channel. This has important implications for surface scattering as well as gate control. Note that in a planar Si MOSFET

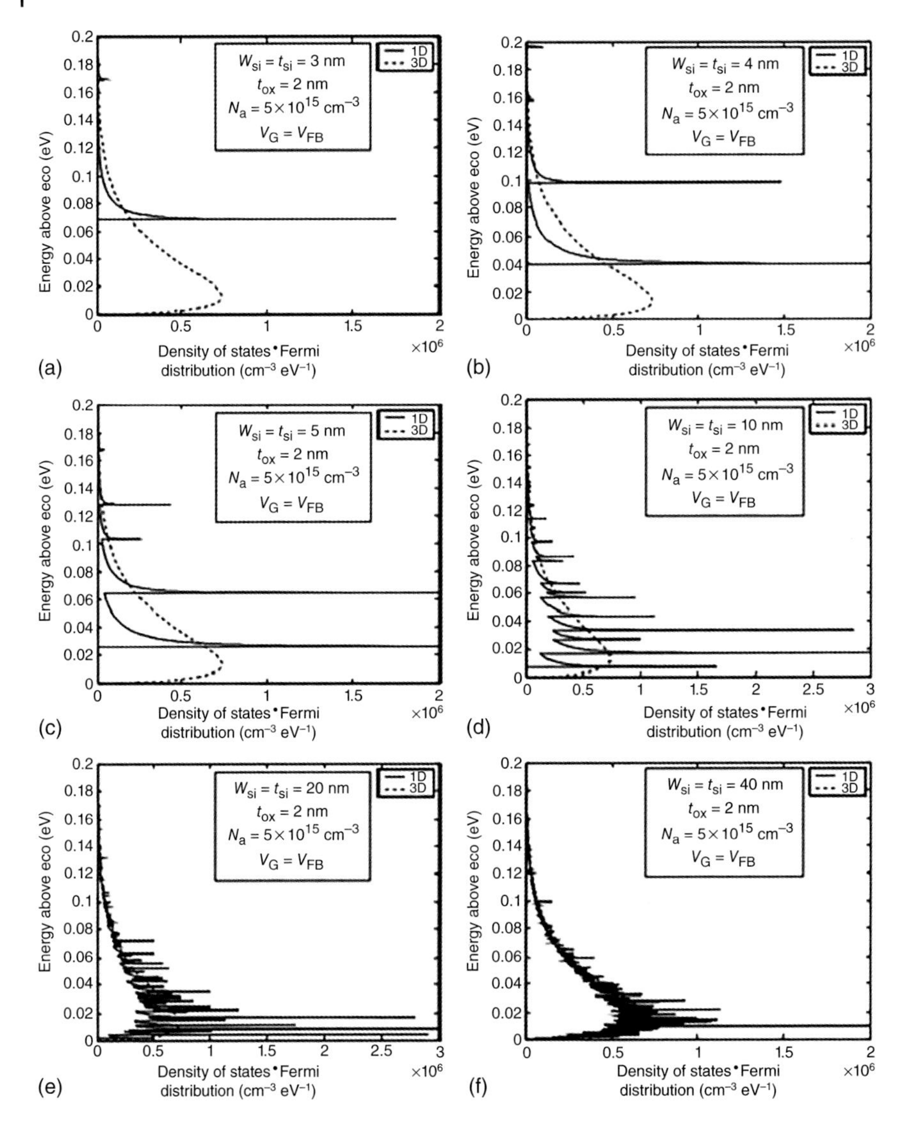

Figure 2.5 Energy spectrum of electron probability density in the conduction band of a square Si nanowire of sides varying from (a) 3 nm through (f) 40 nm. (a) A single sub-band is present for a 2-nm-wide nanowire. (c) Four sub-bands 5 nm wide, while (f) 40-nm nanowire electron occupancy is similar to bulk. Source: Colinge 2007 [36]. Reproduced with permission of Elsevier.

the electrons are typically confined in sub-bands near the surface, and surface scattering plays a significant role in mobility degradation.

Consider the case where the channel lateral dimensions are small enough such that the quantization of energy states is large enough to observe. In the ballistic limit, each discrete energy sub-band can be considered as an independent channel for electron transmission. We can then calculate the electron current as the integrated product of velocity, transmission coefficient, and number of

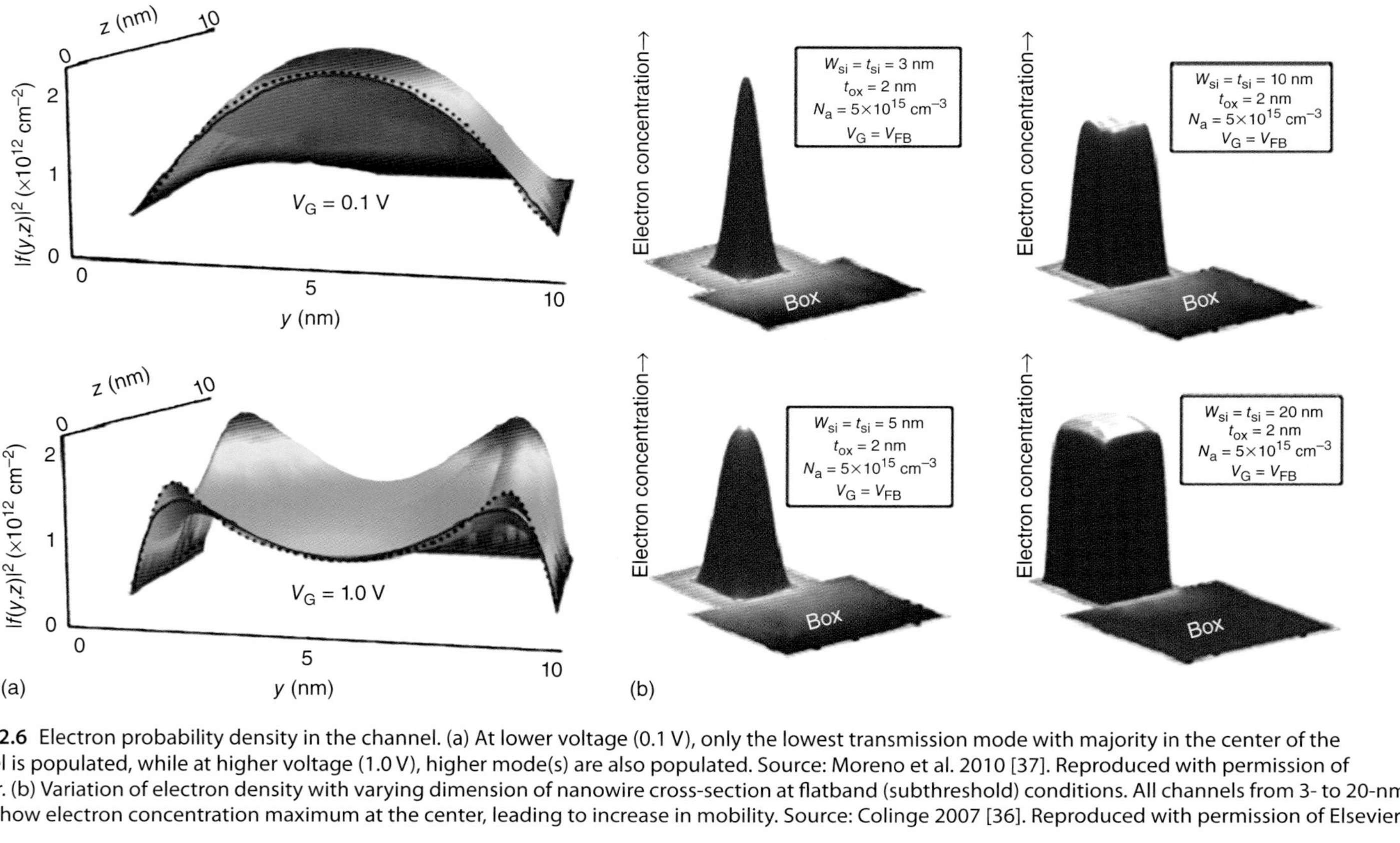

Figure 2.6 Electron probability density in the channel. (a) At lower voltage (0.1 V), only the lowest transmission mode with majority in the center of the channel is populated, while at higher voltage (1.0 V), higher mode(s) are also populated. Source: Moreno et al. 2010 [37]. Reproduced with permission of Elsevier. (b) Variation of electron density with varying dimension of nanowire cross-section at flatband (subthreshold) conditions. All channels from 3- to 20-nm width show electron concentration maximum at the center, leading to increase in mobility. Source: Colinge 2007 [36]. Reproduced with permission of Elsevier.

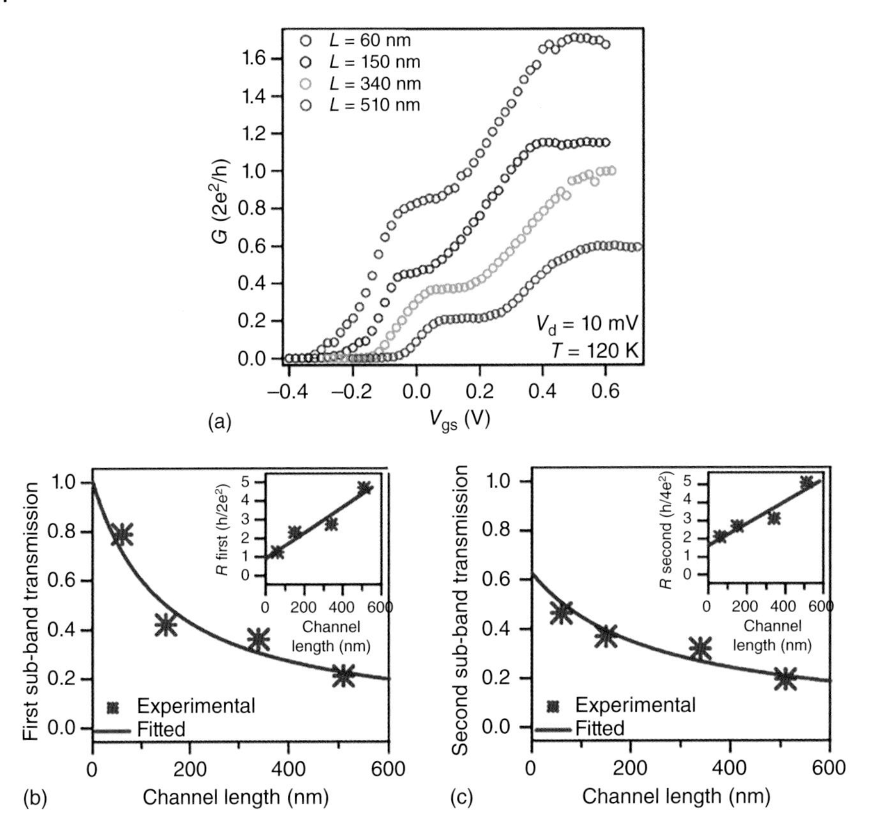

Figure 2.7 Quantum conductance. (a) Variation of conductance with increasing voltage. The staircase feature indicates gradual filling of sub-bands. Transmission probability variation with channel length for (b) first and (c) second sub-band. Insets represent the corresponding resistance variation. Source: Chuang et al. 2013 [39]. Reproduced with permission of ACS.

1D sub-bands. Importantly, for unity transmission, conductance for a single 1D sub-band is constant, irrespective of the traveling wavevector of the electron [38]. Thus, the conductance is quantized, and the maximum conductance per 1D sub-band is $e^2/\pi\hbar$. Several works, such as [39, 40], have reported the experimental observation of the quantum conductance (Figure 2.7). Notably, both these references are based on InAs, which, due to the low effective electron mass, leads to the expression of quantum effects at higher dimensions than a much higher effective electron mass material such as Si.

The presence of such discrete energy sub-bands also gives rise to quantum capacitance. Referring back to our model of filling the nanowire with charges starting from zero, the first sub-band gets filled as the gate voltage is increased. So the charge density in the channel increases until the first sub-band gets filled. Any further increase in gate voltage does not result in addition of more charge until the chemical potential reaches the next sub-band. Thus, in the “sub-bandgap” region, the charge density remains constant, while the voltage still increases. This gives rise to step changes in the capacitance-voltage characteristics, with the rising edge corresponding to charge getting added, and the plateau corresponding to charge remaining constant while the system passes through the sub-bandgap region [41, 42].

As a result of quantum confinement, the lowest occupied energy state by electrons in the conduction band is higher than the energy at the bottom of the conduction band. The same holds true for holes in the valence band, where the ground-state energy is higher than the zero energy of the holes (at the top of the valence band). This leads to an effective increase in the bandgap of the semiconductor. The effective bandgap increases with reducing dimensions of the nanowire. Materials such as InSb or InAs show quantization effects at around 10–20 nm diameter, whereas for Si, we need to go down to about 4 nm to see the effect.

2.1.5 Drive Current

It may be worth mentioning that while GAA nanowires possess much higher gate control than do planar devices, they have a lower drive current. It has been theoretically predicted that nanotubes with core–shell-multigate geometry are poised to bring forth the best of both worlds, combining the superior electrostatic control of the GAA structure with the high drive current of planar MOSFET [43].

2.2 Nanowire Growth and Device Fabrication Approaches

2.2.1 Bottom-up VLS Growth

The most widely used bottom-up growth technique is the vapor–liquid–solid (VLS) method, the controlled experiments and mechanism of which was first explained by Wagner and Ellis in 1964 [44] (Figure 2.8), and several fundamental aspects of which were later developed by Givargizov [45]. While it was described in the previous section that a reduced cross-sectional area would lead to higher electrostatic control, it may be noted that there was thought to be a critical diameter (which varies with growth condition), determined by the Gibbs–Thomson effect, below which the VLS growth rate abruptly vanishes. This diameter was thought to vary anywhere between 200 and 20 nm depending on the supersaturation resulting from the growth condition. However, once it was found by Lieber et al. that nanowires down to the few nanometer scale could be grown [46–55], there was a surge in the number of groups working on these materials. Furthermore, recent developments in the VLS method have shown that lateral nanowires and thin films are also possible [56–63].

Any VLS growth requires the presence of a liquid metal alloy seed which acts as the nucleation site and driver of the growth of the nanowire. Single-element materials like Si or Ge nanowires are grown using metals such as Au, Ag, Pt, and so on, as catalysts. The metal catalyst array is usually pre-patterned on the substrate using common fabrication techniques such as lithography, evaporation, and lift-off. A metal-semiconductor eutectic alloy is formed by heating the

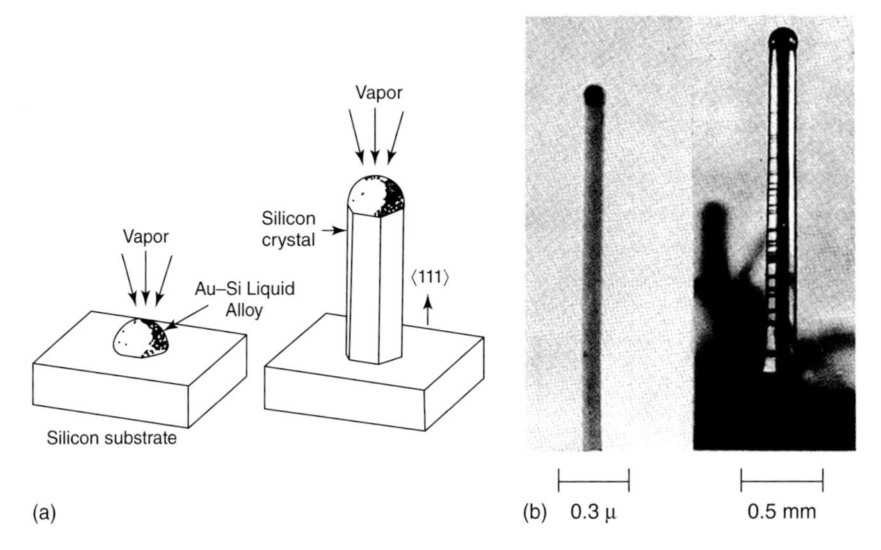

Figure 2.8 Vapor–liquid–solid growth (VLS). (a) Schematic of VLS mechanism as proposed by Wagner and Ellis. (b) SEM micrograph of VLS-grown microwires of different dimensions. Source: Fahad and Hussain 2012 [43]. Reproduced with permission of Springer Nature.

metal-patterned substrate, in the presence of the semiconductor precursor. It may be noted that as the mass of the catalyst metal remains the same, there is only a certain mass of the semiconductor that can get dissolved in the metal for a stable liquid alloy. Any further incorporation of the semiconductor in the melt results in precipitation of the semiconductor from the melt, and the liquid alloy gets back to a stable state, until it gets saturated once again and results in precipitation, and the cycle continues. The precipitation takes place across the bottom surface of the alloy seed, and thus the nanowire grows in size with the alloy seed perched on top.

The VLS growth of compound semiconductors such as InP or GaAs [64, 65] may not need a foreign metal catalyst to form the seed. Instead, one of the elements itself (indium or gallium in this case), works as the catalyst. This is important since some of the catalyst metals, such as gold, introduces deep level traps in the semiconductor.

Axial and radial heterostructures, as well as hybrid structures comprising both axial and radial junctions, can be and have been made using the VLS technique [66] (Figure 2.9).

FET devices made of nanowires from VLS growth summarily revolve around the concept of using the substrate as the drain, the wire as the channel to be surrounded by the gate for maximum electrostatic control, and the top of the wire as the source, contacted to the source electrode. Typical fabrication steps involve growth of the nanowires, surrounding it with the gate dielectric and the gate metal layers, preferentially etching the gate metal from regions away from the central part of the channel (by protecting the central part with resists and etching away the rest), depositing further insulating material as fillers, and then etching back to expose the tops of the wires, source doping if needed, and then depositing drain metal.

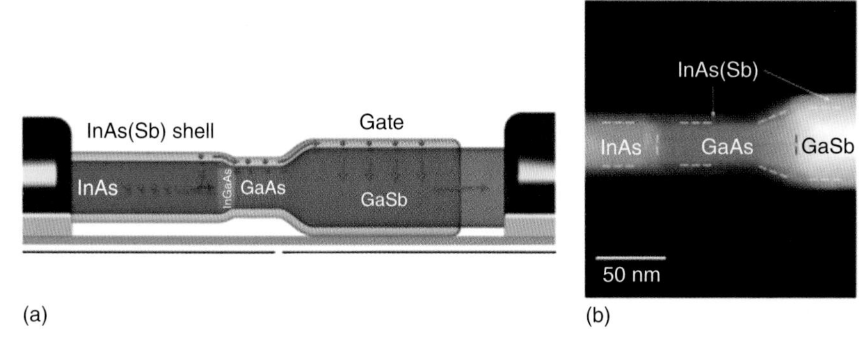

Figure 2.9 Hybrid structure of axial and radial heterostructures. (a) Schematic representation. (b) SEM micrograph. Radial heterostructure around a nanowire leads to the formation of clean 1D electron gas with minimized surface scattering due to absence of unpassivated bonds. Source: Dey et al. 2013 [66]. Reproduced with permission of ACS.

2.2.2 Top-down Oxidation

Another widely used nanowire fabrication technique is the top-down method, which fundamentally involves lithographic patterning and etching down to give rise to nanowire geometries. The material for making the nanowire is the (upper layer of the) starting substrate. For example, to make InAs nanowire on a Si wafer, one may start with an epitaxially grown InAs layer transferred onto Si.

For horizontal or vertical nanowires, the most common procedure would involve lithography of thin rectangular or circular patterns (respectively) so that the nanowire locations are protected by the resist, and the rest is exposed. Plasma etching techniques such as reactive ion etching with or without inductive coupling, using suitable precursor gases such as BCl_3/Cl_2, $CH_4/H_2/Cl_2$, Cl_2/N_2, or HBr are commonly used for anisotropic etching of III–Vs [67].

For horizontal nanowires, several other techniques are employed such as selective substrate etch, mainly to isolate the wire from the bulk of the substrate [68], in order to prevent leakage through the substrate, when the wires are used as channels for FETs, for example. While achieving horizontal nanowires with the advantages of the VLS growth mode (i.e., smooth sidewalls and nanometer geometries) is more challenging than the vertical nanowires, all commercial device fabrication is geared toward planar processing, such as gate dielectric deposition, followed by gate metal, selective patterning and etching the gate metal and dielectric from the source/drain regions, source/drain metal deposition, and annealing.

2.3 State-of-the-Art Nanowire Devices

2.3.1 Silicon Devices

Lieber's research group at Harvard University has been the pioneer in the early Si and Si/Ge nanowire devices, being the early demonstrator of high-performance

Si NWFETs in 2003. Later in 2006 [55], they also demonstrated a Ge/Si core/shell NWFET having scaled transconductance of 3.3 mS cm^{-2}, and on-current of 2.1 mA μm^{-2}, which fared much better than any single-element nanowire because of transport by a well-confined 1D electron gas with much reduced surface scattering from the crystalline Si/Ge interface.

While these were on horizontal nanowire geometries, a group at Berkeley [69] and a European group [70] independently developed a vertical NWFET with a GAA geometry. Both works showed current density per nanowire of similar order of magnitude, and one of them [69] showed an ON/OFF ratio of $\sim 10^5$, and transconductance comparable to high-performance SOI MOSFETs. Multilayer and multifunctional device integration were also demonstrated [50, 71], which showed repeatable performance not degrading with time and being subject to multiple fabrication cycles for the formation of the upper layers.

In 2010, Colinge et al. [72] demonstrated a junctionless nanowire Si FET (electron distribution and transfer characteristics represented in Figure 2.10), the concept of which was first proposed by Lilienfeld [73]. The devices showed excellent transistor performance, with an ON/OFF ratio of $\sim 10^6$, off-current less than 1 fA, and subthreshold slope of 64 mV/decade at room temperature. Being made of a doped Si channel, it is a normally-on device which can be increased by increasing the initial doping concentration, unlike a conventional MOSFET, where the on-current increases with the gate voltage. In a conventional MOSFET, the increase in current to increase speed (for same gate voltage) may be brought about by reducing the equivalent oxide thickness (EOT), which unfortunately also leads to increase of capacitance to effectively slow down the device. In a gated junctionless resistor, however, the current being independent of EOT, the speed actually can be increased by reducing the EOT. It may be noted that such a junctionless FET requires the channel to be of nanowire geometry, so that it can be entirely depleted by application of gate voltage.

2.3.2 III–V Devices

In 2001, Lieber's group demonstrated electronic and optoelectronic device performance of InP nanowires made from laser-assisted catalytic growth [46]. Although the FET device characteristics were not ideal, mainly owing to minimal gate control due to a thick (600 nm) gate dielectric, it set the stage for the widespread research on III–V nanowire device to follow in subsequent years. Later, they also reported GaN nanowire devices (Figure 2.11a) with carrier mobilities ranging from 150 to 650 $cm^2\ V^{-1}\ s^{-1}$ [49], which were statistically larger than the thin-film GaN mobility values between 100 and 300 $cm^2\ V^{-1}\ s^{-1}$.

In 2006, Bryllert et al. [23] addressed the issue of scalability that lay with the previous demonstrations, by following a single-substrate bottom-up approach. VLS-grown InAs nanowires of 80 nm diameter and gate length of 1 μm, were fabricated. A device with 121 channels in parallel (total cross-sectional area of 0.6 μm^2) gave transconductance of 6 mS and drive current of 6 mA, measured at zero gate voltage. In another similar publication [74], they showed high mobility (3000 $cm^2\ V^{-1}\ s^{-1}$) limited by contact resistance, subthreshold slope of

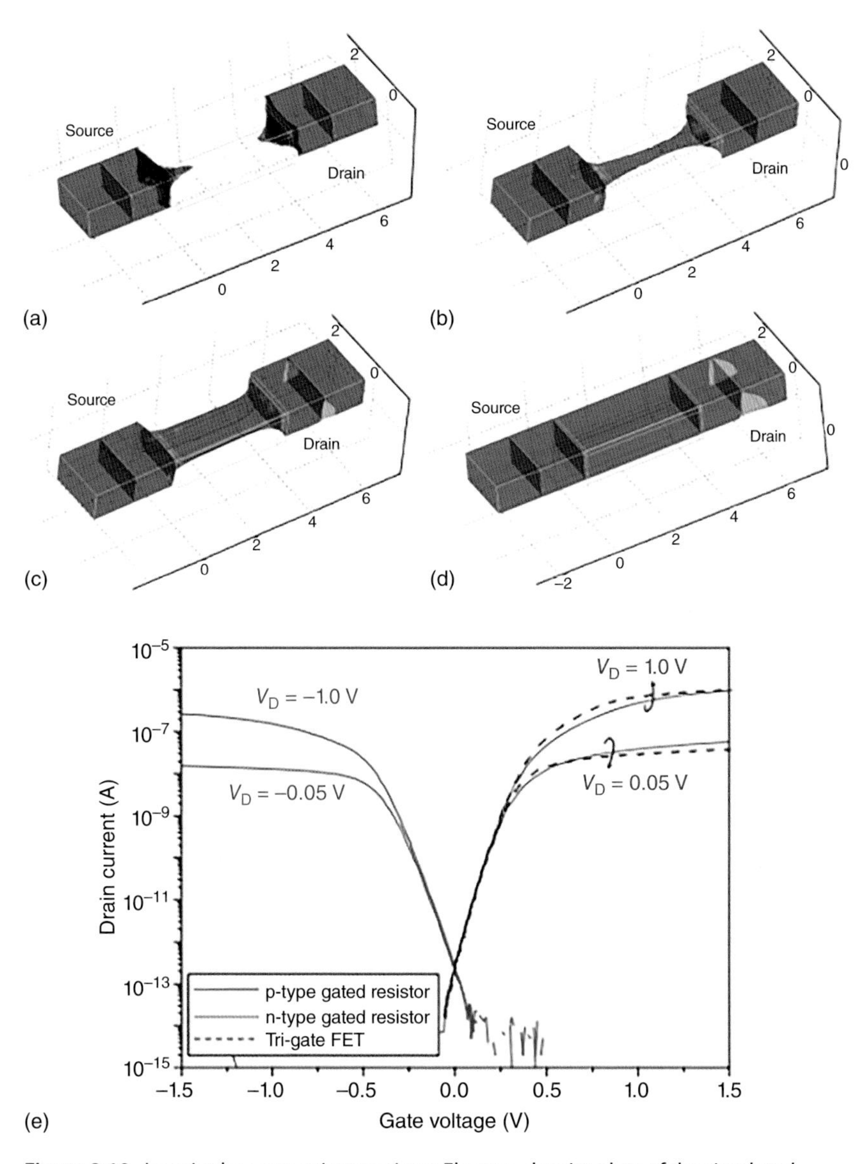

Figure 2.10 Junctionless nanowire transistor. Electron density plots of the simulated nanowire channel at gate voltage (a) below threshold, (b) at threshold, (c) above threshold, and (d) at flat-band potential. (e) I_d–V_g plots of fabricated nanowire gated resistors (both *n*- and *p*-type). Source: Colinge et al. 2010 [72]. Reproduced with permission of Springer Nature.

100 mV/decade, <100 pA gate leakage current, and 100 μA drive current at zero gate voltage (representative device scanning electron microscopy (SEM) shown in Figure 2.11b). In a future work [75] by the same group, improved performance with transconductance 0.5 S mm^{-1}, subthreshold slope around 90 mV/decade, 10^3 ON/OFF ratio, and on-current of 0.2 A mm^{-1} was demonstrated, mainly by introducing high-k dielectric HfO_2 as gate insulator, thus reducing EOT.

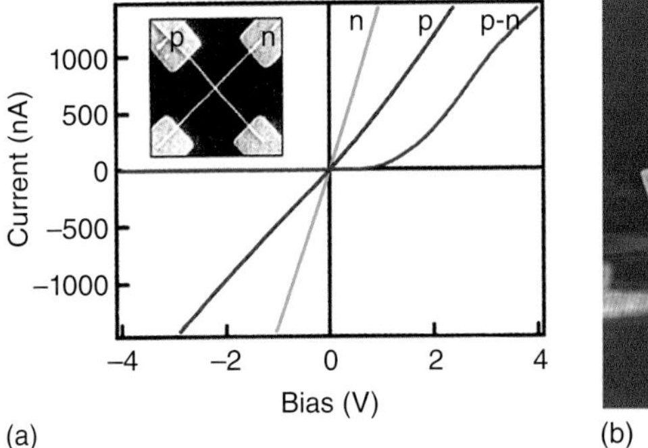

Figure 2.11 III–V nanowire devices. (a) GaN nanowire p–n junction device with *I–V* characteristics. Source: Huang et al. 2002 [49]. Reproduced with permission of ACS. (b) Vertical InAs MOSFET (channels seen as small black lines below air-bridge contact for drain). Source: Bryllert et al. 2006 [74]. Reproduced with permission of IEEE.

GAA horizontal III–V MOSFETs of width 30–50 nm and channel length 50–100 nm obtained by the top-down approach were demonstrated with high-mobility $In_{0.53}Ga_{0.47}As$ channel and Al_2O_3/WN dielectric stack. A representative 50-nm channel length device showed on-current of 0.7 A mm^{-1}, transconductance of 0.5 S mm^{-1}, subthreshold slope 150 mV/decade, and DIBL of 210 mV V^{-1}. Scaling down the EOT from 4.5 to 1.2 nm, for a wire width of 20 nm, improvement in transconductance to 1.74 S mm^{-1}, subthreshold slope of 63 mV/decade, and DIBL of 7 mV V^{-1} was achieved [76].

More recently in 2014 [77], horizontal GAA junctionless NWFETs with gate length 80 nm, width 9 nm, and height 40 nm was fabricated, with an ON/OFF ratio of 10^4, on-current 0.27 A mm^{-1} at Vds = 0.5 V and Vgs = 0.6 V, transconductance ~ 0.3 S mm^{-1} at V_{ds} = 0.4 V, and over a range of 0.2–0.5 V Vgs.

One of the best performing NFET devices so far was from a single top-gated 50-nm diameter n-InAs wire (gate dielectric 30 nm, gate length 2 μm), where the output current reached 3 A mm^{-1} and transconductance greater than 2 S mm^{-1} at Vgs = 0.56 V and Vds = 1 V, corresponding to a low-field mobility of 13 000 cm^2 V^{-1} s^{-1} [78].

The viability of using III–V nanowires in implementing radiofrequency (RF) circuits has also been demonstrated. In 2010, Egard et al. [79] showed vertical InAs wrap-gate devices of gate length 100 nm giving an average unity current gain cutoff frequency of 5.6 GHz, with the best performing device reaching 7.4 GHz and a maximum oscillation frequency above 20 GHz. Later in 2012, Wang et al. [80] made InAs nanomembrane FETs of channel lengths 75 nm, and cutoff frequency of 165 GHz but with a maximum oscillation frequency of 45.4 GHz.

Core–shell structures have also been explored by FET fabrication and characterization [51]. The core–shell structure aids in passivation of surface traps in core, and in formation of a clean electron gas (Figure 2.12).

InGaAs/InP/InAlAs/InGaAs core-multishell structure directly integrated on Si by Tomioka et al. [83] showed enhanced transconductance of 1.42 S mm^{-1}, ON/OFF ratio of 10^8 (with off-current less than 10 pA $μm^{-1}$), compared to 0.28 S mm^{-1} and 10^7, for an InGaAs nanowire without the core–shell structure.

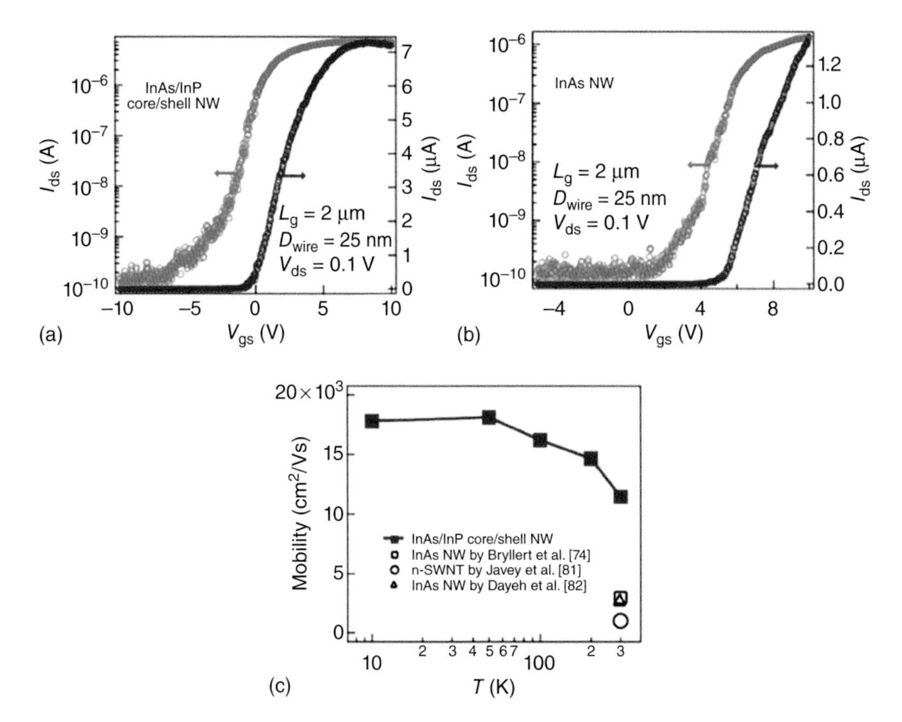

Figure 2.12 Electrical characteristics of an InAs/InP core/shell structure. Log and linear scale transfer characteristics of (a) InAs/InP and (b) InAs nanowire show ~5× increase in saturation current density. (c) Extracted mobility is 2.5× in core/shell structure than in bare nanowire. Increase in mobility with reduction with temperature indicates it is limited by phonon scattering. Source: Jiang et al. 2007 [51]. Reproduced with permission of ACS.

References

1 Hofstein, S.R. and Heiman, F.P. (1963). The silicon insulated-gate field-effect transistor. *Proc. IEEE* 51 (9): 1190–1202.

2 Dennard, R.H., Gaensslen, F.H., Rideout, V.L. et al. (1974). Design of ion-implanted MOSFET's with very small physical dimensions. *IEEE J. Solid-State Circuits* 9 (5): 256–268.

3 Critchlow, D.L., Dennard, R.H., and Schuster, S.E. (1973). Design and characteristics of n-channel insulated-gate field-effect transistors. *IBM J. Res. Dev.* 17 (5): 430–442.

4 Broers, A.N. and Dennard, R.H. (1973). *Impact of Electron-Beam Technology on Silicon Device Fabrication*, C101. Pennington: Electrochemical Soc Inc.

5 Dennard, R.H., Gaensslen, F.H., Kuhn, L., and Yu, H.N. (1972). In: Design of micron MOS switching devices. 1972 International Electron Devices Meeting, 4–6 December 1972, pp. 168–170.

6 John, R. (2006). High dielectric constant gate oxides for metal oxide Si transistors. *Rep. Prog. Phys.* 69 (2): 327.

7 Wilk, G.D., Wallace, R.M., and Anthony, J.M. (2001). High-κ gate dielectrics: current status and materials properties considerations. *J. Appl. Phys.* 89 (10): 5243–5275.

8 Parton, E. and Verheyen, P. (2006). Strained silicon — the key to sub-45 nm CMOS. *III-Vs Review* 19 (3): 28–31.

9 Ghani, T., Armstrong, M., Auth, C. et al. (2003). In: A 90 nm high volume manufacturing logic technology featuring novel 45nm gate length strained silicon CMOS transistors. IEEE International Electron Devices Meeting, 8–10 December 2003, pp. 11.6.1–11.6.3.

10 Young, K.K. (1989). Short-channel effect in fully depleted SOI MOSFETs. *IEEE Trans. Electron Devices* 36 (2): 399–402.

11 Ko, H., Takei, K., Kapadia, R. et al. (2010). Ultrathin compound semiconductor on insulator layers for high-performance nanoscale transistors. *Nature* 468 (7321): 286–289.

12 Subramanian, V., Kedzierski, J., Lindert, N. et al. (1999). In: A bulk-Si-compatible ultrathin-body SOI technology for sub-100 nm MOSFETs. 57th Annual Device Research Conference Digest (Cat. No.99TH8393), 23–23 June 1999, pp. 28–29.

13 Yee Chia, Y., Subramanian, V., Kedzierski, J. et al. (2000). Nanoscale ultra-thin-body silicon-on-insulator P-MOSFET with a SiGe/Si heterostructure channel. *IEEE Electron Device Lett.* 21 (4): 161–163.

14 Takei, K., Chuang, S., Fang, H. et al. (2011). Benchmarking the performance of ultrathin body InAs-on-insulator transistors as a function of body thickness. *Appl. Phys. Lett.* 99: 103507.

15 Takei, K., Madsen, M., Fang, H. et al. (2012). Nanoscale InGaSb heterostructure membranes on Si substrates for high hole mobility transistors. *Nano Lett.* 12 (4): 2060–2066.

16 Madsen, M., Takei, K., Kapadia, R. et al. (2011). Nanoscale semiconductor "X" on substrate "Y"–processes, devices, and applications. *Adv. Mater.* 23: 3115–3127.

17 Takei, K., Fang, H., Kumar, S.B. et al. (2011). Quantum confinement effects in nanoscale-thickness InAs membranes. *Nano Lett.* 11: 5008–5012.

18 Takei, K., Kapadia, R., Fang, H. et al. (2013). High quality interfaces of InAs-on-insulator field-effect transistors with ZrO_2 gate dielectrics. *Appl. Phys. Lett.* 102: 153513.

19 Takei, K., Kapadia, R., Li, Y. et al. (2013). Surface charge transfer doping of III-V nanostructures. *J. Phys. Chem. C* 117: 17845–17849.

20 Takagi, S., Koga, J., Toriumi, A. (1997). In: Subband structure engineering for performance enhancement of Si MOSFETs. International Electron Devices Meeting, IEDM Technical Digest, 10–10 December 1997, pp. 219–222.

21 Hisamoto, D., Kaga, T., Kawamoto, Y., and Takeda, E. (1989). In: A fully depleted lean-channel transistor (DELTA)-a novel vertical ultra thin SOI MOSFET. International Technical Digest on Electron Devices Meeting, 3–6 December 1989, pp. 833–836.

22 Colinge, J.P., Gao, M.H., Romano-Rodriguez, A. et al. (1990). In: Silicon-on-insulator 'gate-all-around device'. International Technical Digest on Electron Devices, 9–12 December 1990, pp. 595–598.

23 Bryllert, T., Wernersson, L.-E., Löwgren, T., and Lars, S. (2006). Vertical wrap-gated nanowire transistors. *Nanotechnology* 17 (11): S227.

24 Natarajan, S., Agostinelli, M., Akbar, S. et al. (2014). In: A 14nm logic technology featuring 2nd-generation FinFET, air-gapped interconnects, self-aligned double patterning and a 0.0588 μm^2 SRAM cell size. 2014 IEEE International Electron Devices Meeting, 15–17 December 2014, pp. 3.7.1–3.7.3.
25 Doyle, B.S. (2003). High performance fully-depleted tri-gate CMOS transistors. *IEEE Electron Device Lett.* 24: 263–265.
26 Xuejue, H., Wen-Chin, L., Charles, K. et al. (1999). In: Sub 50-nm FinFET: PMOS. International Electron Devices Meeting 1999, Technical Digest (Cat. No.99CH36318), 5–8 December 1999, pp. 67–70.
27 Hisamoto, D., Wen-Chin, L., Kedzierski, J. et al. (2000). FinFET-a self-aligned double-gate MOSFET scalable to 20 nm. *IEEE Trans. Electron Devices* 47 (12): 2320–2325.
28 Nam, H. and Shin, C. (2014). Impact of current flow shape in tapered (versus rectangular) FinFET on threshold voltage variation induced by work-function variation. *IEEE Trans. Electron Devices* 61 (6): 2007–2011.
29 Park, T., Choi, S., Lee, D.H. et al. (2003). In: Fabrication of body-tied FinFETs (Omega MOSFETs) using bulk Si wafers. 2003 Symposium on VLSI Technology, Digest of Technical Papers (IEEE Cat. No.03CH37407), 10–12 June 2003, pp. 135–136.
30 Yan, R.H., Ourmazd, A., and Lee, K.F. (1992). Scaling the Si MOSFET: from bulk to SOI to bulk. *IEEE Trans. Electron Devices* 39 (7): 1704–1710.
31 Lee, C.-W., Yun, S.-R.-N., Yu, C.-G. et al. (2007). Device design guidelines for nano-scale MuGFETs. *Solid-State Electronics* 51 (3): 505–510.
32 Auth, C.P. and Plummer, J.D. (1997). Scaling theory for cylindrical, fully-depleted, surrounding-gate MOSFET's. *IEEE Electron Device Lett.* 18 (2): 74–76.
33 Colinge, J.-P. and Greer, J.C. (2016). *Nanowire Transistors: Physics of Devices and Materials in One Dimension*. Cambridge University Press.
34 Colinge, J.P. (2004). *Silicon-on-Insulator Technology: Materials to VLSI: Materials to Vlsi*. Springer Science & Business Media.
35 Colinge, J.P. (2007). *FinFETs and Other Multi-Gate Transistors*. Nature Publishing Group.
36 Colinge, J.-P. (2007). Quantum-wire effects in trigate SOI MOSFETs. *Solid-State Electronics* 51 (9): 1153–1160.
37 Moreno, E., Roldán, J.B., Ruiz, F.G. et al. (2010). An analytical model for square GAA MOSFETs including quantum effects. *Solid-State Electronics* 54 (11): 1463–1469.
38 Levi, A.F.J. (2006). *Applied Quantum Mechanics*. Cambridge University Press.
39 Chuang, S., Gao, Q., Kapadia, R. et al. (2013). Ballistic InAs nanowire transistors. *Nano Lett.* 13 (2): 555–558.
40 Thelander, C., Björk, M.T., Larsson, M.W. et al. (2004). Electron transport in InAs nanowires and heterostructure nanowire devices. *Solid State Commun.* 131 (9–10): 573–579.
41 Afzalian, A., Lee, C.W., Akhavan, N.D. et al. (2011). Quantum confinement effects in capacitance behavior of multigate silicon nanowire MOSFETs. *IEEE Trans. Nanotechnol.* 10 (2): 300–309.

42 Luryi, S. (1988). Quantum capacitance devices. *Appl. Phys. Lett.* 52 (6): 501–503.
43 Fahad, H.M. and Hussain, M.M. (2012). Are nanotube architectures more advantageous than nanowire architectures for field effect transistors? *Scientific Reports* 2: 475.
44 Wagner, R.S. and Ellis, W.C. (1964). Vapor-liquid-solid mechanism of single crystal growth. *Appl. Phys. Lett.* 4 (5): 89–90.
45 Givargizov, E.I. (1975). Fundamental aspects of VLS growth. *J. Crystal Growth* 31: 20–30.
46 Duan, X., Huang, Y., Cui, Y. et al. (2001). Indium phosphide nanowires as building blocks for nanoscale electronic and optoelectronic devices. *Nature* 409 (6816): 66–69.
47 Duan, X. and Lieber, C.M. (2000). Laser-assisted catalytic growth of single crystal GaN nanowires. *J. Am. Chem. Soc.* 122 (1): 188–189.
48 Gudiksen, M.S., Lauhon, L.J., Wang, J. et al. (2002). Growth of nanowire superlattice structures for nanoscale photonics and electronics. *Nature* 415 (6872): 617–620.
49 Huang, Y., Duan, X., Cui, Y., and Lieber, C.M. (2002). Gallium nitride nanowire nanodevices. *Nano Lett.* 2 (2): 101–104.
50 Javey, A., Robin, N., Friedman, R.S. et al. (2007). Layer-by-layer assembly of nanowires for three-dimensional, multifunctional electronics. *Nano Lett.* 7 (3): 773–777.
51 Jiang, X., Xiong, Q., Nam, S. et al. (2007). InAs/InP radial nanowire heterostructures as high electron mobility devices. *Nano Lett.* 7 (10): 3214–3218.
52 McAlpine, M.C., Friedman, R.S., Jin, S. et al. (2003). High-performance nanowire electronics and photonics on glass and plastic substrates. *Nano Lett.* 3 (11): 1531–1535.
53 Morales, A.M. and Lieber, C.M. (1998). A laser ablation method for the synthesis of crystalline semiconductor nanowires. *Science* 279 (5348): 208–211.
54 Tian, B., Zheng, X., Kempa, T.J. et al. (2007). Coaxial silicon nanowires as solar cells and nanoelectronic power sources. *Nature* 449 (7164): 885–889.
55 Xiang, J., Lu, W., Hu, Y. et al. (2006). Ge/Si nanowire heterostructures as high-performance field-effect transistors. *Nature* 441 (7092): 489–493.
56 Chen, K., Kapadia, R., Harker, A. et al. (2016). Direct growth of single-crystalline III-V semiconductors on amorphous substrates. *Nat. Commun.* 7: 10502.
57 Kapadia, R., Yu, Z., Hettick, M. et al. (2014). Deterministic nucleation of InP on metal foils with the thin-film vapor–liquid–solid growth mode. *Chem. Mater.* 26 (3): 1340–1344.
58 Kapadia, R., Yu, Z., Wang, H.-H.H. et al. (2013). A direct thin-film path towards low-cost large-area III-V photovoltaics. *Sci. Rep.* 3: 02275.
59 Fortuna, S.A., Wen, J., Chun, I.S., and Li, X. (2008). Planar GaAs nanowires on GaAs (100) substrates: self-aligned, nearly twin-defect free, and transfer-printable. *Nano Lett.* 8 (12): 4421–4427.
60 Sarkar, D., Tao, J., Wang, W. et al. (2018). Mimicking biological synaptic functionality with an indium phosphide synaptic device on silicon for scalable neuromorphic computing. *ACS Nano* 12 (2): 1656–1663.

61 Lin, Q., Sarkar, D., Lin, Y. et al. (2017). Scalable indium phosphide thin-film nanophotonics platform for photovoltaic and photoelectrochemical devices. *ACS Nano*.

62 Sarkar, D., Wang, W., Lin, Q. et al. (2018). Buffer insensitive optoelectronic quality of InP-on-Si with templated liquid phase growth. *J. Vac. Sci. Technol.* 36 (3): 031204.

63 Sarkar, D., Wang, W., Mecklenburg, M. et al. (2018). Confined liquid phase growth of crystalline compound semiconductors on any substrate. *ACS Nano* doi: 10.1021/acsnano.8b01819.

64 Mårtensson, T., Wagner, J.B., Hilner, E. et al. (2007). Epitaxial growth of indium arsenide nanowires on silicon using nucleation templates formed by self-assembled organic coatings. *Adv. Mater.* 19 (14): 1801–1806.

65 Paek, J.H., Nishiwaki, T., Yamaguchi, M., and Sawaki, N. (2009). Catalyst free MBE-VLS growth of GaAs nanowires on (111)Si substrate. *physica status solidi (c)* 6 (6): 1436–1440.

66 Dey, A.W., Svensson, J., Ek, M. et al. (2013). Combining axial and radial nanowire heterostructures: radial esaki diodes and tunnel field-effect transistors. *Nano Lett.* 13 (12): 5919–5924.

67 Moon, D.I., Choi, S.J., Duarte, J.P., and Choi, Y.K. (2013). Investigation of silicon nanowire gate-all-around junctionless transistors built on a bulk substrate. *IEEE Trans. Electron Devices* 60 (4): 1355–1360.

68 Jurczak, M., Skotnicki, T., Paoli, M. et al. (2000). Silicon-on-Nothing (SON)-an innovative process for advanced CMOS. *IEEE Trans. Electron Devices* 47 (11): 2179–2187.

69 Goldberger, J., Hochbaum, A.I., Fan, R., and Yang, P. (2006). Silicon vertically integrated nanowire field effect transistors. *Nano Lett.* 6 (5): 973–977.

70 Schmidt, V., Riel, H., Senz, S. et al. (2006). Realization of a silicon nanowire vertical surround-gate field-effect transistor. *Small* 2 (1): 85–88.

71 Ahn, J.H., Kim, H.S., Lee, K.J. et al. (2006). Heterogeneous three-dimensional electronics by use of printed semiconductor nanomaterials. *Science* 314: 1754.

72 Colinge, J.-P., Lee, C.-W., Afzalian, A. et al. (2010). Nanowire transistors without junctions. *Nat. Nano.* 5 (3): 225–229.

73 Lilienfeld, J.E. (1925). *Method and Apparatus for Controlling Electric Current*. Nature Publishing Group.

74 Bryllert, T., Wernersson, L.E., Froberg, L.E., and Samuelson, L. (2006). Vertical high-mobility wrap-gated InAs nanowire transistor. *IEEE Electron Device Lett.* 27 (5): 323–325.

75 Thelander, C., FrÖbergFroberg, L.E., Rehnstedt, C. et al. (2008). Vertical enhancement-mode InAs nanowire field-effect transistor with 50-nm wrap gate. *IEEE Electron Device Lett.* 29 (3): 206–208.

76 Gu, J.J., Wang, X. W., Wu, H. et al. (2012). In: 20–80 nm channel length InGaAs gate-all-around nanowire MOSFETs with EOT = 1.2 nm and lowest SS = 63 mV/dec. 2012 International Electron Devices Meeting, 10–13 December 2012, pp. 27.6.1–27.6.4.

77 Song, Y., Zhang, C., Dowdy, R. et al. (2014). III-V junctionless gate-all-around nanowire MOSFETs for high linearity low power applications. *IEEE Electron Device Lett.* 35 (3): 324–326.

78 Do, Q.T., Blekker, K., Regolin, I. et al. (2007). In: Single n-InAs nanowire MIS-field-effect transistor: experimental and simulation results. 2007 IEEE 19th International Conference on Indium Phosphide & Related Materials, 14–18 May 2007, pp. 392–395.

79 Egard, M., Johansson, S., Johansson, A.C. et al. (2010). Vertical InAs nanowire wrap gate transistors with $f_t > 7$ GHz and $f_{max} > 20$ GHz. *Nano Lett.* 10 (3): 809–812.

80 Wang, C., Chien, J.-C., Fang, H. et al. (2012). Self-aligned, extremely high frequency III–V metal-oxide-semiconductor field-effect transistors on rigid and flexible substrates. *Nano Lett.* 12 (8): 4140–4145.

81 Javey, A., Kim, H., Brink, M. et al. (2002). High-κ dielectrics for advanced carbon-nanotube transistors and logic gates. *Nat. Mater.* 1: 241.

82 Dayeh Shadi, A., Aplin David, P.R., Zhou, X. et al. (2007). High electron mobility InAs nanowire field-effect transistors. *Small* 3 (2): 326–332.

83 Tomioka, K., Yoshimura, M., and Fukui, T. (2012). A III-V nanowire channel on silicon for high-performance vertical transistors. *Nature* 488 (7410): 189–192.

3

Two-dimensional Materials for Electronic Applications

Haimeng Zhang and Han Wang

University of Southern California, Ming Hsieh Department of Electrical Engineering, 3740 McClintock Avenue, Los Angeles, CA 90007, USA

The physical dimensionality of crystalline materials plays a critical role in governing their properties. If quantum confinement is in place along one of the physical dimensions, unique properties pertaining to this two-dimensional (2D) geometry will occur; if quantum confinement is experienced in two of the physical dimensions, we will observe wired or one-dimensional (1D) material properties; if quantum confinement is experienced in all the three physical dimensions, typically with physical size of a few nanometers or less for electronic confinement, we usually call it 0D materials. The same compound can exhibit different properties depending on whether it is arranged in a 0D, 1D, 2D or 3D crystal structure. When it comes to the case of sp^2 carbon materials, 0D fullerenes, 1D nanotube, 2D graphene, and 3D graphite exhibit drastically different properties. Although graphite has been known since at least the sixteenth century and has been widely used in the industry for steel making, as brake lining, or as a dry lubricant, it was not until 2004 that a group of scientists from the University of Manchester led by Andre Geim and Konstantin Novoselov succeeded in extracting stable monolayer graphene sheets from graphite and studied their unique properties. Graphene is one of the allotropes of carbon with low dimensionality. It is made of sp^2 hybridized carbon atoms arranged in a hexagonal honeycomb lattice in which three valence electrons form the strong in-plane covalent bonds and the fourth electron remains in the p-orbital and forms the out-of-plane π bond. The electrons in the p-orbitals are easily delocalized among the atoms, leading to the unique linear dispersion relation in graphene.

The isolation of graphene not only triggered intense research activities by physicists and chemists to understand its fundamental properties but also inspired interest from device engineers and circuit designers in 2D-materials-based electronics. Apart from graphene, transition metal dichalcogenides (TMDs), such as MoS_2 and WSe_2, are also important members of the 2D materials family. TMDs can be represented by the general formula MX_2, where M is a transition metal from group 4, 5, or 6, and X is a chalcogen atom (which is sulfur, selenium, or tellurium). The monolayer of TMDs consists of three atomic layers in the form of X-M-X. Adjacent layers of TMDs are held together

Advanced Nanoelectronics: Post-Silicon Materials and Devices,
First Edition. Edited by Muhammad Mustafa Hussain.

by van der Waals forces, which allow them to be easily exfoliated. These 2D materials made it possible to explore the ultimately thin channel transistors and the opportunity for innovations in new device concepts.

In this chapter, we focus on the electronics based on this new class of materials. We first introduce the basic structure and characteristics of 2D-materials-based discrete transistors and discuss their device fabrication technology. Then, we show in detail, with graphene and MoS_2 as examples, how 2D materials transistors perform and how they can be integrated at the circuit level.

3.1 2D Materials Transistor and Device Technology

A field-effect transistor (FET) is a three-terminal device with a semiconducting channel material forming the conducting path between the source and the drain, which can be modulated by the gate electrode. FETs are the primary elements in modern integrated circuits (ICs). Figure 3.1 depicts the basic structure of a top-gate transistor with 2D materials as the channel materials in the classical metal-oxide-semiconductor field-effect transistor (MOSFET) structure.

R_S and R_D are the source and drain access resistance. R_i is the intrinsic resistance. $g_{m,i}$ is the intrinsic transconductance. g_0 is the output transconductance. C_{ds} is the source-drain capacitance. C_{gs} and C_{gd} are the gate-source and gate-drain capacitances. R_G is the resistance in the gate electrode.

Currently, the existing channel materials of commercial FETs are dominated by bulk semiconductors including silicon, GaAs, SiC, GaN, and so on. FETs based on these bulk semiconductors have been successfully scaled down to the micro- and nanometer scale in the 50 years of Moore's law. To keep sustaining Moore's law and meet the technological demand for the next-generation semiconductor devices, new materials and device geometries will be needed to address the challenges in transistor technology.

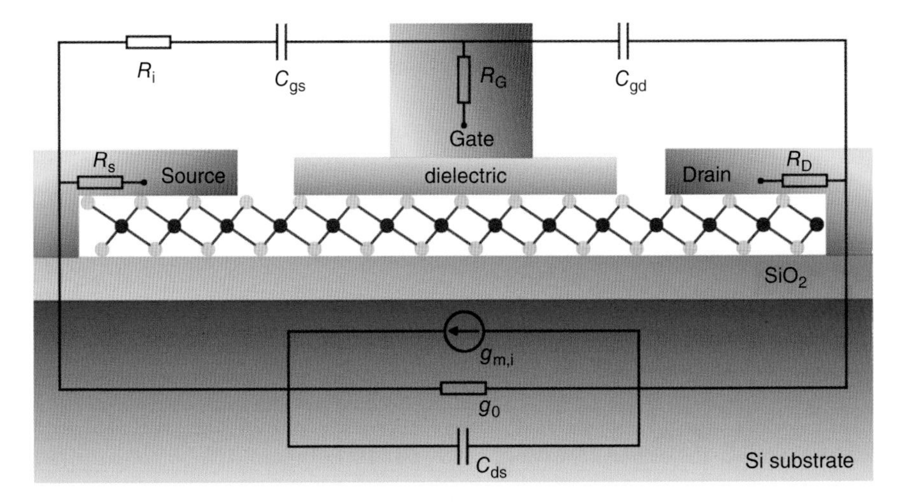

Figure 3.1 Typical structure of an FET based on 2D materials, with the small-signal equivalent circuit labeled.

In the following, we introduce the operation and characteristics of 2D-materials-based FETs to understand the advantage of using 2D materials as the channel materials in FETs. Then, several considerations for FET performance optimization are discussed. Finally, important figures of merit (FOM) are defined and introduced to evaluate their performance and compare them against channel materials of other types.

3.1.1 Operation and Characteristics of 2D-Materials-Based FETs

The effective operation of an FET relies on efficient control of the channel conductivity through the applied voltage between the gate and the source. The transverse electric field induced by the gate voltage can either pinch off the channel of the carrier movement, and thus no current flows between the source and the drain (off-state), or enhance the concentration of the carriers in the channel, leading to higher conductivity in the on-state. This phenomenon where the current through an FET can be tuned by the external applied gate voltage is referred to as the electric field effect.

The performance characteristics of an FET can be most clearly illustrated in its transfer and output characteristics. Transfer characteristics plot the source-to-drain current I_{DS} as a function of the gate voltage, while the output characteristics plot I_{DS} as a function of V_{DS}. The drain current I_{DS} should saturate above a certain V_{DS}.

The ultrathin channel thickness provides improved electrostatic gate control over the channel. The electric field effect in 2D materials was first confirmed by graphene in 2004 [1]. After that, considerable research has been devoted to graphene because of its unique electronic properties. But its lack of bandgap makes graphene unsuitable for applications in digital electronics, in which a sizable bandgap is often essential to suppress the off-state current. Hence, many other 2D semiconducting materials including TMDs, silicone, and phosphorene have been studied as possible channel materials in FETs.

3.1.2 Ambipolar Property of Graphene

Ambipolar conduction, which is characterized by the relatively symmetric electron and hole conduction about the minimum conductivity region, is a clear signature of a graphene-based field-effect transistor (GFET), as shown in Figure 3.2. It also shows that the carrier type of GFET can be selected by the electrical doping owing to this property.

In a GFET, the modulation of V_{GS} shifts the position of Fermi energy in graphene, which is originally at the Dirac point. By controlling the position of the Fermi level, the transport in graphene is ambipolar, i.e. it is performed by either holes or electrons. If V_{GS} is well below the minimum conduction point, the Fermi level in graphene is below its charge neutrality level and the conduction in the channel is dominated by holes. If V_{GS} is well above the minimum conduction point, the Fermi level in graphene is above its charge neutrality level and the conduction in the channel is due to electrons. A mixed state, where the electrons and holes are both present, also exists. In this operating region, electrons and

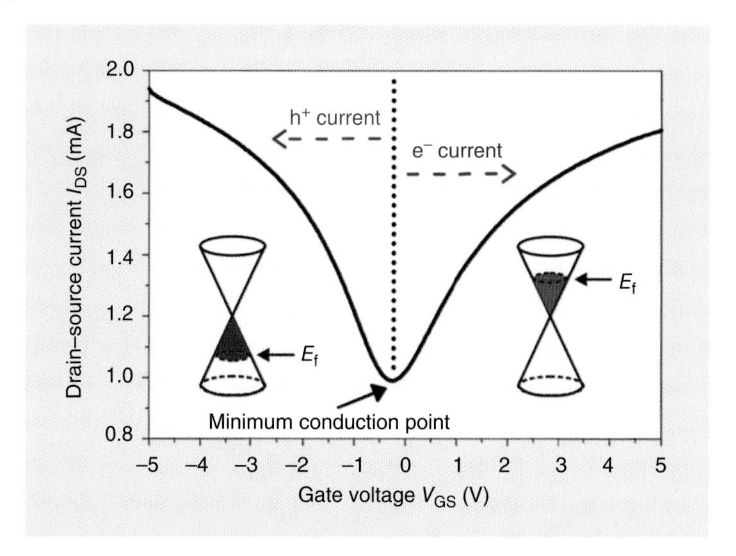

Figure 3.2 V-shape transfer characteristics and ambipolar conduction in GFETs. Source: Wang et al. 2012 [2]. Adapted with permission of IEEE.

holes, both as major carriers, are injected from the source and drain, respectively (assuming $V_{DS} > 0$), and will recombine inside the channel region.

This type of electrical doping is similar to chemical doping, and is commonly used for changing the transport type in semiconductor devices. However, unlike the conventional chemical doping in unipolar silicon MOSFETs where the carrier type cannot be changed once fabricated, the ambipolar behavior of GFET can be utilized to achieve novel applications, and this is discussed in later sections. Ambipolar conduction is also found in other types of materials with a relatively small bandgap, such as WSe_2, carbon nanotubes (CNTs), amorphous silicon, and organic semiconductor heterostructures, among which graphene has the highest mobility and good scalability.

3.1.3 Important Figures of Merit

In this section, some important FOM are defined so that we can compare electronics based on these new 2D materials with the benchmark conventional electronic technology and the requirements of the International Technology Roadmap for Semiconductors (ITRS). The introduced FOM can, for example, be used to evaluate the performance of an FET in digital applications as a switch and in analog/radiofrequency (RF) applications as an amplifier.

3.1.3.1 I_{on}/I_{off} Ratio

When working as a digital switch, the channel of the FET is either in the on-state (where a relatively large I_{on} current can flow) or in the off-state (where only a relatively low I_{off} current flows) depending on whether the voltage applied to the gate is larger than the threshold voltage V_T. During its operation in a circuit, the voltage applied to the FET terminals can only sweep between 0 to the supply voltage V_{DD}. Here, we denote the source-to-drain current as I_{DS}, the gate-to-source voltage as V_{GS}, and the drain-to-source voltage as V_{DS}. I_{on} is defined as the I_{DS} when the largest voltage V_{DD} is applied both between source and drain, and between source and gate ($V_{GS} = V_{DS} = V_{DD}$), while I_{off} is defined as the I_{DS} when V_{DD} is applied between source and drain with no voltage applied at the gate ($V_{DS} = V_{DD}$, $V_{GS} = 0\,V$).

Ideally, one wants to maximize I_{on} since higher I_{on} can provide higher switching speed, and to minimize I_{off} since I_{off} contributes to the standby power consumption. Thus, the I_{on}/I_{off} ratio is widely used to evaluate the switching performance. Integrated logic circuits consist of cascaded logic gates where the output at the drain of a previous transistor serves as the input at the gate of the next transistor. Thus, it is essential for the FETs in such cascaded structures to have the reliability to restore the signal at the output and correctly identify the input signal level ("on" or "off") at the next input. In this regard, the probability of successful operation of such logic circuits is proportional to the I_{on}/I_{off} ratio.

3.1.3.2 Subthreshold Swing

Subthreshold swing (SS) is defined as the change in the gate voltage required to increase the drain current by a factor of 10 (given in mV/decade), and indicates the field-effect switching efficiency of an FET. It is related to FET operation in the subthreshold regime and can be extracted as the inverse slope in the logarithmic scale of the transfer curve which plots I_{DS} versus V_{GS}. SS should be as low as possible, since a lower SS suggests a higher I_{on}/I_{off} ratio at a smaller supply voltage. SS shows a theoretical lower limit of 60 mV/decade at room temperature for thermionic devices. However, this limitation can be overcome in devices utilizing other mechanisms such as in tunneling field-effect transistors (TFETs) [3, 4]. Unlike thermionic devices which use thermal carrier injection, TFET utilizes interband tunneling as a source carrier injection mechanism.

3.1.3.3 Cutoff Frequency and Maximum Frequency of Oscillation

FETs that can operate at RF speed are often used to amplify high-frequency signals. Their amplification characteristics can be described in terms of the small-signal gains, for example, the current gain and the power gain. There are several definitions for the power gain, for example, the unity maximum stable gain (MSG), the maximum available gain (MAG), and the unilateral power gain (U). Closely related to these gains are the two probably most widely used MOFs for RF transistors, namely, the cutoff frequency f_T and the maximum frequency of oscillation f_{max}. All the three power gain quantities should, in principle, lead to the same extracted f_{max}.

The current gain cutoff frequency of a transistor device is the frequency limit of operation where its current gain decreases below unity. The maximum oscillation frequency of a transistor device is the highest operating frequency at which its power gain is still above one. f_T and f_{max} define the upper frequency limits beyond which a transistor loses its ability to amplify. It is important to stress that the power gain and f_{max} are usually more important than the current gain and f_T in practical applications, since in analog applications, a transistor needs to operate at frequencies where it can deliver more power than its input signal.

The operation of such transistors can be described by the small signal-equivalent circuit model, which treats the transistor device as a circuit. From the small-signal equivalent circuit model, we can derive the approximate expressions for the MOF mentioned earlier. These expressions can give us a better understanding of the potential merits and drawbacks of a GFET comparing it with other RF FET types.

Depending on the two-port network theory and from the small-signal equivalent circuit model in Figure 3.1, f_T can be expressed as follows:

$$f_T = \frac{g_m}{2\pi f(C_{gs} + C_{gd})} \tag{3.1}$$

Another expression for f_T that directly depends on the intrinsic parameters of the transistor is given as

$$f_T = \frac{v_{sat}}{2\pi L_g} \tag{3.2}$$

by substituting $g_m = v_{sat}(C_{gs} + C_{gd})/L_g$, where v_{sat} is the saturation velocity, L_g is the gate length, into Eq. (3.1). From this expression, we can see that the f_T value of a transistor is directly related to the intrinsic carrier dynamics and the gate length of the device. In other words, it is directly related to the carrier transit delay, which is the time the carrier takes to move across the channel. However, Eqs. (3.1) and (3.2) only consider the intrinsic components but not the extrinsic circuit elements such as R_S, R_D, and g_0, which are critical in 2D-materials-based FET, especially when the gate length becomes shorter. Taking all the key parasitic components into account for the active region of the device, a more rigorous expression for f_T can be obtained:

$$f_T = \frac{1}{2\pi} \times \frac{g_m}{[C_{gs} + C_{gd}][1 + (R_S + R_D)/R_{ds}] + C_{gd}g_m(R_S + R_D)} \tag{3.3}$$

The power gain cutoff frequency of oscillation f_{max} can be expressed as

$$f_{max,U} = \frac{1}{2\pi[C_{gs} + C_{gd}]} \times \frac{g_m}{\left[4g_0(R_i + R_S + R_G) + 4g_m R_G \frac{C_{gd}}{C_{gs}+C_{gd}}\right]} \tag{3.4}$$

From Eqs. (3.2) and (3.3), we can conclude that, to improve f_T and f_{max} of an FET, we should have a transconductance as high as possible and all other elements of its equivalent circuit as small as possible.

3.1.3.4 Minimum Noise Figure

In practical applications, every device generates noise during its operation, which comes from the fluctuation of voltage and current. Noise is always unwanted, and is critical especially for amplifications of RF signals. The noise factor F is defined as a measure of the noise generated by a transistor:

$$F = \frac{SNR_{in}}{SNR_{out}} \tag{3.5}$$

where SNR_{in} and SNR_{out} are the signal-to-noise ratio at the input and output, respectively. In the RF community, the noise figure is frequently given in the unit of dB as NF:

$$NF = 10\log_{10}F = 10\log_{10}\frac{SNR_{in}}{SNR_{out}} \tag{3.6}$$

On the basis of the equivalent circuit from Figure 3.1, an expression for minimum noise figure NF_{min} can be expressed as

$$NF_{min} = 1 + 2\pi k_f f C_{gs}\sqrt{\frac{R_G + R_S}{g_m}} \tag{3.7}$$

where k_f is an empirical factor and f is the operating frequency of the transistor.

3.1.4 Device Optimization

3.1.4.1 Mobility Engineering

Carrier mobility is one of the key parameters that describes the transport and conduction properties of materials. Figure 3.3 plots the mobility of various nanomaterials that have been proposed for thin-film electronics applications. The high carrier mobility of graphene at room temperature (10 000–15 000 $cm^2V^{-1}s^{-1}$ for chemical vapor deposition (CVD) graphene on SiO_2-covered silicon wafer [6]) has always been its most frequently stated advantage. However, graphene-based FET suffers from a relatively low I_{on}/I_{off} ratio (~10 at room temperature [7, 8])

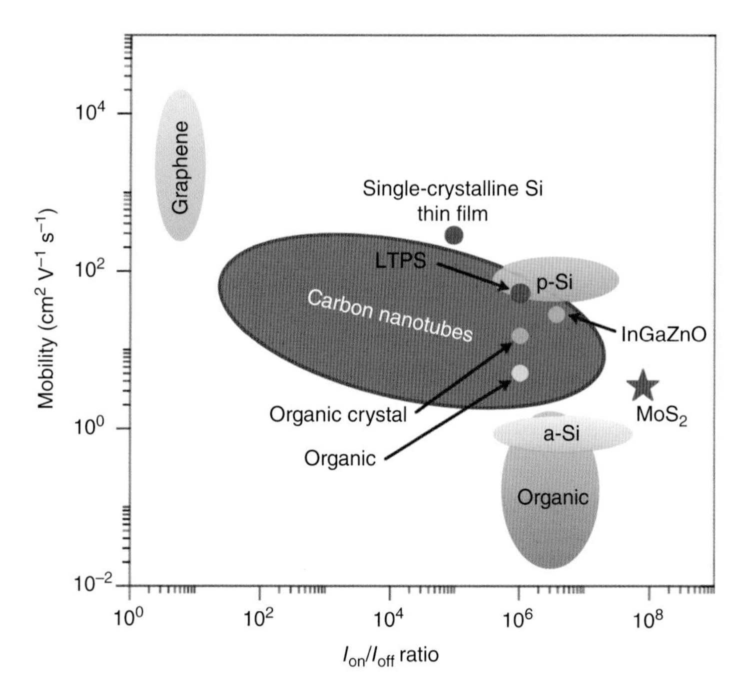

Figure 3.3 Carrier mobility versus I_{on}/I_{off} ratio for various nanomaterials and their transistor devices. Source: Fiori et al. 2014 [5]. Adapted with permission of Springer Nature.

because of its gapless nature. However, in MoS_2, the off current can be suppressed. Hence, the I_{on}/I_{off} ratio is largely improved, which exceeds 10^8 for the first monolayer MoS_2 FET, but its corresponding carrier mobility is only around 200 $cm^2V^{-1}s^{-1}$ [9].

The theoretically predicted values for the carrier mobility of 2D materials are very promising. An intrinsic carrier mobility of 200 000 $cm^2V^{-1}s^{-1}$ is predicted for graphene without charged impurities and extrinsic scattering [10]. Monolayer MoS_2 is predicted by theoretical calculation to have a room-temperature electron mobility in the range of 10–1000 $cm^2V^{-1}s^{-1}$. But the mobility in practice is often limited by disorders and scattering of different kinds. In order to understand the origins of the mobility degradation, time-dependent Hall measurement has been carried out to distinguish between different mechanisms limiting the mobility. Remote optical phonons and Coulomb scattering from the interfacial charge traps are proposed to be the dominant reasons behind the low carrier mobility in atomically thin MoS_2 films [11]. The carrier mobility of 2D materials, in general, is extremely sensitive to the external environment due to their large surface-to-volume ratio. Thus, substrates and dielectrics around 2D materials in the electronic devices need to be carefully engineered to improve their performance.

In this regard, many groups have been searching for a better substrate and dielectric layer for 2D-materials-based devices. hBN, a wide-bandgap 2D material, has been proposed to replace the commonly used SiO_2 substrate. hBN has a hexagonal lattice structure similar to graphene, and can provide 2D-materials-based devices with an ultra-flat substrate without dangling bonds and charge traps. When the graphene channel is encapsulated by hBN, the mobility is reported to be above 60 000 $cm^2V^{-1}s^{-1}$ at $T = 4$ K. Microscale ballistic transport in graphene sandwiched in two hBN layers has also been demonstrated at room temperature. The RF GFET with hBN/graphene/hBN structure has also shown significant improvements in RF performance compared with GFETs on SiO_2 substrate. The encapsulation of MoS_2 in hBN has shown a record-high low-temperature mobility [11]. The enhanced mobility can be attributed to reduced Coulomb scattering due to the high-κ dielectric environment and possible modification of phonon dispersion in 2D materials [12].

Dielectric environment engineering is a promising approach to reduce the extrinsic sources of scattering, and hence can greatly enhance the carrier mobility in the device. Further improvement of carrier mobility would consist of controlling both extrinsic and intrinsic limiting factors such as the crystal defects.

3.1.4.2 Current Saturation

As described in previous sections, a typical FET should demonstrate current saturation of I_{DS} above a certain V_{DS}. Current saturation is important both for discrete device applications and for ICs, for digital applications, and for RF electronics. In an integrated logic circuit, the transistor current saturation is an essential requirement for achieving excellent inverter gain and other logic gate characteristics. Lack of current saturation would also degrade the performance of devices for RF applications. The current saturation directly determines the output conductance, thus affecting both the power gain and current gain FOM of a RF transistor. As for GFET, the drain current reported in the literature does not saturate due to the lack of electronic bandgap in the material. Its often almost-linear output characteristics suggest that the resistance $R_{ds} = 1/g_0$ is very low and is on the same order as R_S and R_D. This can lead to a significant RF loss in this component and degrade the current gain and power gain of the device. Another problem is that since the drain current depends significantly on V_{DS}, the gate-to-drain capacitance C_{gd} is much larger than the C_{gd} for the cases where there is current saturation. On the contrary, for conventional MOSFETs operating in the saturation regime, the amount of charge near the drain has only a weak dependence on the gate voltage, indicating a low C_{gd}. C_{gd}, as the feedback capacitance bridging the input and output would strongly affect the f_T and f_{max} performance of the device. Another problem is that current saturation is an indispensable feature required for building current sources. Without current saturation, it will be impossible to build high-quality current sources, which are critical components for most analog circuits.

Furthermore, a reasonable level of current saturation is needed to have an effective gate control over the channel conductivity. To be more precise, the lack of current saturation can enhance the influence of V_{DS} on the drain current, which can be seen from the strong dependence of the minimum conduction point voltage $V_{G,min}$ on the drain voltage, V_{DS}. The $V_{G,min}$ shifts more positive as V_{DS} increases, causing an almost-linear potential drop along the channel. This is similar to having an FET whose threshold voltage strongly depends on V_{DS}. However, the output of an ideal FET should depend only on the input voltage at the gate, V_{GS}.

The drain current of GFETs cannot saturate since the absence of bandgap in graphene prevents the formation of a depletion region on the drain side of the gate. As a result, drain current in GFETs cannot saturate by drain depletion pitch-off, which is the most common way of saturation in long-channel Si MOSFETs. Many attempts have been made to achieve current saturation in GFETs by achieving saturation of carrier velocity or by introducing a bandgap into graphene, for example, by the use of graphene nanoribbons, which has its own shortcomings and is beyond the scope of this chapter.

3.1.4.3 Metal Contact

Another important issue in the fabrication of 2D-material electronic devices is the contact between the 2D materials and the metals, which need to be ohmic with low resistance. A large contact resistance, owing to the formation of Schottky barrier for charge carrier injection, can significantly degrade the performance of FETs. The contact resistance contributes to the R_S and R_D of the device (Figure 3.1). Large R_S and R_D act as a severe source choke which limits the electron injection from the source to the channel and from the channel to the drain, thus limiting the f_T of the transistors. Large R_S and R_D also reduce the f_{max} of the device, since these resistive components can bring power loss, which degrades the power gain of the device. Furthermore, large contact resistance also weakens the control of the gate over the channel conductivity, since the effective gate voltage at the source injection point is $V_{GS} - I_{DS}R_C$, where R_C is contact resistance for both the source and the drain.

There are four main factors influencing the contact resistance in 2D materials devices: (i) the interface quality, (ii) the sheet resistance of 2D materials underneath the metal, (iii) the electrostatic barrier from the work function mismatch between the 2D materials underneath the metal and the 2D materials in the channel, and (iv) the electrostatic barrier from the work function mismatch between 2D materials and the metal. The last one can be minimized using an electrode material with a work function that best matches the work function of the contact 2D material. Therefore, appropriate metal selection is important in 2D materials devices so that a Schottky barrier is not formed.

Metal deposited on 2D materials usually form high contact resistance (typically 200–2000 Ω μm for graphene and 0.7–10 kΩ μm for semiconducting MoS_2 [13], respectively), which leads to Schottky-limited transport and is much higher than that in Si MOSFET (20–50 Ω μm). With surface contamination being one of the main causes of high contact resistance, much effort has been made to obtain a cleaner 2D materials surface, for example, using high-temperature vacuum annealing, or surface treatment using oxygen plasma, or an Al_2O_3 sacrificial layer. An alternative route of one-dimensional edge contact has also been proposed to improve the contact between graphene and metal [14]. In naturally n-doped 2D MoS_2 FETs, graphene has been shown to be an effective contact electrode. Phase engineering of MoS_2 has also been used to decrease the contact resistance in MoS_2 FET down to 200–300 Ω μm [13]. It is also shown that ferromagnetic contacts, for example, Cobalt (Co), together with the insertion of a thin oxide barrier of MgO between the contact and MoS_2, can reduce the Schottky barrier height significantly [15].

In WS_2 and WSe_2, the contact problem is even more severe. One possible solution is to chemically dope the contact in order to induce high carrier densities and produce ohmic contact [16].

3.2 Graphene Electronics for Radiofrequency Applications

One of the key properties of graphene is that its valence band touches the conduction band at the Dirac point, resulting in its zero bandgap and semi-metallic nature. As a result, GFETs have a relatively large off-state current, which makes it difficult to build logic circuits based on graphene with low standby power dissipation. An approach to solve this problem is to introduce a bandgap in graphene. Bandgaps up to 400 meV have been introduced by patterning graphene into nanoribbons to create quantum-mechanical confinement [17, 18]. But this approach comes with the price of significant mobility reduction [18, 19], or increased off-current [20]. The bandgap can also be introduced by applying a perpendicular electric field in bilayer graphene [21, 22]. The highest reported optical bandgap is 250 meV at a voltage more than 100 V [21]. These facts make GFET unsuitable for digital applications. For RF or analog applications, however, the transistors are continuously operated in the on-state, where it is not necessary to switch off the device. This fact, together with the extraordinarily high carrier mobility and saturation velocity in graphene, makes device researchers believe that graphene can play a relevant role in RF applications.

The state-of-the-art technology in high-frequency electronics is dominated by III–V and SiGe-based semiconductors with the highest reported f_{max} in the range of 1–2 THz [23], enabling simple circuits operating at up to a few hundreds of gigahertz [24, 25]. To some extent, graphene transistors have already fulfilled the expectations by having an intrinsic f_T comparable or close to the best III–V transistors and surpassing Si-based transistors for similar device gate lengths. Moreover, simulation studies [26] have revealed that the limit of f_T is a few tens of THz, which is still well beyond any previous experimental demonstration.

On the other hand, the missing bandgap in graphene can still significantly affect the on-state operation of GFETs and deteriorate its RF performance. In particular, the absence of the bandgap in graphene may prevent proper current saturation, especially at a short gate length. In this section, the performance of discrete RF GFET is discussed and evaluated. Subsequently, the ICs based on GFET are demonstrated to show some unique new electronics concepts that graphene can enable.

3.2.1 Experimental Graphene RF Transistors

In the previous sections, we have shown that high carrier mobility is needed to build an RF transistor. Because of its high carrier mobility, researchers explored the fabrication of GFET specifically designed for GHz operation. Within a relatively short time, the RF performance of GFET, in terms of f_T particularly, was greatly improved. Figure 3.4 summarizes the f_T performance of GFETs together with that of the fastest III–V high electron mobility transistors (HEMTs; i.e. GaAs mHEMTs and InP HEMTs) and silicon-based RF MOSFETs. Clearly,

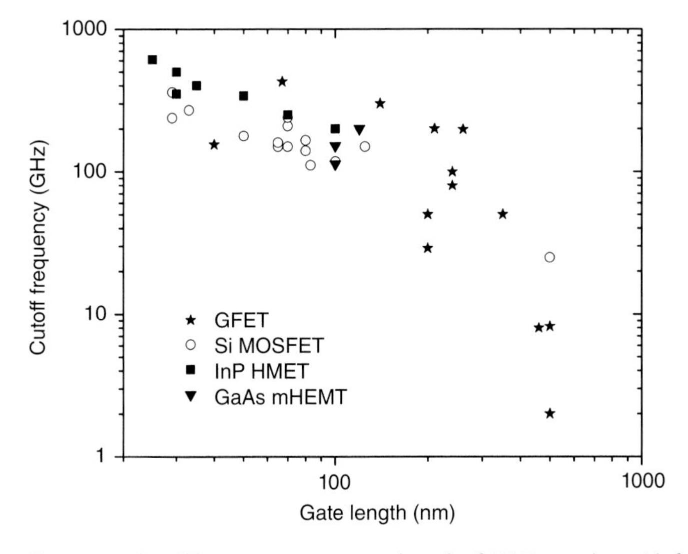

Figure 3.4 Cutoff frequency versus gate length of GFET, together with f_T performance of Si MOSFETs, InP HMETs, and GaAs mHEMTs.

GFETs compete well by delivering intrinsic f_T comparable to the best available III–V transistors, surpassing silicon-based transistors at comparable gate length. Moreover, the simulations have shown that the limitation for the f_T in GFETs can be at tens of THz. Such f_T has not been demonstrated experimentally, yet shows that GFETs may have the potential to surpass other technologies.

However, as mentioned earlier, for many RF applications, it is not the current gain and f_T but rather the power gain and f_{max} that are more relevant. From this point of view, graphene transistors still lag behind III–V and silicon-based MOSFETs, as shown in the f_{max}- L_g plot in Figure 3.5. Compared to the record f_{max} of InP HEMTs and GaAs mHEMTs exceeding 1 THz and that of Si MOSFETs of 420 GHz, the highest reported f_{max} for graphene is only 40 GHz. When comparing the f_T and f_{max} of GFETs with that of other RF FET types, it is also worth mentioning that the measured RF data of GFETs often undergoes a de-embedding procedure to exclude the influence of the RF probe pads and interconnections, which in some cases is different from the one commonly used for conventional RF FETs and usually leads to the reporting of more intrinsic f_T and f_{max} values.

As for the RF noise performance, there are not sufficient data on GFETs so far. Figure 3.6 shows the experimental NF_{min} data from Ref [27] together with the NF_{min} of Si MOSFETs, InP HMETs, and GaAs mHMETs. As can be shown, the noise performance of GFETs is far worse than that of conventional RF transistors. With NF_{min} below 1.6 dB up to 100 GHz, the III–V HMETs show the best noise performance. Si MOSFETs have a NF_{min} below 1 dB up to 20 GHz. For comparison, a Si MOSFET has a NF_{min} of 4.5 dB at 94 GHz, and a GFET shows a comparable NF_{min} (~4.3 dB) but only at 8 GHz. For a fair comparison, it is also important to take into account the gate lengths of the FETs. As for the data collected in Figure 3.6, the GFET noise data are from 1 μm gate length GFETs, the gate lengths of Si MOSFET are between 35 and 65 nm, and the gate lengths for

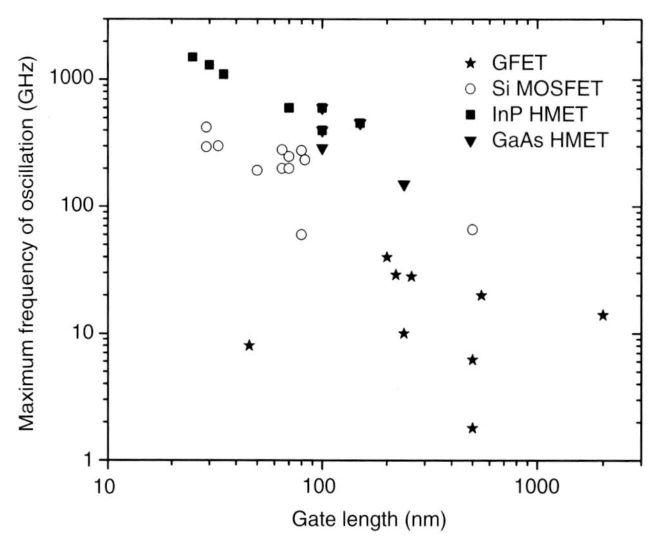

Figure 3.5 Maximum frequency of oscillation versus gate length of GFET, together with f_{max} performance of Si MOSFETs, InP HMETs, and GaAs mHEMTs.

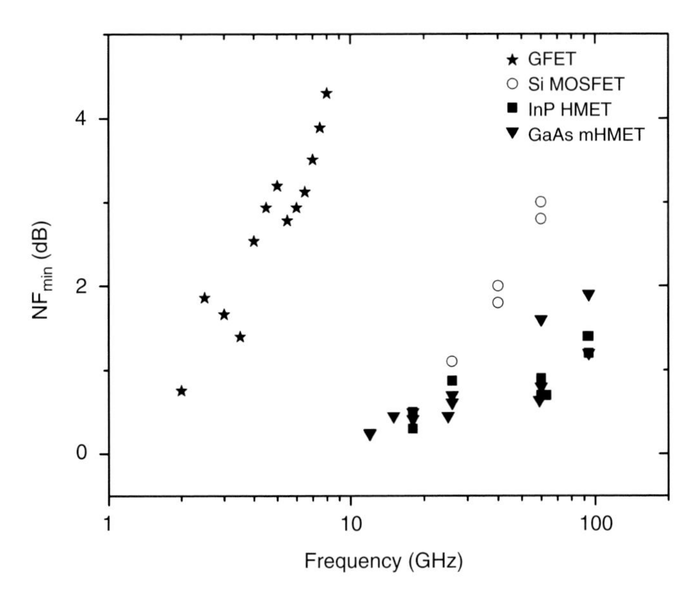

Figure 3.6 Minimum noise figure versus gate length of GFET, together with NF_{min} of Si MOSFETs, InP HMETs, and GaAs mHEMTs.

III–V HEMTs range from 50 to 150 nm. To compare noise performance with different gate lengths, we can use M_{noise} which is defined as

$$M_{noise} = \frac{NF_{min}}{f \cdot L} \tag{3.8}$$

where f is the frequency in GHz and L is the gate length of the FET in μm. Using Eq. (3.8) to estimate the M_{noise} in Figure 3.6, we can obtain M_{noise} ~ 0.5 dB GHz^{-1} μm^{-1} for GFETs, M_{noise} ~ 0.5–2.5 dB GHz^{-1} μm^{-1} for Si MOSFETs, and M_{noise} ~ 0.1 dB GHz^{-1} μm^{-1} for the III–V MOSFETs. Thus, the noise performance in GFETs is worse than that in III–V HMETs, but has comparable noise level with Si MOSFETs. It is expected that the noise performance of GFETs can be better than that of Si MOSFETs as GFETs scale down further.

3.2.2 Graphene-Based Integrated Circuits

The excellent performance of GFET makes it a suitable candidate for the design of high-frequency circuits such as amplifiers, mixers, and demodulators. The following discusses graphene circuits for RF application, starting from the idea of utilizing the ambipolar property of GFET to realize frequency multiplication, frequency mixing, phase-shift keying, and oscillators. Graphene-based RF receiver and electromechanical devices are also introduced.

3.2.2.1 Graphene Ambipolar Devices

Although the carriers in GFET channels may move very fast, the device lacks the current saturation commonly seen in conventional transistors. They also have a relatively low on/off ratio, and their integration in larger scale circuits is still challenging. These limitations hinder the path toward their application at the industrial level. Rather than trying to overcome these limitations (for example, by introducing a bandgap to graphene), some work has shown that we can exploit the unique property of graphene and use it for more advantageous applications. Graphene-based ambipolar electronics are proposed in this regard, relying on the fact that the conduction and valence bands in graphene meet at the Dirac point and thus the Fermi level in graphene can be swept continuously from the conduction band to the valence band by changing the gate voltage of an FET. During this process, the channel carrier changes from electron-type to hole-type and gives the characteristic graphene transfer curve which has been discussed in Section 3.1.2. This ambipolar property, although a drawback for GFETs as amplifiers or logic switches, enables a novel class of nonlinear RF electronics. Graphene frequency multipliers, oscillators, RF switches, high-frequency mixers, and digital modulators have been demonstrated taking advantage of this ambipolar property. The corresponding operating frequency is comparable to the RF ICs based on silicon (from several hundreds of megahertz to tens of gigahertz). However, the above-mentioned circuits are based on the GFETs with long channel length of hundreds of nanometers. Thus, its scalability cannot compete with the latest circuits based on III–V transistors and silicon-based transistors with sub-50-nm gate length.

Graphene Frequency Multipliers Frequency multiplication is one of the most important components for signal generation in radio communication. Frequency multiplication is typically realized by introducing a sinusoidal signal at a lower frequency to a nonlinear element (for example, a diode or a conventional FET) to generate harmonics. For both diode- and FET-based multipliers, the spectral purity of the generated signal is very poor, and additional filters are needed to extract the harmonic component of interest and achieve an integer multiple of the input frequency. To improve the output spectral purity for frequency multiplication, it has been proposed to utilize the ambipolar property of graphene. Figure 3.7 gives a demonstration circuit for a GFET frequency doubler. The inductors and capacitors at the input and output work as chokes and dc-block capacitors. During its operation, the gate voltage of the GFET is biased at its minimum conduction point and a sinusoidal input signal is superimposed to the gate bias voltage. As the conductivity in the GFET channel is modulated between electron and hole transport by the input signal, the current through the GFET remains stable due to the inductor in the bias tee. As a result, the drain voltage swings with the change of the channel conductivity and full-wave rectification can thus be achieved. This process is illustrated in Figure 3.7 to explain the working principle of ambipolar frequency doubling. Points A to E denote the different input and output signal levels at the input and output terminals. The channel conductivity is hole-dominated when the input signal swings from A through B to C, and the channel conductivity is electron-dominated when swinging from C through D to E. Each half-cycle swing of the input signal corresponds to a full-cycle swing at the output; hence, the frequency of the output signal is doubled.

Figure 3.8 shows the output signal spectrum of the first experimental demonstration of the frequency doubling in the K_U-band frequency (12–18 GHz) by

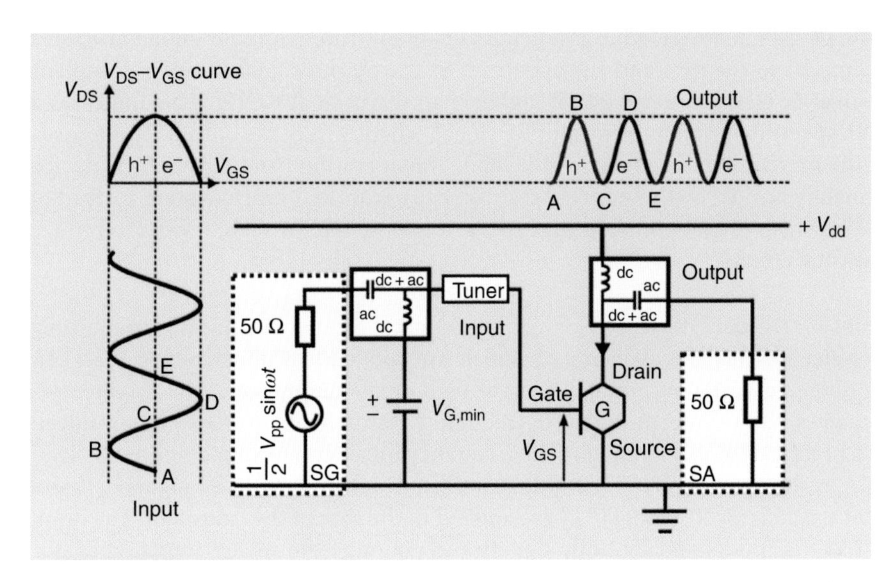

Figure 3.7 Demonstration circuit of a graphene frequency multiplier. Source: Wang et al. 2012 [2]. Adapted with permission of IEEE.

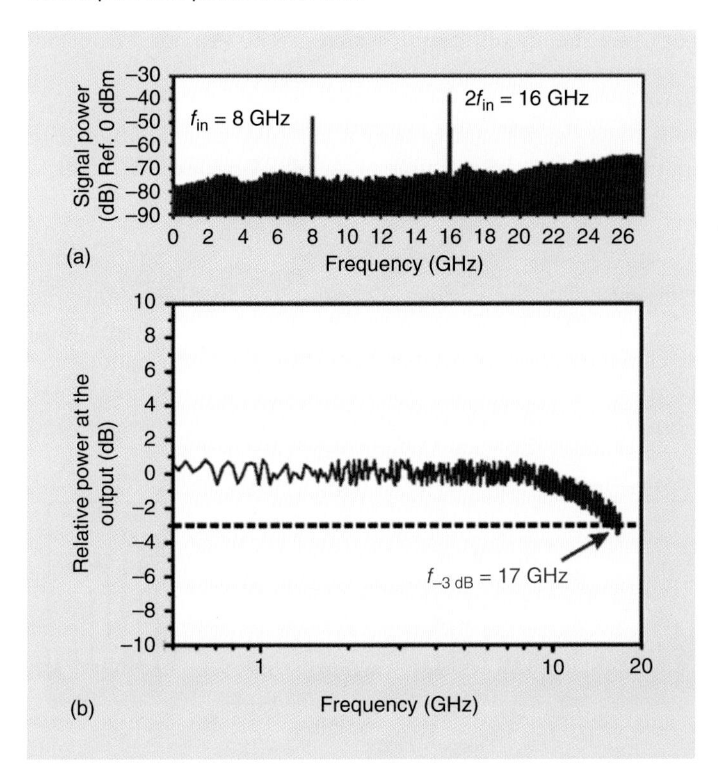

Figure 3.8 (a) Output power spectrum of a graphene ambipolar frequency multiplier; (b) frequency dependence of gain. The −3 dB cutoff frequency is at 17 GHz. Source: Wang et al. 2012 [2]. Adapted with permission of IEEE.

a single GFET device with gate length $L_G = 300\,\text{nm}$. An input signal at 8 GHz is applied to the gate and the spectrum of the output signal shows a dominant peak at 16 GHz, which is 11 dB higher than the peak at 8 GHz. This indicates a high spectral purity in the output RF signal, where 93% of the output RF energy is at the fundamental frequency (16 GHz). The operation bandwidth of the device is mainly limited by the external resistor-capacitor (RC) time constant instead of the intrinsic carrier constant time; thus, it is essential for such devices to reduce parasitic effects.

Graphene Frequency Mixers The main function of frequency mixers is to multiply two electrical signals. When two signals are applied to a mixer, new signals can be obtained at the output at either the sum or the difference of the original frequencies. Other intermodulation frequency components may also be produced in a practical frequency mixer. The frequency mixer can be implemented with an FET, as shown in Figure 3.9. In this structure, a high-frequency RF signal (f_{RF}) and a local oscillator RF signal (f_{LO}) are applied to the gate of the transistor. The drain current is modulated by both signals and contains the mixed frequencies, for example, the sum $f_{RF} + f_{LO}$ and the difference $f_{RF} - f_{LO}$ (referred to as the intermediate frequency, f_{IF}) of the two input frequencies. In practice (for example, in a radio receiver application), input signals f_{RF} and f_{LO} only differ by a small amount. f_{IF} is usually the output frequency of interest, which can be extracted from the drain current by a low-pass filter.

A mixer based on GFET takes advantage of its symmetrical transfer characteristics to produce a very significant quadratic component, as discussed in Section 3.1.2. Assuming that the transfer curve of the GFET is ideally symmetric and infinitely differentiable, its drain current can thus be described as

$$I_D = a_0 + a_1(V_{GS} - V_{G,min})^2 + a_4(V_{GS} - V_{G,min})^4 + \cdots \tag{3.9}$$

where $V_{G,min}$ is the gate voltage at the minimum conduction point and $a_0, a_2, a_4, \ldots$ are constants. This expression shows that when a GFET with ideal symmetric transfer characteristics and is biased at the minimum conduction point, no odd-order intermodulation distortions will appear in the output signal, and all the output power is coupled to the sum and difference frequencies and

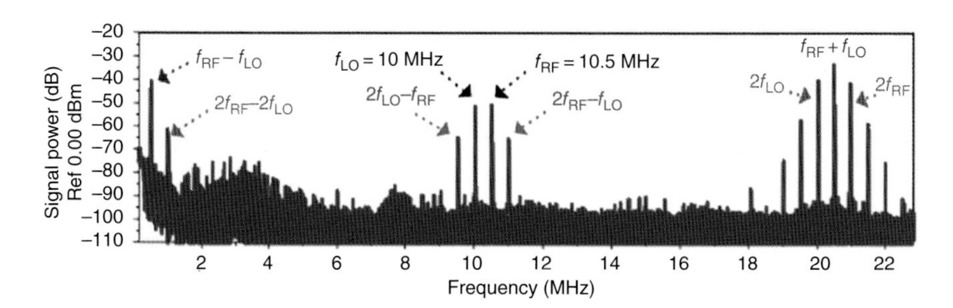

Figure 3.9 Output power spectrum of the first graphene frequency mixer. Source: Wang et al. 2012 [2]. Adapted with permission of IEEE.

other even-order terms. The odd-order intermodulation products, therefore, can be largely suppressed in GFET mixers, which are often present in conventional unipolar mixers and are harmful to circuits operation.

Figure 3.9 shows the power spectrum at the output of a GFET mixer from Ref [2]. When an RF input signal and a local oscillator signal are introduced to the gate, the GFET mixer delivers output signals with a frequency equal to the sum ($f_{RF} + f_{LO} = 20.5$ MHz) and difference ($f_{RF} - f_{LO} = 500$ kHz). Meanwhile, the odd-order intermodulation products are significantly suppressed. GFET mixers, taking advantage of the unique ambipolar transfer characteristics of graphene, have the potential to revolutionize the design of many RF communication circuits. In 2011, researchers at IBM integrated such a GFET mixer with two inductors on a single silicon carbide wafer which operates at frequencies up to 10 GHz (Figure 3.10).

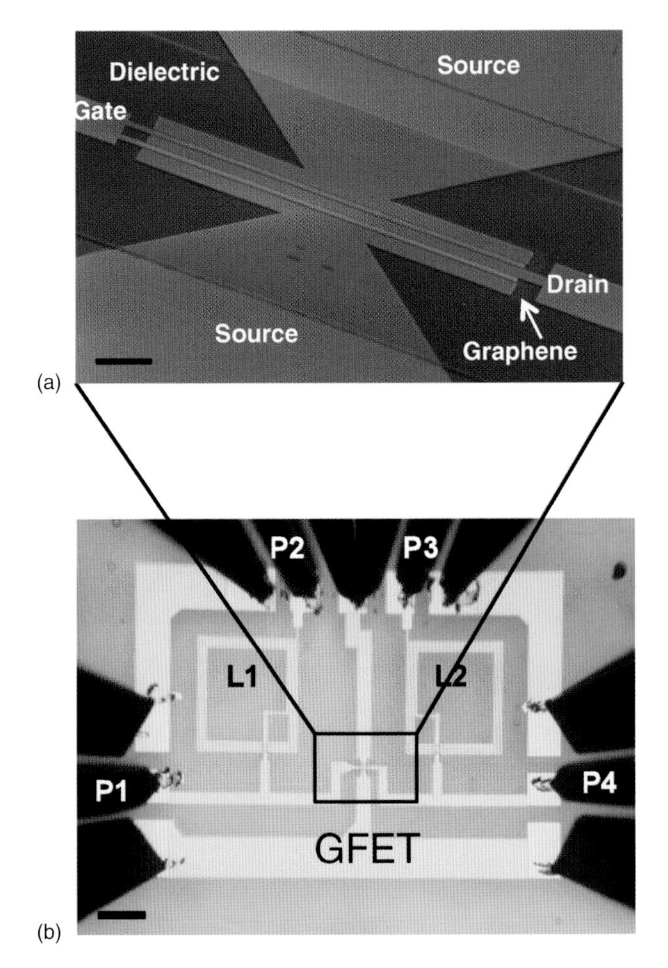

Figure 3.10 An integrated graphene frequency mixer. Source: Lin et al. 2010 [28]. Adapted with permission of AAA.

Binary Phase-Shift Keying Binary phase-shift keying conveys the data by changing the phase of a reference signal. If the phase of the wave does not change, the signal state stays the same (0 or 1). If the phase of the wave changes by 180°, then the signal state changes (from 0 to 1, or from 1 to 0). This type of modulation of digital signal onto an analog signal is essential for wireless digital communication systems such as Bluetooth and Zigbee transceivers.

The ambipolar transfer properties of GFET can also be utilized in such applications. When GFET works as a binary phase-shift keying device, a digital square wave (data signal) and a high-frequency sinusoid (carrier signal) are applied to the gate of the GFET, which is biased at the minimum conduction point by a dc power source. The channel of the GFET is switched between electron and hole conduction by the input digital signal. Due to the negative gain on the hole branch and the positive gain on the electron branch of the GFET transfer characteristics, the output signal equals the carrier wave modulated by the data signal, with a 180° phase shift between "1" and "0."

Figure 3.11 shows the experimental demonstration of a GFET for phase-shift keying application. A carrier signal with $f_{carrier}$ = 500 Hz and a digital data signal with f_{data} = 50 Hz are applied to the gate of GFET, which is biased at the minimum

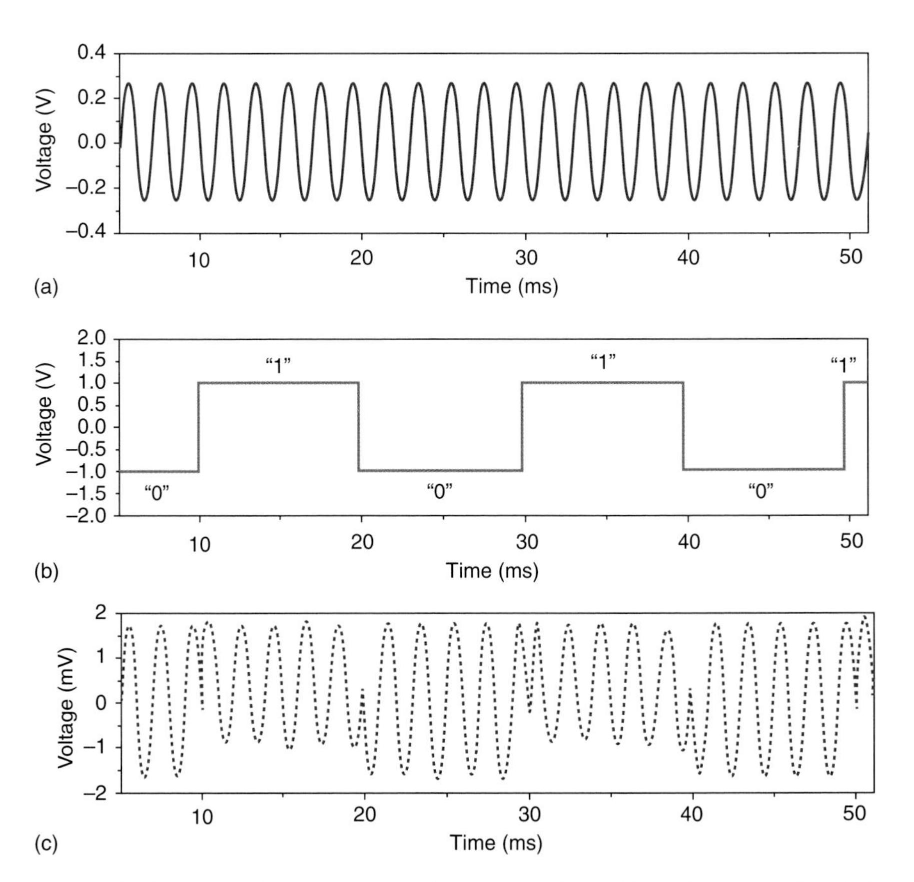

Figure 3.11 Experimental demonstration of a graphene phase-shift keying device. Hsu et al. 2011 [29]. Adapted with permission of Japanese Journal of Applied Physics.

conduction point. The output signal is shown in Figure 3.11 where a 180° phase shift is clearly obtained. Although the experimental demonstration here is at a low frequency, its actual working frequency is only limited by the speed of the GFET.

3.2.2.2 Graphene Oscillators

RF and microwave oscillators can provide signal sources for frequency conversion and carrier generation. They are critical in almost all RF communication systems, including cell phones, satellite transponders, and radio and television transmitters. The basic building blocks of a typical electronic oscillator are the amplifier circuit and the feedback network. In a typical electronic oscillator, the amplifier, which is an FET in this case, amplifies all signals within its bandwidth. The amplifier circuit can be forced to oscillate by feeding some of its output energy back to the input through a positive feedback network. The prerequisites for the oscillation to occur at a particular frequency are, first, the loop gain, which is the gain of the amplifier minus the loss in the feedback network, must be higher than 1; second, the loop phase shift of that particular frequency component must be the integer multiples of 2π.

Figure 3.12 shows the experimental setup which is used to test the first graphene RF electronic oscillator. Picoprobe RF probes are used to contact the drain and gate of the device. Two dc sources bias the drain and gate of the GFET. The oscillation frequency is set by a mechanical tuner. The directional coupler allows a spectrum analyzer to measure the power spectrum of the output signal. Figure 3.12 shows the power spectrum of the first electrical oscillator based on graphene. The tested device is oscillating at a fundamental frequency of 72 MHz, and higher order harmonics are also clearly visible.

3.2.2.3 Graphene RF Receivers

An RF receiver is an indispensable part of wireless communication which receives the modulated RF signal, and demodulates it. In order to realize the graphene RF receiver circuit, researchers at IBM developed a fabrication process that can preserve the GFET quality to the largest extent. Previously, the proof-of-concept ICs we introduced were fabricated by conventional process flow, that is, the active device is fabricated first. This can largely deteriorate the

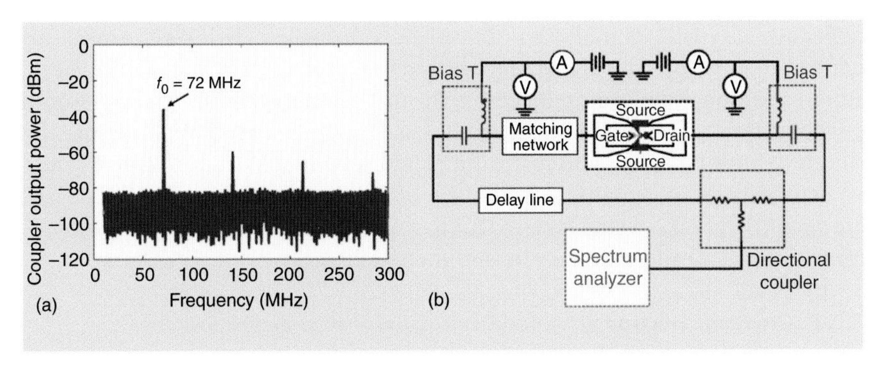

Figure 3.12 (a) Output power spectrum of the first graphene electronic oscillator; (b) demonstration circuit of graphene electronic oscillators. Source: Wang et al. 2012 [2]. Adapted with permission of IEEE.

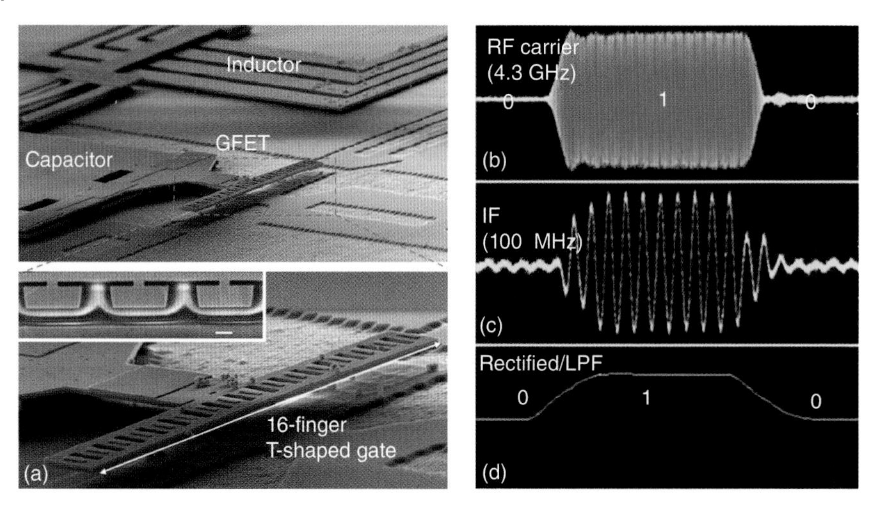

Figure 3.13 (a) Scanning electron microscopic (SEM) image of the key components in an integrated graphene receiver and the enlarged view of the gate; (b) measured waveforms of RF input signal amplitude, f_{IF} output signal and restored binary code after rectifying and low-pass filtering. Source: Han et al. 2014 [30]. Adapted with permission of Springer Nature.

GFET performance due to the delicacy of graphene. To overcome the device degradation problem, a "passive-first active-last" fabrication flow is proposed, which completely reverses the conventional fabrication procedures. In the fabrication process, all passive components together with the device gates are fabricated on silicon wafers before the CVD graphene is transferred. Using the proposed fabrication method, the first RF receiver, a multistage graphene-based IC has been realized in a standard fabrication.

The receiver circuit consists of three stages (Figure 3.13a). The first two stages of the circuit are designed to provide amplification and filtering of the received RF signal. The third stage is designed to provide a down-conversion mixing which converts the GHz RF signal to intermediate frequency (f_{IF}) in the megahertz range.

To demonstrate the function of the IC as an RF receiver, an RF carrier signal of 4.3 GHz is modulated with a bit stream and then sent to the graphene receiver, mimicking a typical digital data transmitted through a wireless carrier. An undistorted IF output signal is achieved from the receiver, shown in Figure 3.13b, without additional signal amplification. The original binary code can be restored by rectification and low-pass filtering. This suggests that receiver function, the abilities of both receiving and restoring digital text, is achieved. The digital text that demonstrates successful reception encoded a three-word message: "I-B-M." This demonstration is a step forward for the applications of graphene in ICs with higher complexity and more advanced functionality.

3.2.2.4 Graphene Electromechanical Devices: Resonators and RF Switches

Nanoelectromechanical devices and systems, such as filters, resonators, and RF switches, are of great interest both for fundamental studies of mechanics

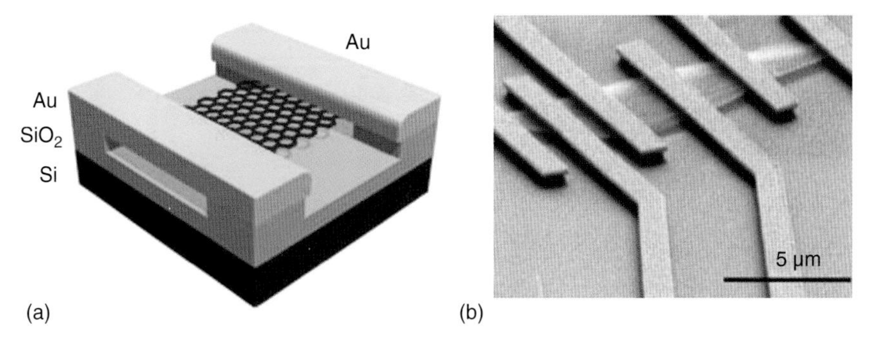

Figure 3.14 (a) Schematic of resonator based on suspended graphene; (b) SEM image of several monolayer graphene resonators. Source: Chen et al. 2009 [31]. Adapted with permission of Springer Nature.

on a nanometer scale, as well as for their novel applications in numerous communication systems. The enormous stiffness and low density of graphene open up the possibility to minimize such devices to one-atom thickness. The first electromechanical device based on graphene was a nanomechanical resonator, which was demonstrated by Chen et al. in 2009. The resonator used suspended monolayer graphene, whose structure is depicted in Figure 3.14. An all-electrical high-frequency mixing approach is implemented to realize the actuation and detection of its mechanical resonance. During the test, a dc gate voltage is applied to introduce a static tension to the device. The motion of the resonator is driven by two RF voltage sources. The first RF voltage at frequency f is fed to the gate and superimposed on the dc bias, while the second RF voltage, at a slightly offset frequency $f + \Delta f$, is applied to the source. The motion of the suspended graphene can be measured as a mixed-down current at the difference frequency Δf since the graphene conductance changes with the distance from the gate. Such a graphene resonator operates at 50–80 MHz with a quality factor of $\sim 1 \times 10^4$ at 5 K.

The resonant frequency can be modeled as

$$f_{\mathrm{res}}(V_{\mathrm{g}}) = \frac{1}{2L}\sqrt{\frac{T_0 + T_{\mathrm{e}}(V_{\mathrm{g}})}{\rho W}} \tag{3.10}$$

where L and W are the length and width of the graphene sheet, respectively. The 2D mass density ρ accounts for the contributions from both the graphene and any surface adsorbates. T_0 is the built-in tension, and T_{e} is the electrostatically induced tension. It is found both theoretically and experimentally that the graphene resonant frequency increases as the length of the graphene membrane decreases (scales approximately as $1/L$). Since graphene is believed to withstand high strains (up to 25% as confirmed by nanoindentation experiments [32]), it is possible to increase the resonant frequency of such devices to the GHz range as the size of the graphene membrane further scales down.

Graphene electromechanical switches have also been demonstrated by Milaninia et al. [33]. A switch, as shown in Figure 3.15, consists of top and

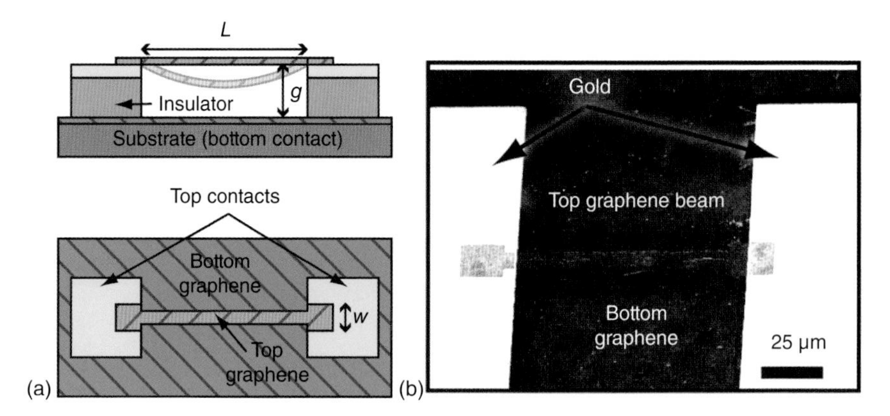

Figure 3.15 (a) Schematic of RF graphene switches, from a cross-sectional view and a top view; (b) the fabricated graphene switch. Source: Milaninia et al. 2009 [33]. Adapted with permission of AIP.

bottom polycrystalline graphene films grown by ambient pressure CVD. The top graphene is brought into electrical contact with the bottom graphene by apply a voltage of 5 V between the layers, and the two films are disconnected by the mechanical restoring force when the bias is removed. The graphene films are able to carry current densities in excess of 7 kA cm^{-2} in switch-on state, which shows they are more robust electrical conductors than traditional metal–metal contacts. Such graphene-based switches are expected to exhibit better reliability if their crystalline and contact resistance can be improved.

3.3 MoS_2 Devices for Digital Application

Despite the high mobility and high saturation velocity graphene possesses, its lack of bandgap leads to a relatively large off current which makes a GFET hard to switch off, hindering its application in digital electronics. TMDs, which have a bandgap in the range 1–2 eV for most of them, can be an alternative.

There are a large number of semiconducting 2D materials in the TMDs family. Past research in electronic devices has been mainly focused on MoS_2, WS_2, WSe_2, and other 2D TMDs. MoS_2 has attracted much attention partly due to its natural abundance. Crystals of MoS_2 are composed of vertical stacked layers held together by van der Waals interactions. A 0.65-nm-thick monolayer can be exfoliated using scotch tape or lithium-based interaction [9]. Its large-area polycrystalline sheets can be synthesized using CVD. Bulk MoS_2 has an indirect bandgap of 1.29 eV, while monolayer MoS_2 has a direct bandgap of 1.8 eV [34]. In this section, single- and thin few-layer MoS_2 is used as an example for constructing transistor devices, and we first show that MoS_2 is promising for the digital application due to its sharp turn-on, decent drive current, and high on/off ratio despite their moderate mobility. Then in the following section, we discuss TMDs-based transistors as basic building blocks for digital and analog circuits.

3.3.1 Experimental MoS_2 Transistors

The first implementation of monolayer MoS_2 transistor was demonstrated by Radisavljevic et al. [9]. The demonstrated monolayer MoS_2 transistor, with HfO_2 as high-κ dielectric layer, exhibits an I_{on}/I_{off} ratio exceeding 10^8 at room temperature with I_{off} as low as 100 fA (25 fA μm^{-1}). Notably, the reported SS value of 74 mV/decade is close to the theoretical estimation of 60 mV/decade for standard FET configuration. Dual-gated transistors based on thin few-layer MoS_2, encapsulated by hBN and with graphene as contacts have also exhibited an I_{on}/I_{off} ratio of 10^6 and low SS of 80 mV/decade [35]. It is worth noting that uniform encapsulation is important for MoS_2 transistors since single-layer MoS_2 is sensitive to adsorbates, according to low-frequency electronic noise measurement [36]. Encapsulation of MoS_2 can not only improve the device mobility but can also help minimize the hysteresis effect due to trapping states induced by absorbed water molecules on the MoS_2 surface and by the photosensitivity of MoS_2 under white illumination [37].

To further improve the switching efficiency of MoS_2 FETs, the FET channel length has been scaled down to 15 nm with minimum SS reported as 66 mV/decade [38], and sub-10 nm with SS reported as 120 mV/decade [39]. The MoS_2 transistor using a single-wall CNT as the gate electrode (with 1-nm physical gate length and ~3.9-nm effective channel length in the on-state) has also been fabricated, which exhibits excellent switching behaviors with SS reported as 65 mV/decade [40]. Negative-capacitance FETs based on MoS_2 have also been proposed to further decrease the SS down to 57 mV/decade, which is below the thermionic limit of 60 mV/decade [41].

Most of the reported MoS_2 transistors appear to be naturally n-doped, which is commonly attributed to Fermi-level pinning at the MoS_2 contact interface [42]. Thus, the large Schottky barrier height for holes at the source/drain contacts limits the hole injection, making it difficult to fabricate *p*-type MoS_2 FETs. Typically, palladium (Pd), with a high work function, has been widely used as contact material to the valence band of various materials, including graphene and CNTs [43]. However, the Fermi level of Pd is slightly above the valence band maximum of MoS_2; thus, it is not sufficient as a hole contact for MoS_2. Although it is suggested that *p*-type transport can be observed in thicker samples [44], many efforts have been made to obtain sufficient hole injection by selecting contact materials. Substoichiometric molybdenum trioxide (MoO_x, $x < 3$), a high work function material, has been shown to be an efficient hole injection layer to MoS_2, which enables fabrication of PFETs with $I_{on}/I_{off} \sim 10^4$ [43].

Although single-layer MoS_2 FETs exhibit high I_{on}/I_{off} ratio and low off-state current, which is important for minimizing the loss in the devices when they are turned off, it can only supply a very limited amount of current when the device is turned on (only 2.5 A m^{-1} at $V_{DS} = 0.5$ V as reported in Ref [9]). Since the speed of a logic circuit is often determined by the ratio between the charge required to change the voltage across the various capacitances in the circuits and the current that can be supported by the transistors, the low on-state current in single-layer MoS_2 may limit the operation speed of any electronic systems constructed from this material. On the other hand, by increasing the number of MoS_2 layers, the on-state current of MoS_2 FETs can be increased significantly with only small degradation in terms of on/off current ratio. For real electronic application, the selection of the number of layers may depend on the type of application. If better frequency performance is needed, then multilayer MoS_2 may be used. If ultralow power performance is necessary, then single-layer MoS_2 may be a better choice. And bilayer and tri-layer MoS_2 thin films may offer good trade-off in between. In short, the capability to control the number of molecular layers in the 2D crystal and the consequent control of the electronic properties enables added flexibility in this material system.

Transistors based on other TMDs materials, including WSe_2, WS_2, and $MoSe_2$, have also been demonstrated. WSe_2 monolayer transistors have shown ambipolar behavior with reported electron and hole mobility of $\sim$100 $cm^2\ V^{-1}\ s^{-1}$[16, 45–47]. The *p*-type FET based on single-layer WSe_2 with chemically p-doped source/drain contacts and high-κ gate dielectrics exhibits an I_{on}/I_{off} of $>10^6$ and an SS of $\sim$60 mV/decade at room temperature [16]. *N*-type WSe_2 FET with indium as contact shows an I_{on}/I_{off} exceeding 10^6 at room temperature and a record high on-current in TMDs-based FETs of 210 μA μm^{-1} [45]. As for WSe_2 FETs, a major challenge is that WSe_2 tends to form a substantial Schottky barrier with most metals commonly used for making electrical contacts. Much work has been focused on lowering the contact resistance for WSe_2 FETs by degenerately surface doping of WSe_2 (e.g. with NO_2 and K) in order to reduce the Schottky barrier thickness or by using low work function metal for n-type FETs or high work function metal for *p*-type semiconductors to lower the Schottky barrier height to the conduction band. It has also been proposed to use ionic-liquid-gated graphene contacts to achieve low-resistance ohmic contacts in WSe_2 [48] since the ionic-liquid gate can greatly modulate the charge carrier density [46]. As for WS_2, liquid-gated FETs based on multilayer and single-layer WS_2 have been fabricated, giving an estimate of the mobility of 40 $cm^2\ V^{-1}\ s^{-1}$ at room temperature [49–51], and can be further optimized to 83 $cm^2\ V^{-1}$ s$^-$1 by interface modification [52]. Standard back-gated FETs based on single-layer WS_2 have shown n-type behavior with a high room-temperature on/off current ratio of $\sim 10^6$ [51, 53]. *P*-type FETs on WS_2 have also been realized by introducing substitutional doping of sulfur with nitrogen atoms, which shows the room-temperature mobility of $\sim$19 $cm^2\ V^{-1}\ s^{-1}$ and an I_{on}/I_{off} of $\sim 10^5$. FETs based on exfoliated multilayer $MoSe_2$ exhibit a room-temperature mobility of 100–160 $cm^2\ V^{-1}\ s^{-1}$ on Parylene-C substrate, and also show highly asymmetric ambipolar behavior, with the I_{on}/I_{off} ratio exceeding 10^6 for electrons and less than 10^3 for holes [54]. Large-grain and highly crystalline multilayer $MoSe_2$ films grown by modified chemical vapor deposition (mCVD) shows n-type FET characteristics with a room-temperature mobility of $\sim$121 $cm^2\ V^{-1}\ s^{-1}$ and an I_{on}/I_{off} of 10^4 [55].

3.3.2 MoS_2-Based Integrated Circuits

3.3.2.1 Direct-Coupled FET Logic Circuits

As shown previously, MoS_2 FETs exhibit n-type characteristics with negative threshold voltage, while *p*-type MoS_2 FET is hard to realize because of the difficulty in chemical doping, the trapping states, and the lack of good metal contact for *p*-type MoS_2. To achieve seamless integration of n-type MoS_2 FETs, direct-coupled transistor logic (DCTL) configuration can be a good choice. In this configuration, both negative and positive threshold voltages can be obtained by selecting gate metals with different work functions; thus, both depletion mode (D-mode) and enhancement mode (E-mode) on a single sheet of MoS_2 can be realized. Moreover, DCFL can offer a trade-off between speed and power consumption with extremely simple circuitry and a minimum number of auxiliary components. With these desirable features, DCFL serves as an excellent architecture for high-speed circuit with low-power dissipation for digital applications.

Figure 3.16 shows the structure of a simple DCTL circuit with D-mode and E-mode MoS_2 transistors sitting side by side on the same chip. To control the gate threshold voltages, working functions of gate metal (w_M) are carefully selected to be either larger than the working function of the semiconductor (w_S) or smaller. Gate metal with a lower working function ($w_M < w_S$) tends to induce electrons in the channel, tuning the channel into the charge accumulation regime; while metal with a higher working function tends to tune the channel into the charge depletion regime, all at zero gate bias. The Al-gate in this structure is selected for the D-mode FET, while the Pd-gate is for the E-mode FET.

Figure 3.17 describes the energy diagram of the transistors in the DCTL circuit. By carefully choosing metal and semiconductor working functions, the channels of the transistors can be designed to be either in the accumulation regime for D-mode FET or in the depletion regime for E-mode FET. The designed gate metal of different working functions can effectively shift the FET threshold voltage. This D/E-mode FET pair, with excellent pinch-off and current saturation behavior, can serve as a building block for digital circuits.

3.3.2.2 Logic Gates

A variety of combinational logic gates (inverter, NAND, NOR, AND, OR, XOR, XNOR) based on CVD-grown MoS_2 have been demonstrated. Such logic gates typically use pull-up networks, which connect the supply and the output node, and pull-down networks, which connect the output node and the ground. In conventional CMOS technology based on silicon, the pull-up network uses pMOS transistors, allowing rail-to-rail operation with lower power consumption. Since

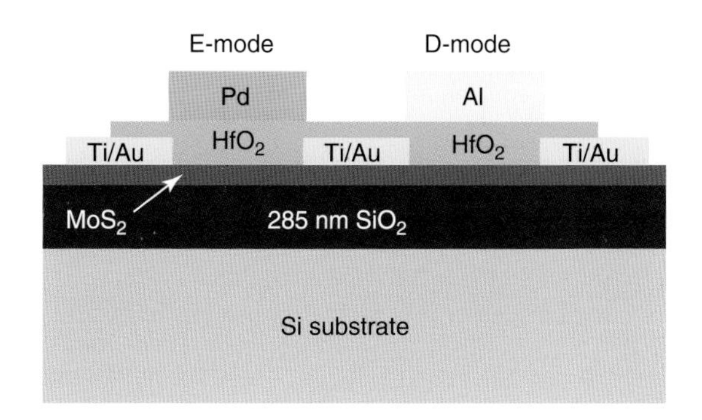

Figure 3.16 Schematic representation of an E-mode and D-mode device. Source: Wang et al. 2012 [56]. Adapted with permission of ACS.

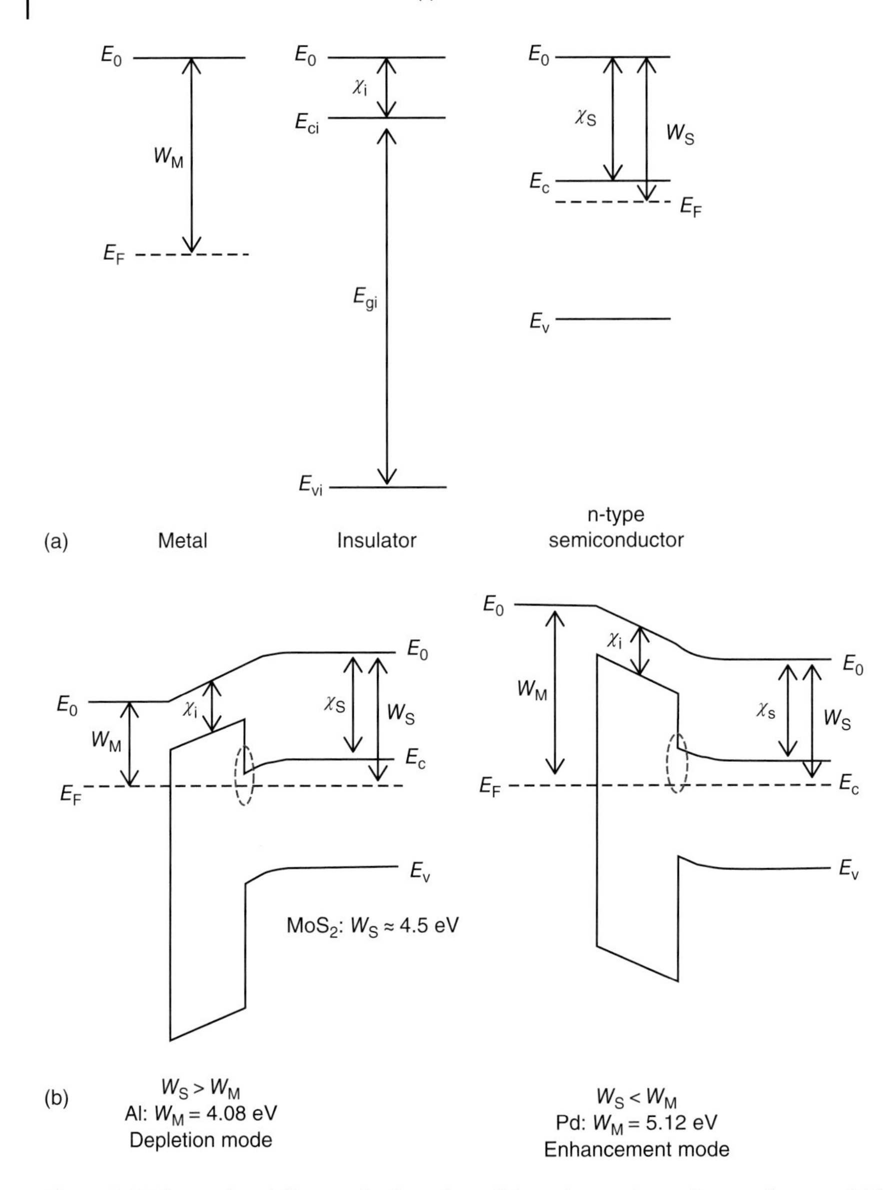

Figure 3.17 Energy band diagram for E-mode and D-mode transistors. Energy diagrams (a) for isolated metal, insulator, and semiconductor, and (b) after bringing them in intimate contact and thermal equilibrium is established.

there are no pMOS transistors in the existing MoS_2 technology, MoS_2-based logic gates have been fabricated with all E-mode FETs or by implementing DCFL technology. The following gives examples of an inverter using all E-mode FETs, and NAND gates using DCFL technology.

A logic inverter is a basic component in digital circuits which can convert logical "1" (high input voltage) into logical "0" (low output voltage), and vice versa.

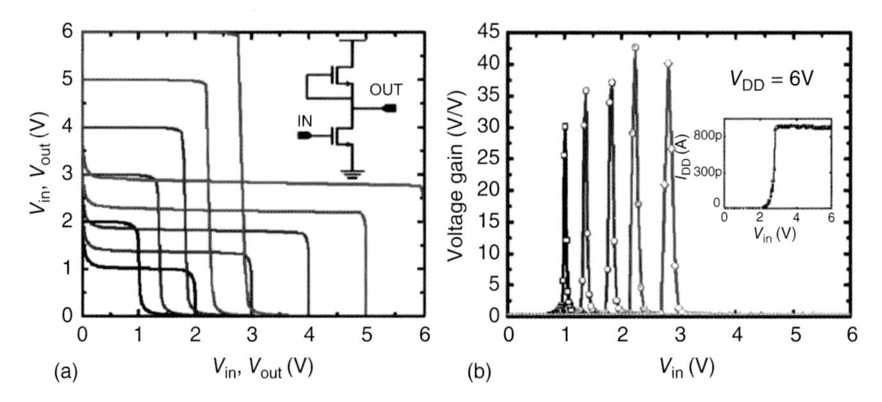

Figure 3.18 Demonstration of an integrated logic inverter. (a) Transfer characteristics and corresponding butterfly curves of MoS_2 logic inverter at different V_{DD}; (b) voltage gain of MoS_2 logic inverter, inset: current changes with input voltage. Source: Yu et al. 2015 [57]. Adapted with permission of IEEE.

An inverter can be constructed from E-mode nMOS FET subthreshold regime where gate-to-source voltage equals zero as depicted in Figure 3.18a (inset) [57]. Figure 3.18 shows an experimental demonstration of such an inverter. The as-fabricated inverter exhibits an excellent inverting performance under a wide range of V_{DD} values. The transition between two logic states "0" and "1" can be characterized by the voltage gain $A_v = -dV_{out}/dV_{in}$, which reaches to 45 for this device at $V_{DD} = 5\ V$. The noise margin of this device is higher than 43% of V_{DD}. This large noise margin, together with its high gain and matched input–output, makes MoS_2-based inverters promising to be integrated into multistage logic circuits.

A NAND gate is a logic gate which outputs false only if all its inputs are true. It is one of the two basic logic gates (the other being the NOR gate) for universal functionality. A NAND gate is significant because any other type logic gates (AND, OR, NOR, XOR, etc.) can be implemented using a combination of NAND gates. Figure 3.19a shows the schematic design of a NAND gate circuit and its experimental demonstration based on bilayer MoS_2. The output of the circuit is close to the supply voltage (V_{DD} = 2 V, logic state 1) when either or inputs are both at logic state 0 ($V_{in} < 0.5$ V). In this case, at least one of the MoS_2 FETs is nonconducting and the output voltage is clamped to the supply voltage V_{DD}. The output is at logic state 0 only when both inputs are at logic state 1. In this case, both MoS_2 FETs are conducting. Figure 3.19c shows the experimental result of such a NAND gate, where the output voltage is measured as a function of time, while two input voltage stages vary across all four possible logic combinations (0,0), (0,1), (1,0), (1,1). This result confirms the stable NAND gate function of the circuit consisting of two active MoS_2 transistor and one MoS_2 load resistor. Hence, it is also possible to fabricate any kind of digital IC with MoS_2 thin films.

Logic gates based on other TMDs materials, for example, WSe_2 [58, 59], ReS_2 [60], and $MoTe_2$ [61], have also been demonstrated with high gain and low power consumption.

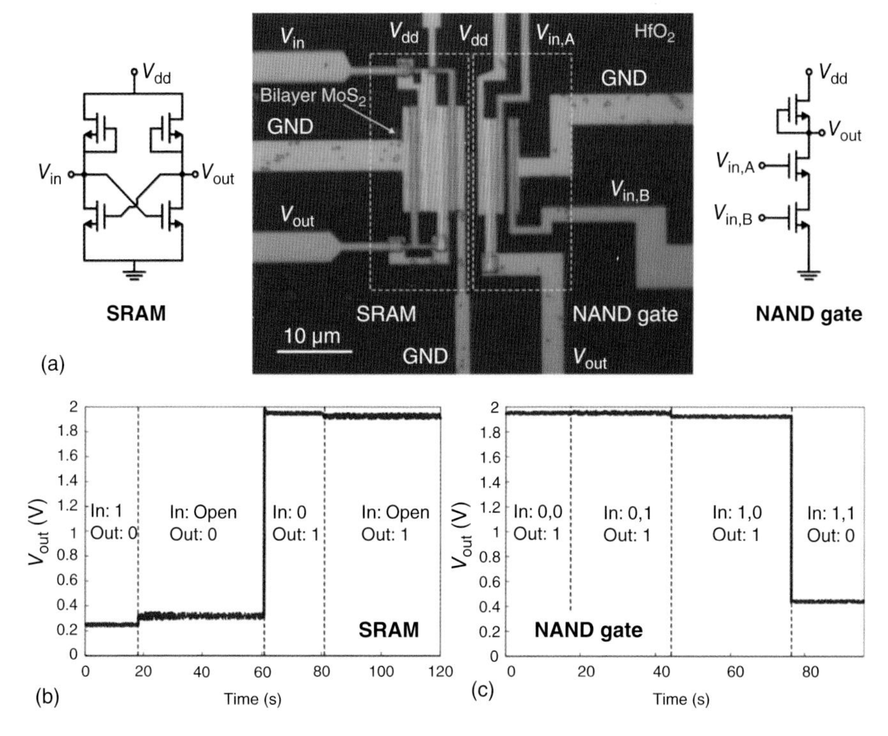

Figure 3.19 Demonstration of an integrated NAND logic gate and a static random access memory (SRAM) cell on bilayer MoS_2. (a) Schematics and optical micrograph of the NAND gate and the SRAM. (b) Output voltage of a flip-flop memory cell (SRAM). (c) Output voltage of the NAND gate. Source: Wang et al. 2012 [56]. Adapted with permission of ACS.

3.3.2.3 A Static Random Access Memory Cell based on MoS_2

Figure 3.19a has also shown a flip-flop memory element (static random access memory (SRAM) constructed from a pair of cross-coupled inverters. This storage cell has two stable states at the output, which are denoted as 0 and 1. The flip-flop cell can be set to logic state 1 (or 0) by applying a low (or high) voltage to the input. Figure 3.19b shows an experimental demonstration of an SRAM based on MoS_2. To verify the functionality of this flip-flop cell, a voltage source is applied to the input to set V_{in} to 2 V at time T = 0 seconds. This drives V_{out} into logic state 0. Then at T = 20 seconds, the switch at V_{in} is opened and the output of the SRAM cell V_{out} remains at logic state 0. At time T = 60 seconds, we apply V_{in} = 0 V at the input to write a logic state 1 into V_{out}. As the switch is opened again at T = 80 second, the output of the SRAM cell remains in the logic state 1, which demonstrates that the flip-flop SRAM circuit fabricated on the bilayer MoS_2 thin film indeed functions as a stable memory cell.

3.3.2.4 Ring Oscillators based on MoS_2

A ring oscillator consists of an odd number of inverters which form a ring structure by connecting the output of each inverter to the next and the last output to the first. It oscillates between high-voltage and low-voltage states. In RF applications and communication systems, oscillators are used for frequency

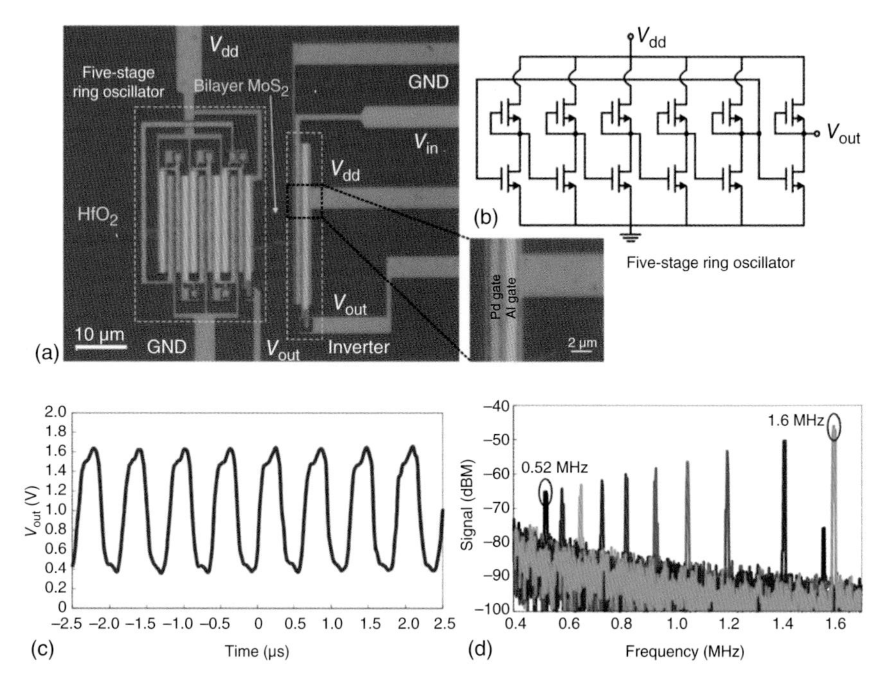

Figure 3.20 Demonstration of a five-stage ring oscillator. (a) Optical micrograph of the ring oscillator. (b) Schematic of the electronic circuit of the five-stage ring oscillator. (c) Output voltage as a function of time for the ring oscillator. (d) The power spectrum of the output signal. Source: Wang et al. 2012 [56]. Adapted with permission of ACS.

translation and channel selection, while they are also important components in digital systems, for example, to provide a clock reference. An example of a five-stage ring oscillator is shown to assess the high-frequency switching capability of MoS_2 and for evaluating the material's ultimate compatibility with conventional circuit architecture (Figure 3.20). The ring oscillator, which integrates 12 bilayer MoS_2 FETs together, was realized by cascading five inverter stages in a closed-loop chain. An extra inverter stage was used to synthesize the output signal by isolating the oscillator operation from the measurement setup to prevent the interference between them. The output of the circuit was connected to either an oscilloscope or a spectrum analyzer for evaluation. For robust ring oscillator performance, it is imperative to have stable operations in all five inverter stages throughout the oscillation cycles, and its tolerance toward noise can be determined from the noise margins for both low and high logic levels, i.e. the shaded regions in Figure 3.18b. The positive feedback loop in the ring oscillator results in a statically unstable system, and the voltage at the output of each inverter stage oscillates as a function of time (Figure 3.20c). The oscillation frequency of a ring oscillator depends on the propagation delay τ_{pd} per stage and the number of stages used in the ring oscillator. For the MoS_2-based ring oscillator, at the supply voltage V_{DD} of 2 V, the fundamental oscillation frequency is at 1.6 MHz, corresponding to a propagation delay of $\tau_{pd} = 1/2nf = 62.5$ ns per stage, where n is the number of stages and f is the fundamental oscillation frequency. The frequency performance of this ring oscillator, while operating at a much lower V_{DD}, is at least an order of magnitude better than the fastest integrated organic semiconductor ring oscillators. It also rivals the speed of ring oscillators constructed from the printed ribbons of single-crystalline silicon reported in the literature. The output voltage swing measured by the oscilloscope is about 1.2 V.

The output signal of the ring oscillator can also be measured in terms of its frequency power spectrum. Figure 3.20d shows the spectrum of the output signal from the ring oscillator as a function of the drain bias voltage V_{DD}. The improvement in frequency performance with increasing V_{DD} can be attributed to the enhancement in the current driving capability of the ring oscillator due to the rise in the drain current I_{DS} in each individual MoS_2 FET with increasing drain and gate voltages. The fundamental frequency of oscillation is currently limited by the parasitic capacitances in various parts of the circuit rather than the intrinsic performance of the MoS_2 devices.

A three-stage ring oscillator based on CVD-grown graphene has also been demonstrated, with the highest oscillation frequency of 1.28 GHz at 2.5 V supply voltage and with output voltage swing of 0.57 V [62]. Graphene ring oscillators can be useful in applications where ultrafast operation is favored over static power dissipation.

Figure 3.21 compares the MoS_2 and graphene ring oscillators with ring oscillators that have been demonstrated in other material systems. For good ring oscillator performance, it is desirable to have high oscillation frequency

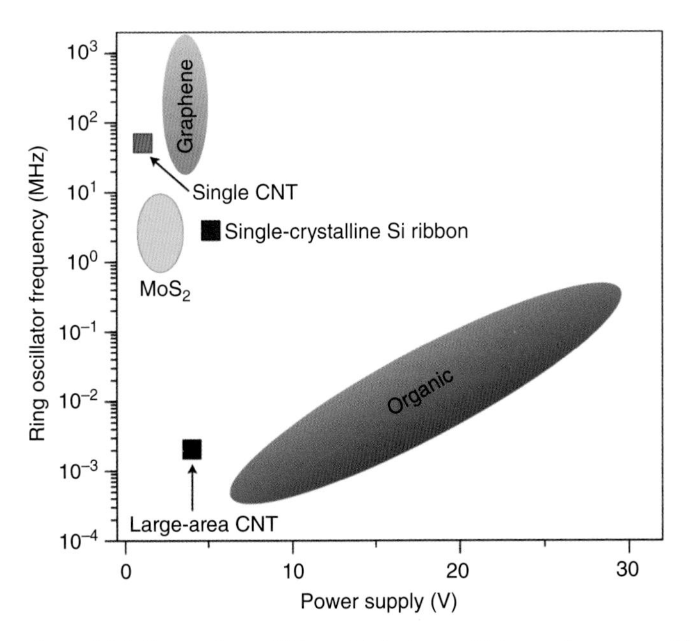

Figure 3.21 Comparison of bilayer MoS_2 ring oscillator performance with ring oscillators demonstrated in other material systems. Source: Fiori et al. 2014 [5]. Adapted with permission Springer Nature.

for performance and low operating power. MoS_2-based and graphene-based ring oscillators clearly have their advantages over ring oscillators demonstrated on other material systems, such as organics and large-area CNTs. The MoS_2 ring oscillator offers similar frequency performance, but with less than half of the operating voltage needed compared to single-crystalline Si ribbon ring oscillators, while a graphene ring oscillator shows the highest frequency in all low-dimensional transistor materials. Although ring oscillators demonstrated on a single CNT show good performance as well, MoS_2 and graphene ring oscillators are more scalable.

3.3.2.5 Microprocessors based on MoS_2

A microprocessor incorporates the functions of a computer's central processing unit (CPU) and serves as a core unit to respond and process the instructions to drive a computer. Important components in a microprocessor, such as a latch and a register [63], have been demonstrated based on CVD-grown MoS_2 by Yu et al. And later, a microprocessor based on MoS_2 has been demonstrated by Wachter et al. showing the feasibility of large-scale integration of MoS_2 in complex logical circuits.

Figure 3.22A shows the architectural block diagram of a microprocessor which contains all the basic building blocks of a common microprocessor: an arithmetic logic unit (ALU), an accumulator (AC), a control unit (CU), an instruction register (IR), an output register (OR) and a program counter (PC). The clock signal and memory are implemented off-chip. The as-fabricated circuit can execute predefined programs stored in the external memory, perform logical operation, and transfer the calculation results to the output port.

The microprocessor was fabricated on the basis of the CVD-grown bilayer MoS_2 films on a silicon chip using gate-first technology. The MoS_2 FETs exhibit a mobility of 3 $cm^2\ V^{-1}\ s^{-1}$, an I_{on}/I_{off} ratio of $\sim 10^8$, and uniform behavior over a $\sim 50\ mm^2$ area on the wafer. The circuit is designed on the basis of the NMOS logic family, which uses n-type enhancement-mode FETs to realize pull-up and pull-down networks. To match voltage levels and achieve high noise margin in cascaded logic stages, the ratio of width and length of the FETs is carefully designed, as it determines the switching threshold voltage. The microscopic image of the as-fabricated microprocessor is shown in Figure 3.22B which consists of 115 MoS_2 transistors and is 0.6 mm^2 in size. The static power consumption is estimated to be 1.4 μW per logic gate, and 60 μW for the whole circuit consisting of 41 stages.

The functionality of the circuit has been verified by logical conjunction and disjunction operations on 1-bit data. Although the speed of operation performed in this work is limited by the frequency of the clock signal, its intrinsic speed is ultimately limited by the current-driving capacity of the pull-up transistor. The maximum operating frequency is estimated to be 2–20 kHz and can be further improved by increasing the drain current of the pull-up transistor in the subthreshold regime. This can be achieved by employing depletion-mode load FETs, controlled chemical doping, improving the carrier mobility of 2D channel material of the transistor, or by reducing the transistor channel lengths.

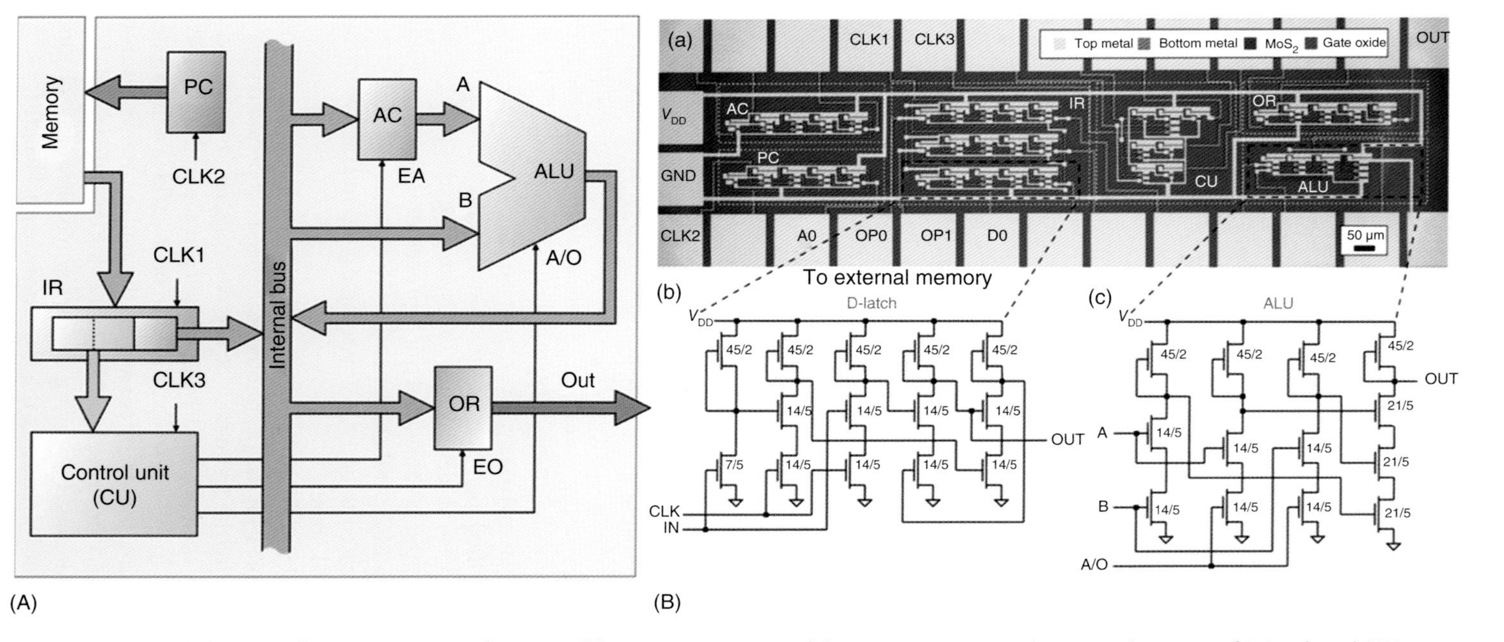

Figure 3.22 (A) Block diagram of microprocessor architecture; (B) microscopic image of the microprocessor with circuit schematics of D-Latch and ALU. Source: Wachter et al. 2017 [64]. Licensed under https://creativecommons.org/licenses/

References

1 Novoselov, K.S., Geim, A.K., Morozov, S.V. et al. (2004). Electric field effect in atomically thin carbon films. *Science* 306 (5696): 666–669.

2 Wang, H., Hsu, A.L., and Palacios, T. (2012). Graphene electronics for RF applications. *IEEE Microwave Magazine* 13 (4): 114–125.

3 Ionescu, A.M. and Riel, H. (2011). Tunnel field-effect transistors as energy-efficient electronic switches. *Nature* 479 (7373): 329–337.

4 Seabaugh, A.C. and Zhang, Q. (2010). Low-voltage tunnel transistors for beyond CMOS logic. *Proceedings of the IEEE* 98 (12): 2095–2110.

5 Fiori, G., Bonaccorso, F., Iannaccone, G. et al. (2014). Electronics based on two-dimensional materials. *Nat. Nanotechnol.* 9 (10): 768–779.

6 Hao, Y.F., Bharathi, M.S., Wang, L. et al. (2013). The role of surface oxygen in the growth of large single-crystal graphene on copper. *Science* 342 (6159): 720–723.

7 Kedzierski, J., Hsu, P.L., Healey, P. et al. (2008). Epitaxial graphene transistors on SIC substrates. *IEEE Trans. Electron Devices* 55 (8): 2078–2085.

8 Lemme, M.C., Echtermeyer, T.J., Baus, M., and Kurz, H. (2007). A graphene field-effect device. *IEEE Electron Device Lett.* 28 (4): 282–284.

9 Radisavljevic, B., Radenovic, A., Brivio, J. et al. (2011). Single-layer MoS_2 transistors. *Nat. Nanotechnol.* 6 (3): 147–150.

10 Chen, J.H., Jang, C., Xiao, S.D. et al. (2008). Intrinsic and extrinsic performance limits of graphene devices on SiO_2. *Nat. Nanotechnol.* 3 (4): 206–209.

11 Radisavljevic, B. and Kis, A. (2013). Mobility engineering and a metal-insulator transition in monolayer MoS(2). *Nat. Mater.* 12 (9): 815–820.

12 Jena, D. and Konar, A. (2007). Enhancement of carrier mobility in semiconductor nanostructures by dielectric engineering. *Phys. Rev. Lett.* 98 (13): 136805.

13 Kappera, R., Voiry, D., Yalcin, S.E. et al. (2014). Phase-engineered low-resistance contacts for ultrathin MoS_2 transistors. *Nat. Mater.* 13 (12): 1128–1134.

14 Wang, L., Meric, I., Huang, P.Y. et al. (2013). Onedimensional electrical contact to a two-dimensional material. *Science* 342 (6158): 614–617.

15 Chen, J.R., Odenthal, P.M., Swartz, A.G. et al. (2013). Control of Schottky barriers in single layer MoS_2 transistors with ferromagnetic contacts. *Nano Lett.* 13 (7): 3106–3110.

16 Fang, H., Chuang, S., Chang, T.C. et al. (2012). High-performance single layered WSe_2 p-FETs with chemically doped contacts. *Nano Lett.* 12 (7): 3788–3792.

17 Han, M.Y., Ozyilmaz, B., Zhang, Y., and Kim, P. (2007). Energy band-gap engineering of graphene nanoribbons. *Phys. Rev. Lett.* 98 (20): 206805.

18 Li, X., Wang, X., Zhang, L. et al. (2008). Chemically derived, ultrasmooth graphene nanoribbon semiconductors. *Science* 319 (5867): 1229–1232.

19 Jiao, L., Zhang, L., Wang, X. et al. (2009). Narrow graphene nanoribbons from carbon nanotubes. *Nature* 458 (7240): 877–880.

20 Yoon, Y. and Guo, J. (2007). Effect of edge roughness in graphene nanoribbon transistors. *Appl. Phys. Lett.* 91 (7): 073103.

21 Zhang, Y., Tang, T.T., Girit, C. et al. (2009). Direct observation of a widely tunable bandgap in bilayer graphene. *Nature* 459 (7248): 820–823.

22 Xia, F., Farmer, D.B., Lin, Y.M., and Avouris, P. (2010). Graphene field-effect transistors with high on/off current ratio and large transport band gap at room temperature. *Nano Lett.* 10 (2): 715–718.

23 Lai, R., Mei, X.B., Deal, W.R. et al. (2007). Sub 50 nm InP HEMT device with F_{max} greater than 1 THz. 2007 IEEE International Electron Devices Meeting, Washington, DC, pp. 609–611.

24 Forstner, H.P., Knapp, H., Jager, H. et al. (2008). A 77GHz 4-channel automotive radar transceiver in SiGe. 2008 IEEE Radio Frequency Integrated Circuits Symposium, Atlanta, GA, pp. 233–236.

25 Long, J.R. (2005). SiGe radio frequency ICs for low-power portable communication. *Proceedings of the IEEE* 93 (9): 1598–1623.

26 Fiori, G. and Iannaccone, G. (2012). Insights on radio frequency bilayer graphene FETs. 2012 International Electron Devices Meeting, San Francisco, CA, pp. 17.3.1–17.3.4.

27 Tanzid, M., Andersson, M.A., Sun, J., and Stake, J. (2014). Microwave noise characterization of graphene field effect transistors. *Appl. Phys. Lett.* 104 (1).

28 Lin, Y.M., Valdes-Garcia, A., Han, S.J. et al. (2011). Wafer-scale graphene integrated circuit. *Science* 332 (6035): 1294–1297.

29 Hsu, A., Wang, H., Kim, K.K. et al. (2011). High frequency performance of graphene transistors grown by chemical vapor deposition for mixed signal applications. *Jpn. J. Appl. Phys.* 50 (7): 070114.

30 Han, S.J., Garcia, A.V., Oida, S. et al. (2014). Graphene radio frequency receiver integrated circuit. *Nat. Commun.* 5: 3086.

31 Chen, C.Y., Rosenblatt, S., Bolotin, K.I. et al. (2009). Performance of monolayer graphene nanomechanical resonators with electrical readout. *Nat. Nanotechnol.* 4 (12): 861–867.

32 Lee, C., Wei, X.D., Kysar, J.W., and Hone, J. (2008). Measurement of the elastic properties and intrinsic strength of monolayer graphene. *Science* 321 (5887): 385–388.

33 Milaninia, K.M., Baldo, M.A., Reina, A., and Kong, J. (2009). All graphene electromechanical switch fabricated by chemical vapor deposition. *Appl. Phys. Lett.* 95 (18).

34 Mak, K.F., Lee, C., Hone, J. et al. (2010). Atomically thin MoS_2: a new direct-gap semiconductor. *Phys. Rev. Lett.* 105 (13).

35 Lee, G.H., Cui, X., Kim, Y.D. et al. (2015). Highly stable, dual-gated MoS_2 transistors encapsulated by hexagonal boron nitride with gate-controllable contact, resistance, and threshold voltage. *ACS Nano.* 9 (7): 7019–7026.

36 Sangwan, V.K., Arnold, H.N., Jariwala, D. et al. (2013). Low-frequency electronic noise in single-layer MoS_2 transistors. *Nano Lett.* 13 (9): 4351–4355.

37 Late, D.J., Liu, B., Matte, H.S.S.R. et al. (2012). Hysteresis in single-layer MoS_2 field effect transistors. *ACS Nano.* 6 (6): 5635–5641.

38 Nourbakhsh, A., Zubair, A., Huang, S. et al. (2015). 15-nm channel length MoS_2 FETs with single- and double-gate structures. 2015 Symposium on VLSI Technology (VLSI Technology).

39 Shih, C.H. and Chien, N.D. (2011). Sub-10-nm tunnel field-effect transistor with graded Si/Ge heterojunction. *IEEE Electron Device Lett.* 32 (11): 1498–1500.

40 Desai, S.B., Madhvapathy, S.R., Sachid, A.B. et al. (2016). MoS_2 transistors with 1-nanometer gate lengths. *Science* 354 (6308): 99–102.

41 Nourbakhsh, A., Zubair, A., Joglekar, S. et al. (2017). Subthreshold swing improvement in MoS_2 transistors by the negative-capacitance effect in a ferroelectric Al-doped-HfO_2/HfO_2 gate dielectric stack. *Nanoscale* 9 (18): 6122–6127.

42 Neal, A.T., Liu, H., Gu, J.J., and Ye, P.D. (2012). Metal contacts to MoS_2: A two-dimensional semiconductor. 70th Device Research Conference, University Park, TX, pp. 65–66.

43 Chuang, S., Battaglia, C., Azcatl, A. et al. (2014). MoS_2 p-type transistors and diodes enabled by high work function MoO_x contacts. *Nano Lett.* 14 (3): 1337–1342.

44 Zhang, Y., Ye, J., Matsuhashi, Y., and Iwasa, Y. (2012). Ambipolar MoS2 thin flake transistors. *Nano Lett.* 12 (3): 1136–1140.

45 Liu, W., Kang, J., Sarkar, D. et al. (2013). Role of metal contacts in designing high-performance monolayer n-type WSe_2 field effect transistors. *Nano Lett.* 13 (5): 1983–1990.

46 Allain, A. and Kis, A. (2014). Electron and hole mobilities in single-layer WSe_2. *ACS Nano* 8 (7): 7180–7185.

47 Fang, H., Tosun, M., Seol, G. et al. (2013). Degenerate n-doping of few-layer transition metal dichalcogenides by potassium. *Nano Lett.* 13 (5): 1991–1995.

48 Chuang, H.J., Tan, X., Ghimire, N.J. et al. (2014). High mobility WSe_2 p- and n-type field-effect transistors contacted by highly doped graphene for low-resistance contacts. *Nano Lett.* 14 (6): 3594–3601.

49 Braga, D., Gutierrez Lezama, I., Berger, H., and Morpurgo, A.F. (2012). Quantitative determination of the band gap of WS_2 with ambipolar ionic liquid-gated transistors. *Nano Lett.* 12 (10): 5218–5223.

50 Jo, S., Ubrig, N., Berger, H. et al. (2014). Mono- and bilayer WS_2 light-emitting transistors. *Nano Lett.* 14 (4): 2019–2025.

51 Khalil, H.M., Khan, M.F., Eom, J., and Noh, H. (2015). Highly stable and tunable chemical doping of multilayer WS_2 field effect transistor: reduction in contact resistance. *ACS Appl. Mater. Interfaces* 7 (42): 23589–23596.

52 Cui, Y., Xin, R., Yu, Z. et al. (2015). High-performance monolayer WS_2 field-effect transistors on high-kappa dielectrics. *Adv. Mater.* 27 (35): 5230–5234.

53 Ovchinnikov, D., Allain, A., Huang, Y.S. et al. (2014). Electrical transport properties of single-layer WS_2. *ACS Nano.* 8 (8): 8174–8181.

54 Chamlagain, B., Li, Q., Ghimire, N.J. et al. (2014). Mobility improvement and temperature dependence in $MoSe_2$ field-effect transistors on parylene-C substrate. *ACS Nano* 8 (5): 5079–5088.

55 Rhyee, J.S., Kwon, J., Dak, P. et al. (2016). High-mobility transistors based on large-area and highly crystalline CVD-grown $MoSe_2$ films on insulating substrates. *Adv. Mater.* 28 (12): 2316–2321.

56 Wang, H., Yu, L.L., Lee, Y.H. et al. (2012). Integrated circuits based on bilayer MoS_2 transistors. *Nano Lett.* 12 (9): 4674–4680.

57 Yu, L., El-Damak, D., Ha, S. et al. (2015). Enhancement-mode single-layer CVD MoS_2 FET technology for digital electronics. 2015 IEEE International Electron Devices Meeting (IEDM).

58 Tosun, M., Chuang, S., Fang, H. et al. (2014). High-gain inverters based on WSe_2 complementary field-effect transistors. *ACS Nano* 8 (5): 4948–4953.

59 Resta, G.V., Balaji, Y., Lin, D. et al. (2018). Doping-free complementary logic gates enabled by two-dimensional polarity-controllable transistors. *ACS Nano* doi: 10.1021/acsnano.8b02739.

60 Dathbun, A., Kim, Y., Kim, S. et al. (2017). Large-area CVD-grown sub-2 V ReS_2 transistors and logic gates. *Nano Lett.* 17 (5): 2999–3005.

61 Larentis, S., Fallahazad, B., Movva, H.C.P. et al. (2017). Reconfigurable complementary monolayer $MoTe_2$ field-effect transistors for integrated circuits. *ACS Nano* 11 (5): 4832–4839.

62 Guerriero, E., Polloni, L., Bianchi, M. et al. (2013). Gigahertz integrated graphene ring oscillators. *ACS Nano* 7 (6): 5588–5594.

63 Yu, L.L., El-Damak, D., Radhakrishna, U. et al. (2016). Design, modeling, and fabrication of chemical vapor deposition grown MoS_2 circuits with E-mode FETs for large-area electronics. *Nano Lett.* 16 (10): 6349–6356.

64 Wachter, S., Polyushkin, D.K., Bethge, O., and Mueller, T. (2017). A microprocessor based on a two-dimensional semiconductor. *Nat. Commun.* 8.

4

Integration of Germanium into Modern CMOS: Challenges and Breakthroughs

Wonil Chung, Heng Wu, and Peide D. Ye

Purdue University, School of Electrical and Computer Engineering, 465 Northwestern Ave., West Lafayette, IN 47907, USA

4.1 Introduction

The introduction of the transistor has completely revolutionized the modern world over the past few decades and still is playing a fundamental role in almost all aspects of human lives. Thriving microelectronic industries presented humans with the unprecedented ability of computation, all of which stemmed from the very first transistor developed back in 1948 at Bell Labs based on germanium [1].

Equipped with intrinsically superior carrier mobility than silicon, germanium was the popular material as substrate back in the early days of the semiconductor industry. However, it was overwhelmed by silicon-based devices owing to inferior surface quality and difficulties in fabrication processes. Ever since the introduction of Moore's law, the golden rule has been abided by the industry for several decades [2]. Now the leading companies strive for the deep sub-10-nm regime and the law is currently being questioned and challenged. Recently "More than Moore" (MtM) strategies [3–5] are actively being discussed in the field. Nearing the physical limit in dimensions, scaling alone seems to be not enough as the only winning strategy.

Silicon still stands firm as the substrate in the major electronics market, alternative channel materials such as III–V and germanium have been widely studied aiming for the position of substitute materials in the post-silicon era. Specifically, germanium boasts much higher electron and hole (electron: $1900\,cm^2\,V^{-1}\,s^{-1}$, hole: $3900\,cm^2\,V^{-1}\,s^{-1}$) mobility than does silicon (electron: $1500\,cm^2\,V^{-1}\,s^{-1}$, hole: $450\,cm^2\,V^{-1}\,s^{-1}$). Therefore, employing the advantages of both Ge and III–V, complementary metal-oxide semiconductor (CMOS) structures with III–V/Ge hybrid CMOS have also been proposed [6].

In addition, both silicon and germanium are categorized in the group IV, enabling higher process compatibility within the current mainstream silicon-based CMOS process technology. With the help of state-of-the-art fabrication equipment, germanium devices are regaining much attention as one of the possible solutions for the future electronics.

Advanced Nanoelectronics: Post-Silicon Materials and Devices,
First Edition. Edited by Muhammad Mustafa Hussain.

However, there are many issues that should be addressed and solved in order for germanium to join the mainstream industry. The most critical challenges for germanium devices can be summarized into three major parts: (i) high-quality contacts, (ii) stable gate stack formation for good interface quality, and (iii) integration of the aforementioned processes into a complete working device in various structures with minimum dimensions.

In the following sections, various research efforts to implement the leading-edge germanium transistors are introduced and discussed in detail, including issues with making good metal/n-type contacts, development of various surface treatments for stable interface, novel process schemes for minimized equivalent oxide thickness (EOT) and finally the full integration into high-performance CMOS circuits.

4.2 Junction Formation for Germanium MOS Devices

4.2.1 Charge Neutrality Level and Fermi Level Pinning

Unlike the blessed silicon source and drain, the most challenging issue for the germanium devices would be metal-germanium contact for the n-type substrate. Without solving the metal-nGe contact issue, implementation of germanium CMOS devices would be impossible. Factors that govern the nature of metal-semiconductor (MS) contact include the Fermi level pinning (FLP) and the location of charge neutrality level (CNL). In the case of germanium, CNL is located at only 0.1 eV above its valence band edge and the pinning factor (S) is extracted to be 0.05 [7–9]. Figure 4.1 shows the experimentally obtained Schottky barrier heights (SBHs) on n-Si and n-Ge along with reported metal work functions.

Very low pinning factor (S) near the Bardeen limit ($S = 0$) extracted from the interface of metal and nGe contact translates to strong pinning of the Fermi level. Unable to modulate the SBH of the metal-nGe contact, metals in contact with n-type germanium suffer from intrinsically high contact resistivity. On

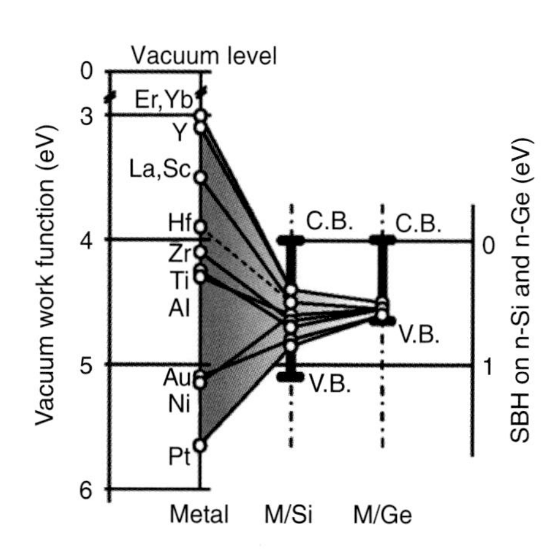

Figure 4.1 Extracted Schottky barrier heights (SBH) on n-Si and n-Ge of various reported metal work functions. Source: Nishimura et al. 2007 [8]. Reproduced with permission of AIP.

top of that, occupied acceptor type traps above the CNL form a large negative charge buildup at the interface, dramatically pulling down the performance of Ge nMOSFET (n-channel metal-oxide semiconductor field-effect transistor) [7, 9].

Various researches have been conducted to address the reasons behind it and have presented possible novel process strategies to alleviate the issue [10–13], enabling the breakthroughs in Ge nMOSFET [14–16].

4.2.2 Metal/Ge Contacts

4.2.2.1 Alleviation of FLP

FLP at the metal/n-Ge contact interface was studied and discussed to explain the physics behind it. Although the exact mechanism causing the FLP on germanium surface has not been clarified, metal-induced gap states (MIGS) or effect of electric dipole at the interface of metal/n-Ge are suggested as the possible reasons for the phenomenon. MIGS theory claims that the FLP arises from the decay of the metal's electron wavefunction tail into the substrate [8, 10, 17]. If this is the case, alleviation of FLP caused by MIGS can be solved by inserting an insulating layer in between the metal and the germanium substrate which will push away the metal's electron wavefunction further from the interface. Therefore, the metal/n-Ge contact property would be significantly affected by the thickness of the inserted layer and the intrinsic property of the bulk germanium [8]. Also, FLP should be very weakly related or even irrelevant to the interfacial characteristics as it should only be caused by the incoming metal electron wavefunction.

SBH and ideality factor is found to be modulated according to the number of single-layer graphene (SLG) inserted between the metal and n-Ge [10]. With two layers of SLG, SBH shows the minimum value; but when the number of layers increases beyond 3, both forward and reverse current decreases significantly implying a large increase in the tunneling resistance. Graphene is known to be conductive, but it exhibits poor conductivity perpendicular to the plane which acts as an insulator in this case. It clearly exhibits significant modulation of both parameters, suggesting the MIGS is behind the FLP of metal/n-Ge system.

On the other hand, the presence of various electric dipoles is also considered as another origin of the strong FLP in the metal/n-Ge interface [11, 18]. Researchers have tested changing the interface characteristics so that the dipole formed at the interface varies from sample to sample. It was reported that addition of higher nitrogen concentration into the TaN metal in contact with n-Ge effectively reduces the SBH for electrons. Experimental SBH for electrons in TaN/n-Ge stack was extracted to be 0.552 eV with N_2 flow rate of 0 sccm, which decreased to 0.22 eV when the N_2 flow rate was increased to 12 sccm [11]. This suggests that increasing the nitrogen compositional ratio (confirmed by X-ray photoelectron spectroscopy (XPS) analysis) by altering the nitrogen flow rate during TaN sputtering results in alleviation of FLP.

4.2.2.2 Metal/n-Ge Contact

As already mentioned, it requires careful chemical and physical analysis on metal/n-Ge contact area to acquire the optimal n-Ge contact. Fortunately p-type germanium has much lower SBH ($q\Phi_{B,h}$) than the strongly pinned counterpart for n-Ge ($q\Phi_{B,e}$). Moreover, low diffusivity and high dopant activation efficiency in case of p-type dopants such as B, BF_2, and Ga facilitates the fabrication of Ge pMOSFET.

Experimental data showing such high specific contact resistivity (ρ_c) can be found in multiple references. Ni/n-Ge shows few orders of magnitude larger ρ_c when compared to that of Ni/n-Si for all annealing temperature conditions [19]. This suggests that careful process is needed to solve the resistivity issue in metal/n-Ge contact.

Experiments aiming to solve such issues were conducted using some novel processes developed to effectively fabricate well-behaved metal/n-Ge contacts. Ohmic contacts and very low specific resistivity on n-Ge have been reported by inserting a thin layer of Si between the metal and n-Ge [20]. In situ doped (phosphorous, 1×10^{20} cm^{-3}) silicon was epi-grown selectively on n-Ge and Ti/TiN stack was deposited for the contact. Ten and sixteen nanometers of silicon showed $\rho_c = 1.4 \times 10^{-6} \Omega\ cm^2$ and $\rho_c = 1.7 \times 10^{-6} \Omega\ cm^2$, respectively.

Another passivation method using Ge_3N_4 layer formed by plasma nitridation on a clean Ge surface prepared by annealing in vacuum ($\sim 1 \times 10^{-9}$ Torr) was tested, which was also reported to effectively reduce the barrier height when Al, Cr, and Co were used as the metal [21]. Au and Pt in contact with n-Ge were found to be not affected by the inserted Ge_3N_4 layer and still showed strong rectification. Such thin layers of Si or Ge_3N_4 films help reduce the FLP and therefore alleviate the strongly rectifying I–V nature of metal/n-Ge contact.

4.2.2.3 Recessed Contact Formation

Using effective material and fabrication process for metal/n-Ge contact is favorable, but when it comes to realization of both nMOS and pMOS on the same wafer for CMOS operation, single contact formation with the same metal would be more preferred. However, using single-metal and contact formation recipe might benefit in metal/n-Ge contact, but usually increases the barrier for the holes. A novel recessed source/drain (S/D) formation process was proposed and it has helped fabricate breakthroughs in high-performance Ge nMOSFET and pMOSFET in various 3D structures (planar [14], Fin field-effect transistors (FinFET) [22] and nanowires [23]) and in the realization of the first Ge CMOS Circuits [23].

The principles in improving the current properties in metal/semiconductor consist of two different strategies: (i) lowering the Schottky barrier height ($q\Phi_B$) and (ii) reducing the barrier width (W_{SB}). As discussed previously, the first strategy requires difficult and complex processes which are not favorable.

The motivation behind the recessed contact formation falls into the second strategy, to reduce the barrier width, as depicted in Figure 4.2a. Etching was done with carefully optimized BCl_3/Ar-based recipe using inductive-coupled plasma (ICP) etching to make the Ge contact. The scanning electron microscopic (SEM) image of the etched region is shown in Figure 4.2b,c. The germanium surface is etched down approximately 12 nm for the best contact result.

The recessed S/D formation method significantly improves the n-Ge contact quality, as seen from Figure 4.3a,b. After the recessed S/D process, both Al (red) and Ni (blue) contact on n-Ge show better contact property. Contact

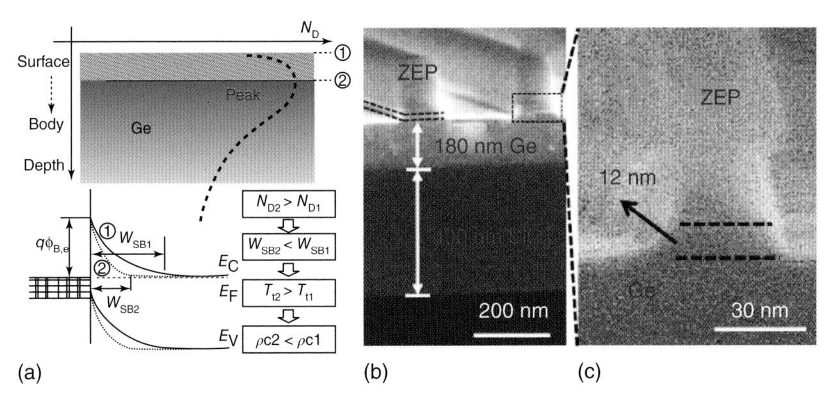

Figure 4.2 (a) Recessed S/D structure enhances the electron tunneling efficiency by reducing W_{SB} and thus lowering the resistivity. (b,c) SEM images showing the cross-section of the test recessed S/D structure [14]. With single S/D recess and deposition of Ni, both high-performance Ge nMOSFET and pMOSFET can be fabricated.

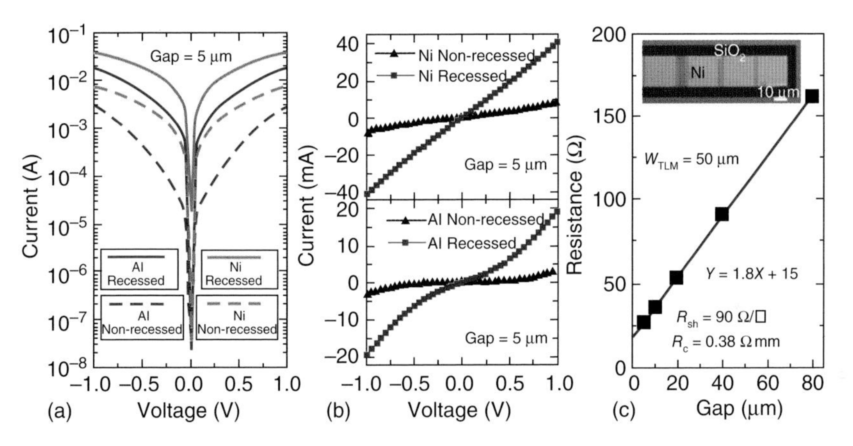

Figure 4.3 (a) I–V curves between two TLM metal pads with gap distance of 5 μm are shown before and after the S/D recess etching on n-Ge. (b) Linear scaled representation of (a). (c) Measured resistance from the TLM pattern shown in the inset figure with Ni metal pads separated by SiO_2 mesa isolation. Source: Wu et al. 2014 [14]. Reproduced with permission of IEEE.

resistance (R_c) and sheet resistance (R_{sh}) are extracted from Figure 4.3c, which were 0.38 Ω mm and 90 Ω/□, respectively.

Recessed S/D contact on p-Ge was also studied and it also showed improvement, as seen from Figure 4.4. With similar mechanism as n-Ge contact improvement after recessed S/D, p-Ge dopants implanted into the S/D Ge area also follow the near-Gaussian distribution and therefore etching away the top surface is helpful in making better p-Ge contact as well. Note that R_{sh} stays similar, whereas the R_c is reduced by approximately 80%.

Finding the optimal recess etch depth is also critical in forming good metal/n-Ge contact. As found from recess etch time experiment, summarized in Figure 4.5, etching time strongly affects the contact property. Four different

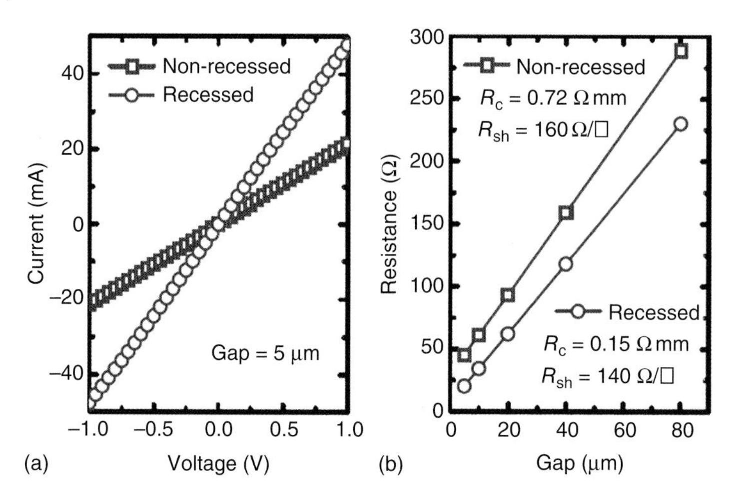

Figure 4.4 (a) I–V curves on TLM pads that are 5 μm apart on p-Ge before and after the S/D recess etching. From (b) the measured resistance versus gap, recessed S/D proved to be also effective on p-Ge contact [24].

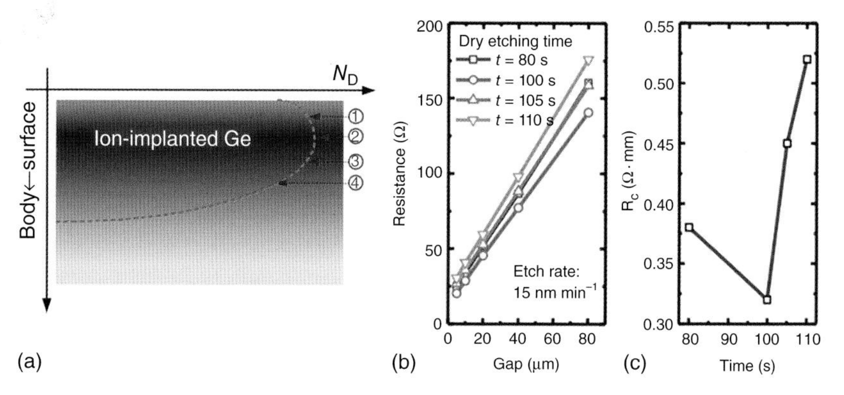

Figure 4.5 (a) Diagram representing the Gaussian distribution of doping ions inside the Ge bulk. (b) Resistance data from TLM pattern for four different etching times and (c) extracted R_c. Source: (Panels b and c) Wu et al. 2015 [25]. Reproduced with permission of IEEE.

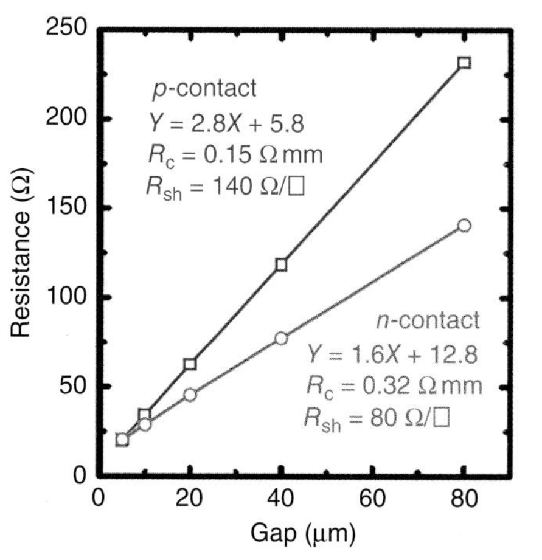

Figure 4.6 The best TLM contact results for n-Ge and p-Ge from recessed S/D process [24].

etching times (80, 100, 105, and 110 s) using the aforementioned BCl_3/Ar-based RIE recipe were carried out resulting in four different etch profiles, respectively.

The best TLM contact results on n-Ge and p-Ge using recessed S/D etch recipes can be found in Figure 4.6. From this result, the optimal etching recipe can be found and implemented into the actual state-of-the-art nMOSFET extensively discussed in the later section.

4.3 Process Integration for Ge MOS Devices

4.3.1 Interface Engineering Issues

High-quality SiO_2 film (excellent insulator, high break down field) and its perfect interface (low interface trap density) with silicon made it possible for silicon to thrive as the base material for the modern microelectronics. However, in case of germanium, GeO_2 is known for its volatility above ~430 °C [26, 27]. GeO_2 reacts with Ge in the interface through redox reaction ($GeO_2 + Ge \leftrightarrow 2GeO$) and GeO (g) diffuses through GeO_2 [26]. GeO_2 is not only volatile but also reacts with water [28] showing hygroscopic property. The disadvantage of GeO_2 continues with low-k value (~4–6 [29, 30]), which suggests that it is not suitable to use GeO_2 layer as the only gate oxide for state-of-the-art Ge transistors.

Therefore, it is indispensable to not only include high-κ oxides such as HfO_2 or ZrO_2 within the gate oxide stacks for Ge MOSFETs but also carefully engineer the passivation process to yield low leakage current, interface trap density (D_{it}), and high dielectric constant (k). Various gate stacks used for high-performance devices are dealt with in the following section.

It is known that the MOSFET performance not only is dependent on the gate stack but also on the Ge wafer itself, in terms of atomically flat surface and impurities within the layer [31]. It implies that even with the same process flow, different Ge wafers give different results due to physical and chemical variations among the wafers. It was experimentally extracted that different Ge wafers with the same process yielded in two distinctive trends in electron mobility. However, both wafers showed similar D_{it} profiles, suggesting that the scattering source for these two wafers is electrically inactive. It was concluded that the mobility degradation came from different oxygen concentrations in the wafer, thus resulting in different amounts of neutral impurities in bulk Ge. A wafer with lower oxygen concentration gave higher electron mobility and H_2 annealing found to be helpful in improving the mobility because it extracts the oxygen out of the wafer [31].

4.3.2 Various Gate Stack Combinations for Ge MOSFET

The most effective way to obey Moore's law has been scaling down various dimensions of MOSFET. Among them, oxide thickness is one of the key factors. By decreasing the oxide thickness, the oxide capacitance increases, which results in higher channel carrier charge and thus better output performance. However, decreasing the oxide thickness had its own limit caused by exponentially increasing leakage current. Therefore, higher dielectric constant (high-κ) oxides were introduced to acquire the same amount of high oxide capacitance even with thicker oxides. Whenever gate oxides with different dielectric constants are used, the EOT concept is needed to assess the performance of the device and compare it with the conventional SiO_2 process. The same high-κ oxide strategy applies to germanium devices. To keep the EOT as small as possible, various combinations of interfacial layers comprised of high-κ materials such as HfO_2, ZrO_2, Y_2O_3, and $LaLuO_3$ were widely investigated [16, 32–34].

Many different interface passivation methods were studied, such as vacuum annealing [34], Si interfacial layer passivation [35–37], Ge (oxy)nitridation [38, 39], sulfur-based passivation [40, 41], high-pressure O_2 oxidation [35, 42, 43] and plasma (or ozone) post oxidation (PPO or OPO) [16, 44]. Among them, germanium-oxide-based gate stacks showed stable and more promising properties when integrated into Ge metal-oxide semiconductor (MOS) devices.

4.3.2.1 GeO*x*-Free Gate Stack with ALD High-κ

As stated earlier, GeO_2 (or non-stoichiometric GeO_x) suffers from unstable desorption and hygroscopic properties. Furthermore, the low dielectric constant makes it difficult to push the EOT down to deep sub-1-nm range. It is plausible to expect minimized EOT without non-stoichiometric germanium sub-oxides (GeO_x). Annealing the HF-cleaned Ge wafer in a high-vacuum chamber (low to mid 10^{-7} Torr) removes the native germanium oxide effectively as confirmed in Ge 3d XPS spectra with low Ge sub-oxide peaks, as found in Figure 4.7a. Ultrathin in situ metal (Hf or Zr) capping protects the clean Ge surface from forming further GeO_x. Deposition of atomic layer deposition (ALD) HfO_2 or ZrO_2 with O_3 as oxidant naturally oxidizes the capping metal into high-κ oxide (HfO_2, ZrO_2). This process is found to be effective in reducing the gate leakage current and pushing the EOT into deep-sub-1 nm. EOT was extracted to be 0.62 nm for Hf-cap/HfO_2 [45] and 0.606 nm for Zr-cap/ZrO_2, as presented in Figure 4.7b [34].

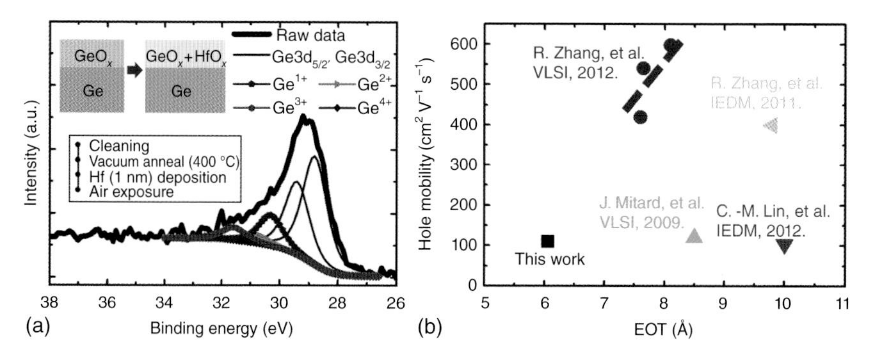

Figure 4.7 (a) XPS spectra showing Ge 3d peaks measured from air-exposed Hf-capped/vacuum-annealed Ge. (b) Peak hole mobility of Ge pMOSFET in comparison with other reported works. Source: Shin et al. 2014 [34]. Reproduced with permission of IEEE.

4.3.2.2 Silicon Interfacial Layer Passivation

A very thin silicon layer can be inserted in between the main gain oxide and germanium substrate to act as an interfacial layer [37]. After typical HF wet cleaning, H_2 annealing was done at 650 °C for 60 s followed by Si growth using SiH_4 in N_2 ambient. Room-temperature N_2O plasma oxidation was done and the HfO_2 was deposited with ALD. It was observed that the number of Si monolayers deposited onto the germanium wafer determines the electrical performance of the MOS gate stack. With more than four monolayers of Si, TEM images revealed periodical contrast variation along the epi-Si and Ge, suggesting the existence of dislocations due to lattice mismatch of Ge and Si.

It was also confirmed that in case of Si passivation, the thickness of the Si layer typically affects the interface trap density in the upper half of the energy bandgap which increases proportionally with Si thickness, whereas the lower half stays constant. Si 2p peak in XPS spectra graph shows that as the Si thickness increases above 0.5 nm, Si—Si bonding peak became clearer and the surface roughness probed with AFM showed no dependence of roughness with the thickness [35].

4.3.2.3 Germanium (Oxy)Nitridation

Direct formation of Ge nitride or Ge oxynitride is also studied as one of the possible interfacial layer stacks. Advantages of incorporation of nitrogen include higher thermal stability [46] and higher k value than GeO_x [38, 39]. Germanium surface nitridation through direct irradiation of nitrogen radical is one way of inserting thin oxygen-free Ge nitride (Ge_3N_4) between the Ge bulk and the main dielectric layer. Such Ge nitride in stack with HfO_2 on top shows an atomically flat profile and was found to be stable in oxygen ambient at high temperatures [38]. Ge nitride along with additional forming gas annealing (FGA) at 400 °C for 30 minutes further improves the interface quality which fixes the severe stretch in measured C–V curve as a result of significant reduction in D_{it}, pulling it down to $\sim 1.8 \times 10^{11}$ cm^{-2} eV^{-1}.

Apart from formation of Ge_3N_4 via direct nitridation of the germanium surface, nitridation of GeO_2 has been studied aiming for better interface quality [39]. Since ultrathin GeO_2 was reported to degrade after deposition of ALD high-κ films, the plasma-oxidized GeO_2 layer was transformed into GeON layer. After forming the GeO_2 on p-Ge through plasma oxidation, in situ nitridation with Ar and N_2 was carried out to form GeON at various microwave power levels (100–600 W). GeON film (2.7 nm) was confirmed with XPS analysis, showing clear N1s and Ge-N peaks after nitridation. By adjusting the nitrogen to Ge ratio, it was concluded that appropriate nitrogen ratio within pre-oxidized GeO_2 layer is effective in reducing the D_{it} which pulled it down to approximately $\sim 4 \times 10^{11}$ cm^{-2} eV^{-1} at N/Ge atomic ratio of 0.34. However, increasing the ratio over 90% gave inferior interface quality, which implies that GeON/Ge with optimized N content should be used for thin EOT and low D_{it} stack.

4.3.2.4 GeO_2-Based Gate Stacks

Among many different combinations of gate formation strategies, GeO_2-based gate stacks show superior results in terms of electrical properties [6, 14, 16, 23, 33, 47]. Obviously due to low k value, the GeO_x-based gate stacks suffer from difficulties in scaling the EOT. However, recently it was experimentally proved that GeO_x-based stacks can also be scaled down to EOT of sub-1-nm regime [16]. EOT was brought down to ~0.6 nm [47]. In the following sections, three experimental GeO_2-based gate stacks that demonstrated devices with good performance are listed and summarized.

High-Pressure Oxidation (HPO) As discussed earlier, GeO_2 suffers from a desorption nature and the interface state density deteriorates substantially. Therefore, it is favorable to suppress the GeO desorption as much as possible using high-κ materials or a Si capping layer to block the GeO [48] or oxidize at high pressure [42]. Bi-directional $C-V$ sweep of Au/GeO_2/p-Ge MOS capacitor which was oxidized at 70 atm (HPO, high-pressure oxidation) exhibits significantly better $C-V$ curves than the one oxidized at atmospheric pressure (APO). It clearly suggests that HPO is significantly effective in reducing voltage hysteresis and alleviating frequency dispersion of capacitance. This is attributed to suppression of GeO desorption from the Ge/GeO_2 interface causing self-passivation of the surface. The minimum D_{it} extracted near the midgap is 2×10^{11} cm^{-2} eV^{-1}. In combination with this HPO process, low-temperature oxygen annealing (LOA) and introduction of Y_2O_3 boosts the electron and hole mobility well above the silicon universal mobility. With successful interface passivation through HPO and LOA, electron and hole mobility that is 2.5 and 3.5 times larger than Silicon mobility were acquired [35].

Plasma (or Ozone) Post Oxidation (PPO or OPO) Recently, oxidation using electron cyclotron resonance (ECR) oxygen plasma through a thin layer of Al_2O_3 as control layer called "plasma post oxidation" (PPO) has proved to be very successful in realizing high-performance Ge CMOS devices, pushing the EOT down to deep-sub-1-nm region (0.7 nm) [16, 44]. To further reduce the EOT to 0.6 nm, HfO_2 can be used along with ozone-oxidized GeO_x [47]. Post oxidation does not have to be done only with ECR oxygen plasma. GeO_x underneath AlO_x was formed after deposition of AlO_x (0.3 nm) and HfO_2 (2 nm) using ALD. In situ post oxidation with ozone (O_3/O_2 ambient at the pressure of ~100 Pa) was carried out, yielding a total of a 2.7-nm-thick crystallized $HfO_2/AlO_x/GeO_x$ gate. A sample without OPO was also fabricated to compare electrical properties, which shows worse (SS = 155 mV/decade, peak effective hole mobility <100 cm^2 V s^{-1}) data. The fabricated gate stack with OPO showed a subthreshold swing (SS) of 85 mV/decade and peak effective hole mobility = 417 cm^2 V s^{-1}. The extracted effective mobility is the highest value ever reported at EOT = 0.6 nm regime. When compared to the device without GeO_x (ZrO_2/Zr-cap/vacuum-annealed Ge discussed in Section 3.2.1 [34]), the OPO GeO_x-based gate stack shows better interface quality, resulting in 4.1 times larger peak hole mobility.

4.3.2.5 Rare-Earth Oxides Integrated into Germanium MOSFETs

Various high-κ dielectric materials have been introduced for integration into high-performance germanium transistors aiming for higher performance. Along with the relatively widely studied HfO_2, ZrO_2, and Al_2O_3, incorporation of rare-earth oxides and elements such as Y [33], Y_2O_3 [35, 48, 49], $LaLuO_3$ [32, 50], and La_2O_3 [51] were also investigated for further EOT scaling and improvement in interface quality.

Yttrium-Based Oxide on Germanium Incorporation of the rare-earth element such as yttrium can be done using Y_2O_3 directly [49], doping yttrium into gate oxide layers forming Y-GeO_2 [33] or forming composite oxide films such as $YScO_3$ films [33].

As discussed in the previous section, germanium oxide shows a thermally unstable property and intermixing with high-κ might act as a deteriorating factor for germanium devices. Therefore, it is favorable to effectively block aggressive desorption of GeO into high-κ oxides. When compared with HfO_2, Y_2O_3 was found to reduce GeO volatilization due to formation of amorphous $YGeO_x$ layer between GeO_2 and Y_2O_3, resulting in better $C-V$ characteristics than between the HfO_2/GeO_2 counterpart [48]. Such a Y_2O_3/Ge system was found to exhibit a dielectric constant value of 12.

Another work that used yttrium oxide as the gate stack in germanium nMOS utilized additional high-pressure O_2 annealing [49]. Relatively low interface trap density (D_{it}) of $\sim 10^{11}\ cm^{-2}\ eV^{-1}$ and stable electron mobility of $\sim 900\ cm^2\ V^{-1}\ s^{-1}$ ($N_s = 10^{12}\ cm^{-2}$) were acquired, which were analyzed to be due to reduced GeO desorption and Y passivation of the oxygen-deficient GeO_2 layer.

It was also recently demonstrated that the $YScO_3$/Y-doped GeO_2/Ge stack could be used for EOT of 0.5 nm and a low D_{it} level [33]. Y-doped GeO_2 (denoted as Y-GeO_2) was found to be a very robust interfacial layer which acts as a strong oxygen diffusion barrier among various trivalent metal-oxide-doped GeO_2 (M-GeO_2) [52]. On top of such a Y-GeO_2 layer, additional gate oxide ($YScO_3$) was deposited to show EOT of 0.5 nm. $YScO_3$ showed a dielectric constant of 17 and a bandgap of 6 eV, serving as an amorphous oxide which does not react with germanium up to 600 °C. This gate scheme successfully exhibited the highest electron mobility of $1057\ cm^2\ V^{-1}\ s^{-1}$, maintaining sub-1-nm EOT (0.8 nm) and low leakage current density (J_G).

Lanthanum-Based Oxide on Germanium Another rare-earth element studied for germanium gate stack is lanthanum [32, 50, 51]. Ternary rare-earth oxide of ALD-$LaLuO_3$ and thermally grown thin GeO_2 (dry oxidation at 350 °C) stack were formed on a germanium-on-insulator (GeOI) wafer with Ni/Au gate metallization [32]. Such a stack resulted in increase in on-current, transconductance, and effective hole mobility. A similar $La_{2x}Lu_{2(1-x)}O_3$ ($x = 0.6$, $k = 23$, energy gap = 5.5–5.8 eV)/GeO_2/Ge stack which was annealed at high pressure (50 atm) resulted in more stable $C-V$ curves with negligible frequency dispersion, proving the effectiveness of the stack [50]. Furthermore, an effective leakage current suppression at 0.5 nm EOT was also reported using a combination of La_2O_3 and thin ZrO_2 capping layer. As reported earlier, lanthanum could also serve as an effective element that could improve the overall performance of germanium devices when optimized with various novel methods and gate stacks.

4.3.3 Stress and Relaxation of Ge Layer on an Si-Based Substrate

Integrating a germanium channel into conventional Si-based substrates raises another important issue that is related to the physical nature of the layer. Lattice mismatch of germanium and silicon (Ge's lattice constant being approximately 4% higher than Si [53]) causes severe misfits at the interface and thus hinders formation of a high-quality germanium layer on the silicon platform. Several techniques have been studied experimentally to form a high-quality germanium layer on conventional silicon-based technology, including the germanium condensation method [54–56] and wafer bonding [57–59]. GeOI wafers are now widely employed for active researches in the field of germanium MOS devices.

The germanium condensation method utilizes the thermal oxidation of an epitaxially grown $Si_{1-x}Ge_x$ layer on a silicon-on-insulator (SOI) wafer. Such a SiGe layer can be grown with chemical vapor deposition (CVD) on a typical SOI wafer. Through thermal oxidation at temperatures below the melting point of $Si_{1-x}Ge_x$, Ge atoms within the SiGe layer are rejected from the SiO_2 layer and accumulate within the SiGe layer ($Si_{1-y}Ge_y$, $y > x$), which results in an increase of the Ge percentage [56]. Using this method, 7 nm of a single-crystal germanium layer could be formed with Ge fraction above 99.5% and 1.1% compressive strain. Furthermore, local Ge condensation can be applied to add higher strain to the channel by increasing the Ge mole fraction locally at the selectively grown SiGe source and drain regions [60].

The wafer bonding method is already used widely in commercially available SOI wafers. Using the similar method (SmartCut™) used for SOI wafer fabrication, a germanium layer with plasma-enhanced chemical vapor deposition (PECVD) tetraethoxysilane (TEOS) oxide on top can be implanted with H ions and then detached upside down onto the surface of a base silicon substrate. The detached surface can be further polished by chemical mechanical polishing (CMP). The resulting Ge layer showed a very smooth surface with an atomic force microscopy (AFM) roughness root mean square (RMS) value below 0.2 nm for 1 μm × 1 μm scan [57]. These GeOI wafers were then processed to show their possibilities for application toward development of high-performance germanium transistors [14, 22, 23, 61], which is dealt in great detail in Chapter 5.

4.4 State-of-the-Art Ge CMOS with Recessed Channel and S/D

4.4.1 Germanium CMOS Devices

Recently, a series of germanium complementary metal-oxide semiconductor (Ge CMOS) circuits were demonstrated on a GeOI substrate using different structures (Figure 4.8a planar ultrathin body (UTB) field-effect transistor (FET) [62], Figure 4.8b FinFET [22], and Figure 4.8c nanowire FET [23]). As mentioned earlier, germanium nMOS devices usually suffer from inferior contact property when compared to pMOS due to strong FLP and activation issues.

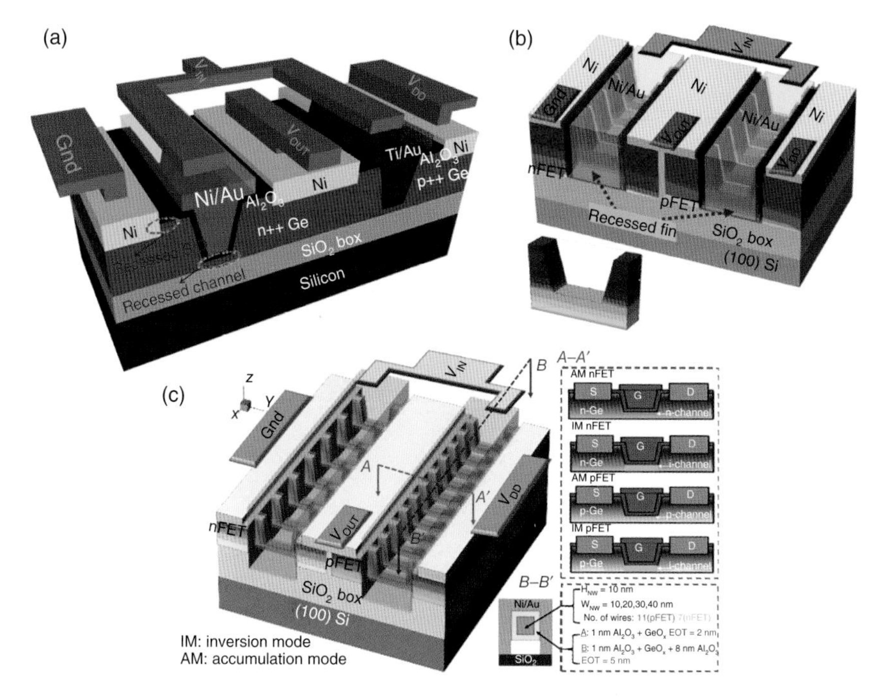

Figure 4.8 3D schematics of Ge CMOS devices using recessed channel and S/D process in (a) planar UTB, (b) fin, and (c) nanowire structures. Inset diagrams in (c) shows the cross-section view along A–A′ and B–B′. Source: (Panel a) Wu et al. 2014 [62]. Reproduced with permission of IEEE; (Panel b) Wu et al. 2015 [22]. Reproduced with permission of IEEE; (Panel c) Wu et al. 2015 [23]. Reproduced with permission of IEEE.

However, employing "recessed S/D formation" and "recessed channel" processes, high-performance Ge CMOS circuits were made possible. In addition, the whole fabrication process was developed considering the manufacturer's point of view, which favors the single S/D formation method and the same material for both nMOS and pMOS devices. The key benefit of this scheme lies in the low contact resistivity and excellent gate electrostatic control realized by recessed S/D and fully depleted ultrathin body (FD-UTB) recessed channel, respectively. All three structures (planar UTB, FinFET, and nanowire) share the same process except for the channel area definition steps. Channel thickness is controlled with optimized power, time, and gas pressure using SF_6 ICP dry etching. The precise control of the dry etching step allows the formation of ultrathin recessed channel thicknesses down to the 10-nm range. UTB devices enjoy many advantages related to electrostatic gate controllability and were extensively studied both experimentally and theoretically [56, 63–66].

The formation of the recessed channel is followed by fin etch using the same dry etch recipe used for the channel recess. Lithography patterns and etch conditions were calibrated to precisely fabricate fin widths and lengths down to 10 and 40 nm, respectively, confirmed with scanning electron microscopy (SEM)

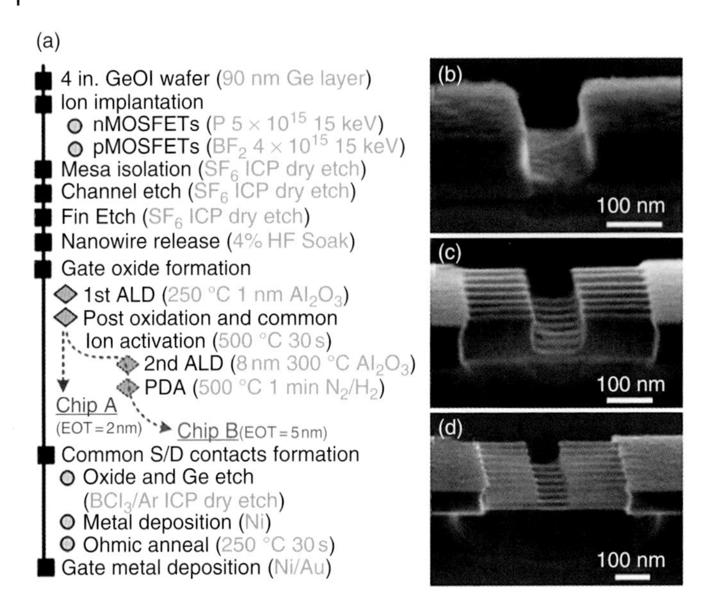

Figure 4.9 (a) Process flow of the germanium nanowire CMOS devices. (b) Planar, (c) fin, and (d) nanowire structures are clearly seen from tilted SEM images. Source: Wu et al. 2016 [67]. Reproduced with permission of IEEE.

images [67]. All patterns were written with the E-beam lithography tool. After channel structure definition, nanowires were formed by wet etching the underlying SiO_2 (BOX) with HF solution. Source and drain contacts were then made with the BCl_3/Ar ICP dry etching recipe followed by S/D and gate metal deposition (Ni/Au). The whole process and SEM images for the process are shown in Figure 4.9a–d.

In the case of nFET, accumulation mode (AM) was used; but for pFET, the inversion mode (IM) was used. This is related to the trap neutrality level (TNL) located near the valence band of germanium. In the case of IM pFET, to induce channel inversion of n-type Ge, negative gate bias was applied which will pull down the Fermi level (E_F) with the aid of large negative acceptor traps, as seen in Figure 4.10a. However, in the case of IM nFET, it is difficult to turn the device on easily by inverting the p-type channel (pulling up E_F toward the conduction band edge) due to increasing negative acceptor traps as well as the large number of acceptor traps consuming the inverted electrons from the channel [7, 9, 24]. Therefore, instead of utilizing the IM nFET, AM nFET was fabricated.

Electrical data of the recessed channel and S/D devices are summarized in Figure 4.11a–f. The transfer curves (I_d–V_{gs}) of AM Ge nanowire nFET and IM Ge nanowire pFET show SS of 64.1 mV/decade for nMOS and 113 mV/decade for pMOS, as presented in Figure 4.11a and b, respectively. More than 800 devices were measured and the On/Off ratio (I_{ON}/I_{OFF}) data were plotted in BOX and histogram plots as well. From Figure 4.11d, the mean value of I_{ON}/I_{OFF} was found to be 2×10^5. Ge hybrid CMOS was fabricated with IM pFET and AM nFET, with carefully balanced currents as seen in Figure 4.11c. Seven and 11 nanowires were

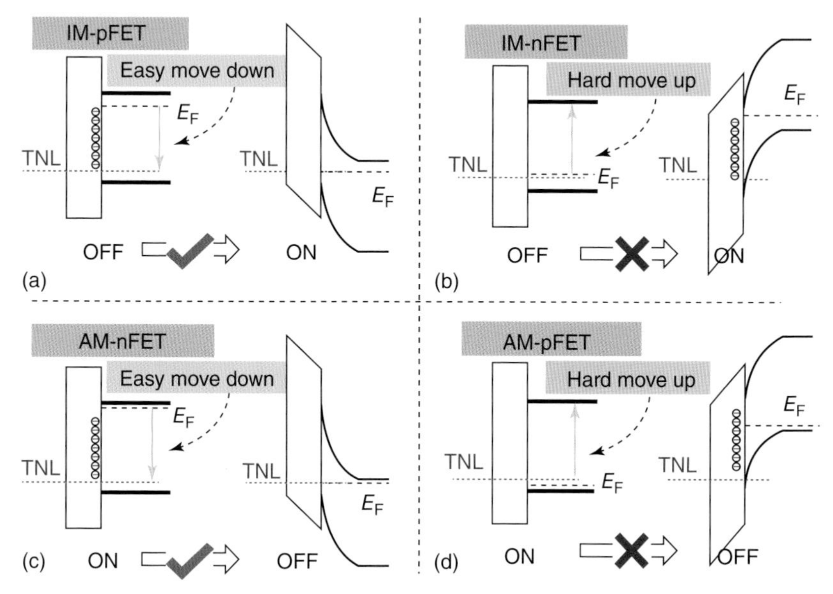

Figure 4.10 (a–d) Band diagram showing the location of TNL (or charge neutrality level, CNL, as discussed in the section on junction formation) and respective type (n/p) and mode (IM/AM) of the MOSFET [24].

used for nMOS and pMOS. Output characteristics of the hybrid CMOS devices (Figure 4.11c) show balanced output currents.

4.4.2 Germanium CMOS Circuits

Seven AM parallel nanowire nFETs and 11 IM parallel nanowire pFETs form a simple Ge CMOS inverter, as depicted in Figure 4.8c to balance their asymmetric electrical performance. The balanced transfer curve is found in Figure 4.11c. These inverters were measured with supply voltages V_{DD} ranging from 0.2 to 1.6 V and from V_{OUT} versus V_{IN} graphs in Figure 4.12a–c; voltage gains were extracted as well. Extracted maximum voltage gains (V_{DD} = 1.2 V) from planar, fin, and nanowire structures are 36 V/V [62], 34 V/V [22], and 54 V/V [23], respectively.

A nine-stage Ge CMOS ring oscillator has also been demonstrated using ultra-thin channel thickness of 10 nm fabricated through the recess channel and S/D process [68]. Figure 4.13a–c briefly introduces the 3D structure views and equivalent circuit diagram of the nine-stage oscillator. Figure 4.13d,e shows the output characteristics of the oscillator from which an oscillation frequency of 88.3 MHz and propagation delay ($\tau_{pd} = 1/2nf_0, f_0$ = oscillation frequency, n = number of stages) of 629 ps could be extracted. With channel length scaling, f_0 increases (reduced τ_{pd}) due to increasing drive current.

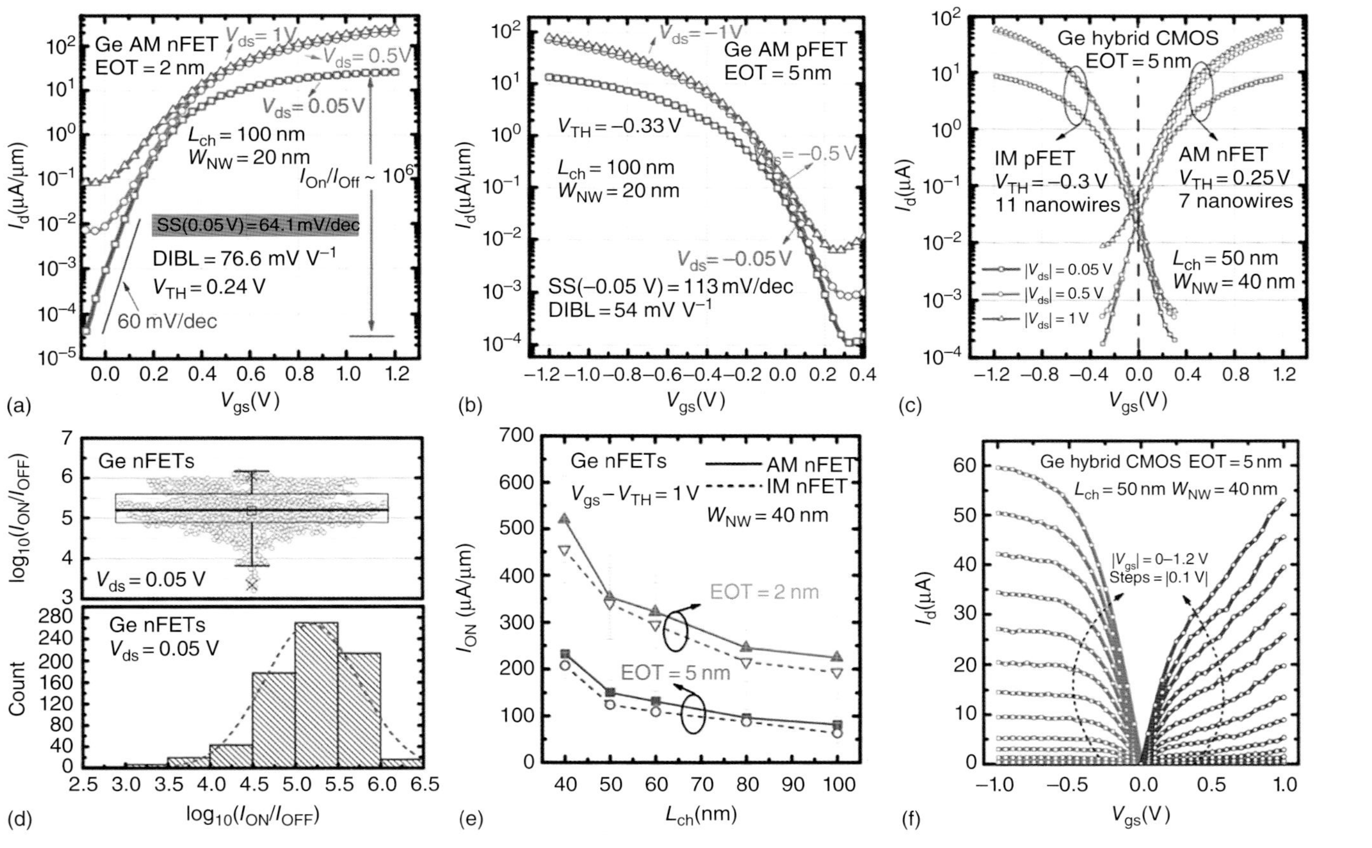

Figure 4.11 I_d–V_{gs} graphs of (a) AM Ge nanowire nFET and (b) IM Ge nanowire pFET. (c) Balanced I_d–V_{gs} curve consisting of 7 AM nanowire nFETs and 11 IM nanowire pFETs that were used to form Ge hybrid CMOS circuits. (d) On/off ratios were extracted from more than 800 Ge nanowire nFETs and were plotted in BOX plot and histogram. (e) On current trend of Ge nFETs (both AM and IM) at V_{gs}–V_{th} = 1V with channel length scaling. (f) Output characteristic graph (I_d–V_d) of Ge hybrid CMOS shown in (c). Note that $V_{IN} = V_g$ and $V_{OUT} = V_d$. Source: Wu et al. 2015 [23]. Reproduced with permission of IEEE.

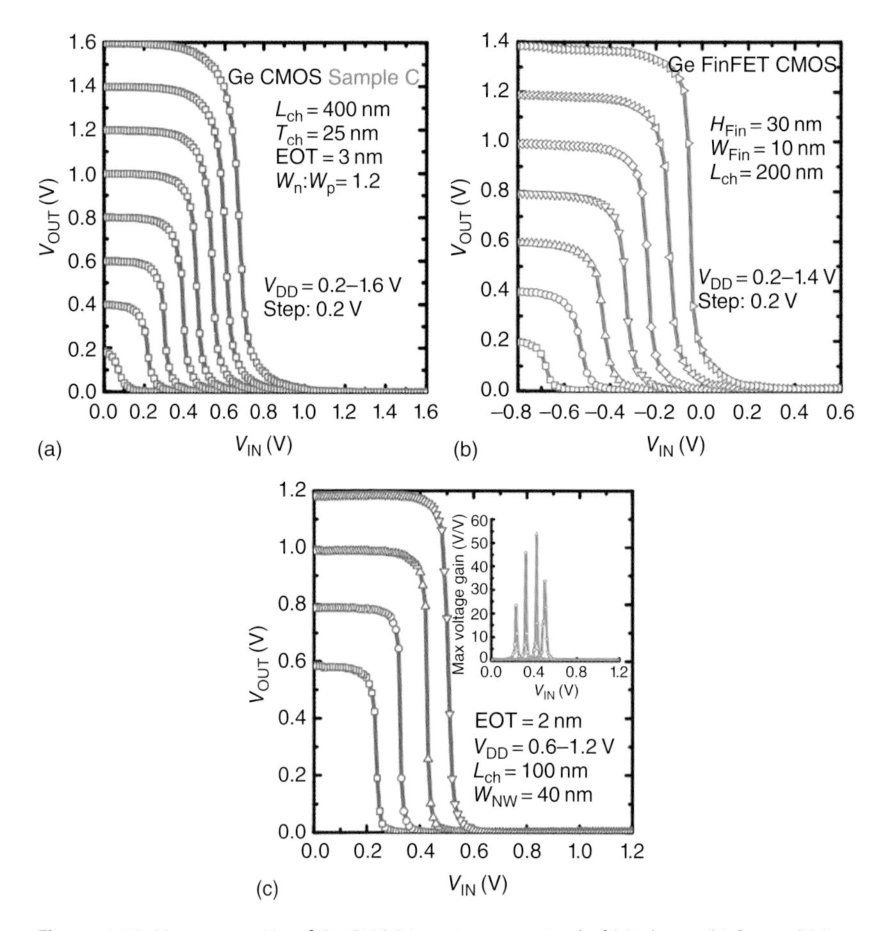

Figure 4.12 V_{OUT} versus V_{IN} of Ge CMOS inverter comprised of (a) planar, (b) fin, and (c) nanowire FETs. Source: (Panel a) Wu et al. 2014 [62]. Reproduced with permission of IEEE; (Panel b) Wu et al. 2015 [22]. Reproduced with permission of IEEE; (Panel c) Wu et al. 2015 [23]. Reproduced with permission of IEEE.

4.5 Steep-Slope Device: NCFET

Device dimension scaling has successfully served as an effective way to continue the Moore's law by increasing the density of devices on a single chip. With the number of mobile devices increasing along with advent of IoT (Internet of Things), demands for low-power devices are growing faster than ever. For lower power devices, reducing the operation voltage is the most effective solution. However, due to the lower limit (60 mV/decade at room temperature) of SS for conventional thermionic-emission-based MOS devices, reduction in supply voltage naturally causes an increase in off-state current.

When the gate voltage (V_G) is applied, only a partial fraction is delivered to the surface potential (ψ_s) which modulates I_D. Current modulation with respect to surface potential is fixed to a constant $n = 60$ mV/decade and the voltage dividing

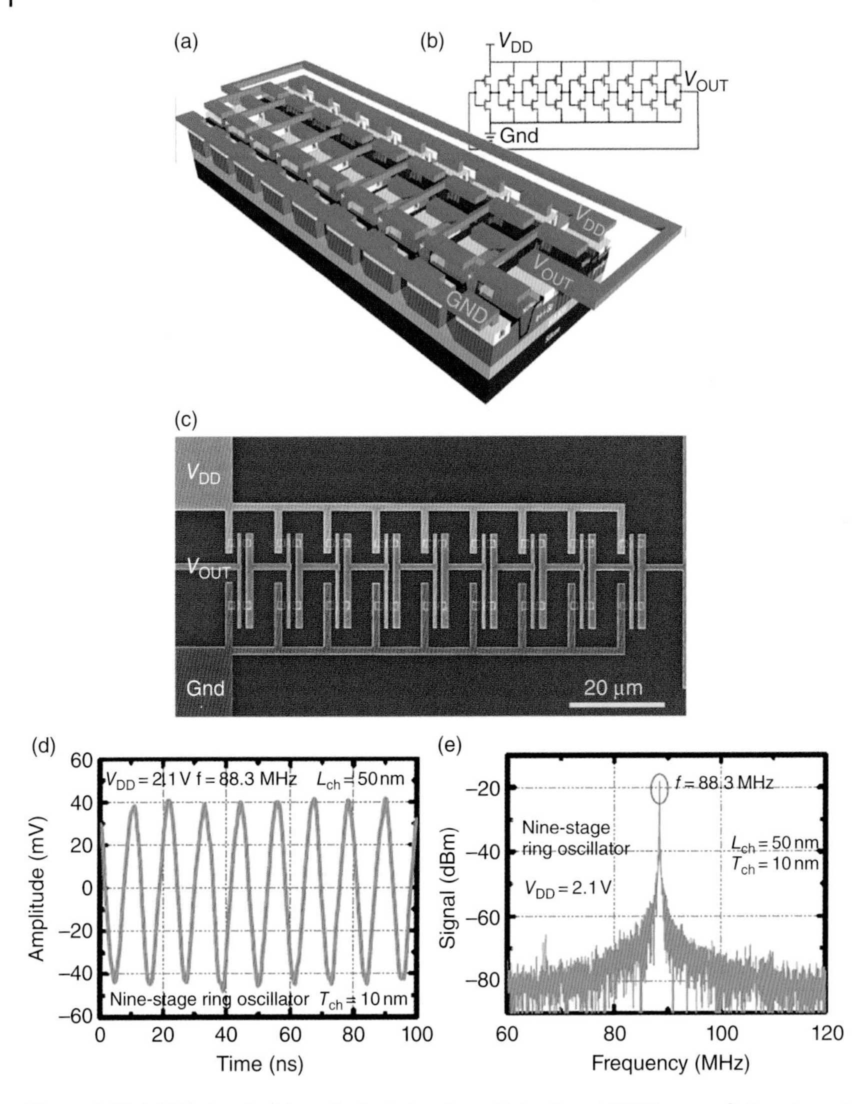

Figure 4.13 (a) 3D sketch, (b) equivalent circuit, and (c) colored SEM image of nine-stage ring oscillator. (d) Output characteristic of the ring oscillator and (e) power spectrum showing oscillating frequency of 88.3 MHz at V_{DD} = 2.1 V. Source: Wu 2016 [24]. Reproduced with permission of Purdue University.

factor is $m = 1 + C_s/C_{ox} > 1$, yielding overall SS to be larger than 60 mV/decade. This limit serves as the absolute minimum value for SS as long as the MOS device uses conventional gate oxides and carrier transport mechanism. To overcome this fundamental limit holding back the SS reduction, steep-slope devices with different current mechanisms should be developed. Tunnel field-effect transistor (TFET) is one of them, in which the term "n" is different from Figure 4.14b since the current is dependent on the carrier's band-to-band tunneling probability. However, typically, TFETs suffer from low on-current. Devices taking advantage

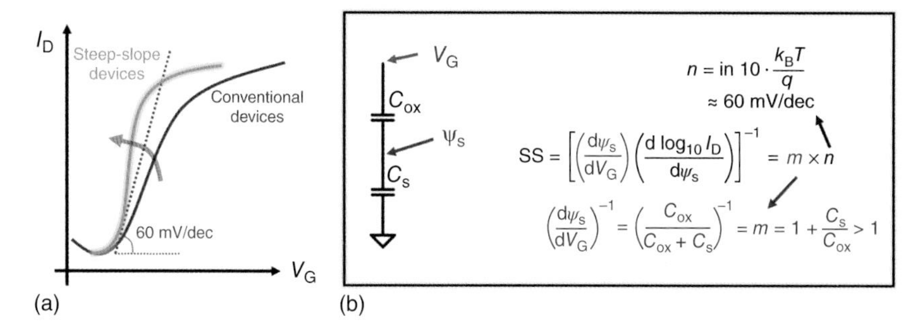

Figure 4.14 (a) SS for conventional MOS devices and steep-slope devices with respect to the Boltzmann limit of 60 mV/decade. (b) Gate oxide capacitance (C_{ox}) and semiconductor capacitance (C_s) in a conventional MOS device and terms leading to calculation of SS.

of impact ionization usually require high supply voltage, which is against the initial motivation of developing steep-slope devices.

Recently, negative capacitance field-effect transistor (NCFET) has been gaining a great deal of attention due to its possibility to pull down SS under 60 mV/decade and its compatibility with conventional fabrication processes. Multiple studies aiming to demonstrate sub-60 with the conventional CMOS platform were done [69–74]. NCFET uses capacitors that show negative capacitance over a certain range of voltage [70]. An ideal negative capacitor does not exist as it is not stable. Therefore, ferroelectric (FE) material can be inserted as the gate oxide material in the MOS structure. With negative capacitance, body factor "m" in Figure 4.14b can be smaller than 1 with negative C_{ox}. In this case, surface potential $d\psi_s$ changes more than the applied dV_G implying voltage amplification, and the FE oxide serves as the voltage step-up transformer which helps the surface potential to see a larger voltage value than what was applied [75].

Figure 4.15a,b shows the energy landscape of a regular capacitor ($C > 0$) and FE material (partially $C < 0$), respectively. Taking the derivative of energy U by charge Q (dU/dQ) gives Q–V graphs plotted in dotted lines (red). The dotted box (blue) in FE material near $Q = 0$ specifies the negative capacitance region ($C = \left(\frac{d^2U}{dQ^2}\right)^{-1} < 0$), where the energy curve is concave downward.

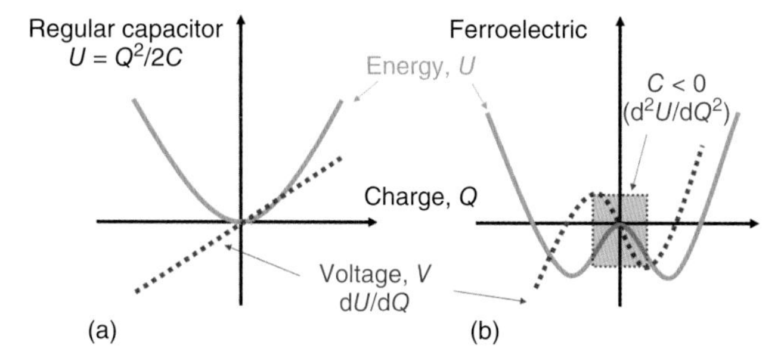

Figure 4.15 Energy, voltage (dU/dQ), and capacitance ($C = d^2U/dQ^2$) as a function of Q. Graphs for (a) regular capacitor and (b) ferroelectric are shown for comparison.

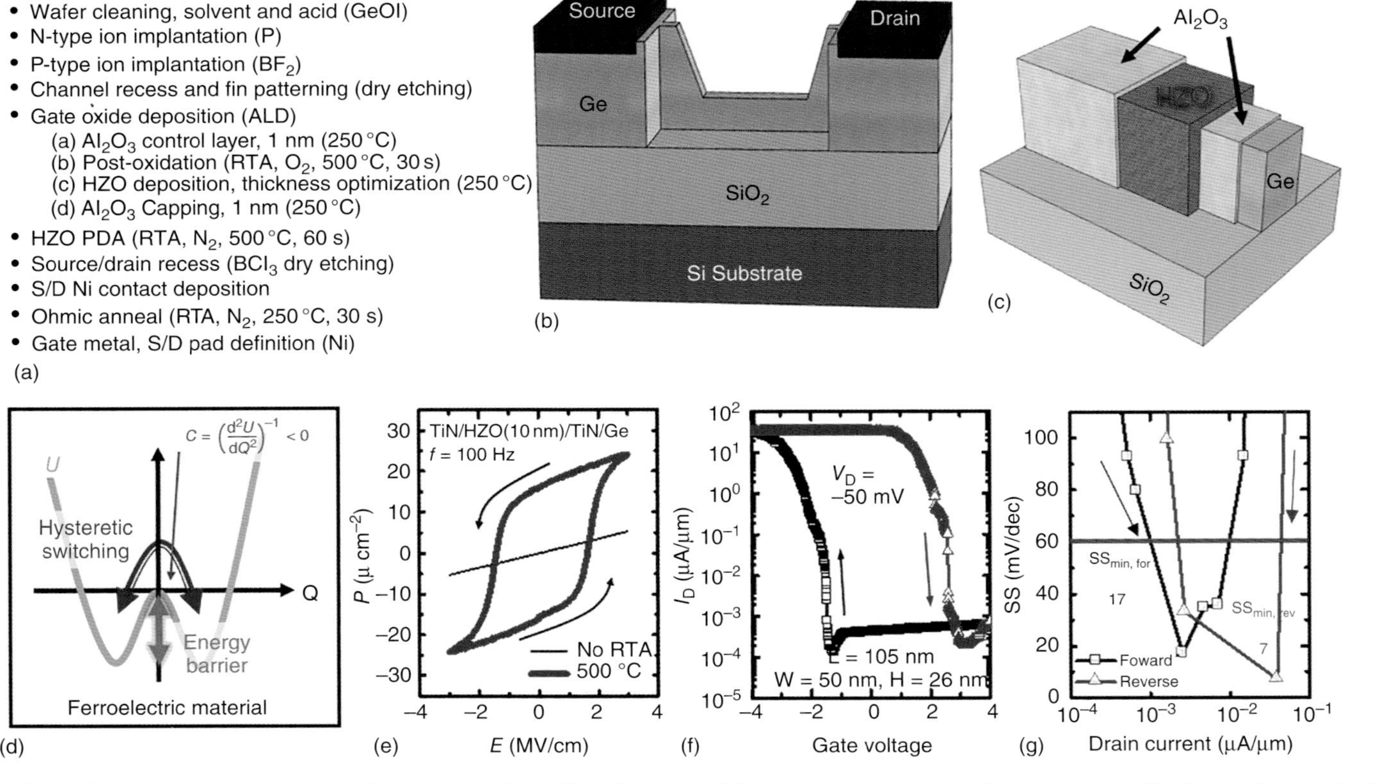

Figure 4.16 (a) The key process steps used to fabricate Ge NC FinFET. (b) 3D figure of the fabricated fin structure before gate metallization. (c) Gate oxide stack used in Ge NC FinFET. (d) Conceptual idea of origin of hysteresis in FEFET. (e) P–E graph of TiN/HZO/TiN on Ge substrate before (linear line) and after (closed loop) annealing at 500 °C showing clear ferroelectric behavior. (f) Ge FE pFinFET fabricated with 10-nm HZO demonstrates steep switching properties in both forward and reverse sweep. (g) SS extracted from (f). Chung et al. 2017 [78]. Reproduced with permission of IEEE.

Germanium MOS devices showing promising results with help from the FE gate stacks are also demonstrated [72, 73]. Among different FE materials, $Hf_{0.5}Zr_{0.5}O_2$ (HZO) deposition using ALD is popular since it shows decent ferroelectricity and has high compatibility with the conventional CMOS platform [71–73, 76, 77].

Figure 4.16a–g shows process flow and electrical data from an experimental sub-60 mV/decade Ge NC FinFET device [78]. FE HZO was utilized to integrate NC into the gate stack. Figure 4.16b,c shows the 3D structure of the demonstrated device. The first thin Al_2O_3 layer on a Ge channel is used for the control layer, preventing aggressive oxidation of the Ge channel. Rapid thermal annealing (RTA) was done in O_2 atmosphere to oxidize the Ge underneath the control layer to form thin GeO_x to enhance the interface quality [16, 67]. Al_2O_3 on top of HZO is used to prevent HZO exposure to air during the fabrication process.

Figure 4.16d explains briefly about the origin of unstable hysteresis in typical ferroelectric FET (FeFET). Energy (U) of the FE material exhibits negative capacitance values in a specific charge region ($C = (d^2U/dQ^2)^{-1} < 0$) and the hysteresis comes from the energy barrier between the local minimums of U [79]. Such hysteresis can be alleviated by carefully optimizing device parameters such as the FE oxide's thickness, compositional ratio, and annealing temperature or device dimensions, altering existing positive capacitances within the device itself which results in demonstration of NCFETs [78, 80].

As clearly seen from the linear line in Figure 4.16e, as-dep HZO shows no polarization-electric field (P-E) loop, implying there is no FE property. However, with 500 °C annealing, the FE property is activated showing a clear P-E loop with coercive field (E_C) of 1.6 MV cm^{-1} and remnant polarization (P_r) of 16 μC cm^{-2}. Integrating 10 nm of HZO into the Ge FinFET structure with recessed channel, source and drain structure, a steep slope of 7 and 17 mV/decade in reverse and forward sweep were demonstrated, as shown in Figure 4.16f,g.

It is reported that there is a trade-off between SS and I–V hysteresis, which requires careful optimization of the FE film and the process steps to minimize both SS and the hysteresis simultaneously [79, 81]. With optimized gate stack, negligible hysteresis and sub-60 mV/decade switching can be realized [71, 78]. Moreover, ferroelectric gate oxide's capabilities related to polarization time response and speed should be addressed and studied for more implementation of such devices into the mainstream industry [82–84].

4.6 Conclusion

Although germanium is equipped with much higher electron and hole mobility than silicon, it has not been the material of choice for the mainstream microelectronic products. Its inferior interface properties with various dielectric materials (including the native oxide, GeO_2), strong FLP, and lack of process integration research on germanium were the key factors that hindered active researches on it. However, recently, high-performance germanium MOS devices with various materials and structures were announced and studied, mainly owing to the introduction of novel materials and state-of-the-art fabrication equipment.

Better interface quality with various interfacial layers was demonstrated. Lower resistivity in metal/n-Ge contacts were made possible with precise control of the fabrication processes. With these strategies, both high-performance nMOS and pMOS devices in planar, fin, and nanowire structures were successfully demonstrated, leading to more active researches on germanium. The conventional silicon industry already incorporates germanium into its process flow in the form of SiGe, which increases the demand for a more systematical understanding of germanium-based technologies. With several more issues (process simplicity, reliability issues, and so on) to be solved, germanium could expected to present itself as a promising alternative for the future electronics.

References

1 Bardeen, J. and Brattain, W.H. (1948). The transistor, a semi-conductor triode. *Phys. Rev.* 74 (2): 230–231.

2 Moore, G.E. (2006). Cramming more components onto integrated circuits, Reprinted from Electronics, volume 38, number 8, April 19, 1965, pp. 114 ff. *IEEE Solid-State Circuits Newsl.* 20 (3): 33–35.

3 Clavelier, L., Deguet, C., Di Cioccio, L. et al. (2010). Engineered substrates for future More Moore and More than Moore integrated devices. 2010 International Electron Devices Meeting, 2.6.1–2.6.4.

4 Kazior, T.E. (2013). More than Moore: III–V devices and Si CMOS get it together. 2013 IEEE International Electron Devices Meeting, 28.5.1–28.5.4.

5 Yeric, G. (2015). Moore's law at 50: are we planning for retirement? 2015 IEEE International Electron Devices Meeting (IEDM), 1.1.1–1.1.8.

6 Takagi, S., Zhang, R., Kim, S.-H. et al. (2012). MOS interface and channel engineering for high-mobility Ge/III–V CMOS. 2012 International Electron Devices Meeting, 23.1.1–23.1.4.

7 Dimoulas, A., Tsipas, P., Sotiropoulos, A., and Evangelou, E.K. (2006). Fermi-level pinning and charge neutrality level in germanium. *Appl. Phys. Lett.* 89 (25): 252110.

8 Nishimura, T., Kita, K., and Toriumi, A. (2007). Evidence for strong Fermi-level pinning due to metal-induced gap states at metal/germanium interface. *Appl. Phys. Lett.* 91 (12): 123123.

9 Kuzum, D., Park, J.-H., Krishnamohan, T. et al. (2011). The effect of donor/acceptor nature of interface traps on Ge MOSFET characteristics. *IEEE Trans. Electron Devices* 58 (4): 1015–1022.

10 Baek, S.C., Seo, Y.-J., Oh, J.G. et al. (2014). Alleviation of fermi-level pinning effect at metal/germanium interface by the insertion of graphene layers. *Appl. Phys. Lett.* 105 (7): 73508.

11 Seo, Y., Lee, S., Baek, S.C. et al. (2015). The mechanism of schottky barrier modulation of tantalum nitride/Ge contacts. *IEEE Electron Device Lett.* 36 (10): 997–1000.

12 Ahn, H.J., Moon, J., Seo, Y. et al. (2017). Formation of low-resistivity nickel germanide using atomic layer deposited nickel thin film. *IEEE Trans. Electron Devices* 64 (6): 2599–2603.

13 Miyoshi, H., Ueno, T., Hirota, Y. et al. (2014). Low nickel germanide contact resistances by carrier activation enhancement techniques for germanium CMOS application. *Jpn. J. Appl. Phys.* 53 (4S): 04EA05.
14 Wu, H., Si, M., Dong, L. et al. (2014). Ge CMOS: Breakthroughs of nFETs (I_{max} = 714 mA/mm, g_{max} = 590 mS/mm) by recessed channel and S/D. 2014 Symposium on VLSI Technology (VLSI-Technology): Digest of Technical Papers, 1–2.
15 Kuzum, D., Krishnamohan, T., Nainani, A. et al. (2009). Experimental demonstration of high mobility Ge NMOS. 2009 IEEE International Electron Devices Meeting (IEDM), 1–4.
16 Zhang, R., Huang, P.-C., Lin, J.-C. et al. (2013). High-mobility Ge p- and n-MOSFETs With 0.7-nm EOT using $HfO_2/Al_2O_3/GeO_x/Ge$ gate stacks fabricated by plasma postoxidation. *IEEE Trans. Electron Devices* 60 (3): 927–934.
17 Tersoff, J. (1984). Schottky barrier heights and the continuum of gap states. *Phys. Rev. Lett.* 52 (6): 465–468.
18 Tung, R.T. (2001). Formation of an electric dipole at metal-semiconductor interfaces. *Phys. Rev. B* 64 (20): 205310.
19 Oh, J., Huang, J., Chen, Y.-T. et al. (2011). Comparison of Ohmic contact resistances of n- and p-type Ge source/drain and their impact on transport characteristics of Ge metal oxide semiconductor field effect transistors. *Thin Solid Films* 520 (1): 442–444.
20 Martens, K., Rooyackers, R., Firrincieli, A. et al. (2011). Contact resistivity and Fermi-level pinning in n-type Ge contacts with epitaxial Si-passivation. *Appl. Phys. Lett.* 98 (1): 13504.
21 Lieten, R.R., Degroote, S., Kuijk, M., and Borghs, G. (2008). Ohmic contact formation on n-type Ge. *Appl. Phys. Lett.* 92 (2): 22106.
22 Wu, H., Luo, W., Zhou, H. et al. (2015). First experimental demonstration of Ge 3D FinFET CMOS circuits. 2015 Symposium on VLSI Technology (VLSI Technology), T58–T59.
23 Wu, H., Wu, W., Si, M., and Ye, P.D. (2015). First demonstration of Ge nanowire CMOS circuits: Lowest SS of 64 mV/dec, highest gmax of 1057 μS/μm in Ge nFETs and highest maximum voltage gain of 54 V/V in Ge CMOS inverters. 2015 IEEE International Electron Devices Meeting (IEDM), 3, 2.1.1–2.1.4.
24 Wu, H. (2016). *Non-Silicon CMOS Devices and Circuits on High Mobility Channel Materials: Germanium and III–V*. Purdue University.
25 Wu, H., Si, M., Dong, L. et al. (2015). Germanium nMOSFETs with recessed channel and S/D: contact, scalability, interface, and drain current exceeding 1 A/mm. *IEEE Trans. Electron Devices* 62 (5): 1419–1426.
26 Kita, K., Lee, C.H., Tabata, T. et al. (2010). Desorption kinetics of GeO from GeO_2/Ge structure. *J. Appl. Phys.* 108 (5): 54104.
27 Prabhakaran, K., Maeda, F., Watanabe, Y., and Ogino, T. (2000). Distinctly different thermal decomposition pathways of ultrathin oxide layer on Ge and Si surfaces. *Appl. Phys. Lett.* 76 (16): 2244–2246.
28 Kamata, Y., Takashima, A., and Tezuka, T. (2009). Material properties, thermal stabilities and electrical characteristics of Ge MOS devices, depending

on oxidation states of Ge oxide: monoxide [GeO(II)] and dioxide [GeO_2(IV)]. *MRS Proc.* 1155: 1155-C02-4.

29 Hurley, P.K., Cherkaoui, K., O'Connor, E. et al. (2008). Interface defects in HfO_2, LaSiOx, and Gd_2O_3 high-k/metal-gate structures on silicon. *J. Electrochem. Soc.* 155 (2): G13.

30 Murad, S.N.A., Baine, P.T., McNeill, D.W. et al. (2012). Optimisation and scaling of interfacial GeO_2 layers for high-k gate stacks on germanium and extraction of dielectric constant of GeO_2. *Solid State Electron.* 78: 136–140.

31 Lee, C.H., Nishimura, T., Tabata, T. et al. (2013). Reconsideration of electron mobility in Ge n-MOSFETs from Ge substrate side – atomically flat surface formation, layer-by-layer oxidation, and dissolved oxygen extraction. 2013 IEEE International Electron Devices Meeting, 2.3.1–2.3.4.

32 Gu, J.J., Liu, Y.Q., Xu, M. et al. (2010). High performance atomic-layer-deposited $LaLuO_3$/Ge-on-insulator p-channel metal-oxide-semiconductor field-effect transistor with thermally grown GeO_2 as interfacial passivation layer. *Appl. Phys. Lett.* 97 (1): 12106.

33 Lu, C., Lee, C.H., Nishimura, T., and Toriumi, A. (2015). Design and demonstration of reliability-aware Ge gate stacks with 0.5 nm EOT. 2015 Symposium on VLSI Technology (VLSI Technology), T18–T19.

34 Shin, Y., Chung, W., Seo, Y. et al. (2014). Demonstration of Ge pMOSFETs with 6 Å EOT using TaN/ZrO_2/Zr-cap/n-Ge(100) gate stack fabricated by novel vacuum annealing and in-situ metal capping method. 2014 Symposium on VLSI Technology (VLSI-Technology): Digest of Technical Papers, 1–2.

35 Lee, C.H., Nishimura, T., Tabata, T. et al. (2010). Ge MOSFETs performance: Impact of Ge interface passivation. 2010 International Electron Devices Meeting, 18.1.1–18.1.4.

36 Bai, W.P., Lu, N., and Kwong, D.-L. (2005). Si interlayer passivation on germanium MOS capacitors with high-k dielectric and metal gate. *IEEE Electron Device Lett.* 26 (6): 378–380.

37 De Jaeger, B., Bonzom, R., Leys, F. et al. (2005). Optimisation of a thin epitaxial Si layer as Ge passivation layer to demonstrate deep sub-micron n- and p-FETs on Ge-On-Insulator substrates. *Microelectron. Eng.* 80: 26–29.

38 Maeda, T., Nishizawa, M., Morita, Y., and Takagi, S. (2007). Role of germanium nitride interfacial layers in HfO_2/germanium nitride/germanium metal-insulator-semiconductor structures. *Appl. Phys. Lett.* 90 (7): 72911.

39 Zhang, R., Iwasaki, T., Taoka, N. et al. (2011). Suppression of ALD-induced degradation of Ge MOS interface properties by low power plasma nitridation of GeO_2. *J. Electrochem. Soc.* 158 (8): G178.

40 Kim, G.-S., Kim, S.-H., Kim, J.-K. et al. (2015). Surface passivation of germanium using SF6 plasma to reduce source/drain contact resistance in Germanium n-FET. *IEEE Electron Device Lett.* 36 (8): 745–747.

41 Xie, R. and Zhu, C. (2007). Effects of sulfur passivation on germanium MOS capacitors with HfON gate dielectric. *IEEE Electron Device Lett.* 28 (11): 976–979.

42 Lee, C.H., Tabata, T., Nishimura, T. et al. (2009). Ge/GeO_2 interface control with high-pressure oxidation for improving electrical characteristics. *Appl. Phys. Express* 2: 71404.

43 Lee, C.H., Nishimura, T., Nagashio, K. et al. (2011). High-electron-mobility Ge/GeO_2 n-MOSFETs with two-step oxidation. *IEEE Trans. Electron Devices* 58 (5): 1295–1301.

44 Zhang, R., Iwasaki, T., Taoka, N. et al. (2012). High-mobility Ge pMOSFET with 1-nm EOT Al_2O_3/GeO_x/Ge gate stack fabricated by plasma post oxidation. *IEEE Trans. Electron Devices* 59 (2): 335–341.

45 Chung, W. (2014). *Research on Improvement of Electrical Properties of Ge pMOS Devices Using Vacuum Annealing and Ultrathin Hf Layer with sub-1nm EOT*. Korea Advanced Institute of Science and Technology.

46 Kim, H., McIntyre, P.C., Chui, C.O. et al. (2004). Interfacial characteristics of HfO_2 grown on nitrided Ge (100) substrates by atomic-layer deposition. *Appl. Phys. Lett.* 85 (14): 2902–2904.

47 Zhang, R., Tang, X., Yu, X. et al. (2016). Aggressive EOT scaling of Ge pMOSFETs with $HfO_2/AlO_x/GeO_x$ Gate-Stacks Fabricated by Ozone Postoxidation. *IEEE Electron Device Lett.* 37 (7): 831–834.

48 Kita, K., Takahashi, T., Nomura, H. et al. (2008). Control of high-k/germanium interface properties through selection of high-k materials and suppression of GeO volatilization. *Appl. Surf. Sci.* 254 (19): 6100–6105.

49 Nishimura, T., Lee, C.H., Tabata, T. et al. (2011). High-electron-mobility Ge n-channel metal–oxide–semiconductor field-effect transistors with high-pressure oxidized Y_2O_3. *Appl. Phys. Express* 4 (6): 64201.

50 Tabata, T., Lee, C.H., Kita, K., and Toriumi, A. (2008). Impact of high pressure O_2 annealing on amorphous $LaLuO_3$/Ge MIS capacitors. *ECS Trans.* 16: 479–486.

51 Bethge, O., Zimmermann, C., Lutzer, B. et al. (2014). ALD grown rare-earth high-k oxides on Ge: lowering of the interface trap density and EOT scalability. *ECS Trans.* 64 (8): 69–76.

52 Lu, C., Hyun Lee, C., Zhang, W. et al. (2014). Enhancement of thermal stability and water resistance in yttrium-doped GeO_2/Ge gate stack. *Appl. Phys. Lett.* 104 (9): 92909.

53 Dismukes, J.P., Ekstrom, L., and Paff, R.J. (1964). Lattice parameter and density in germanium-silicon alloys 1. *J. Phys. Chem.* 68 (10): 3021–3027.

54 Tezuka, T., Nakaharai, S., Moriyama, Y. et al. (2004). Selectively-formed high mobility SiGe-on-Insulator pMOSFETs with Ge-rich strained surface channels using local condensation technique. Digest of Technical Papers. 2004 Symposium on VLSI Technology, 198–199.

55 Hashemi, P., Ando, T., Balakrishnan, K. et al. (2015). High-mobility high-Ge-content $Si_{1-x}Ge_x$-OI PMOS FinFETs with fins formed using 3D germanium condensation with Ge fraction up to x ~0.7, scaled EOT ~8.5 Å and ~10 nm fin width. 2015 Symposium on VLSI Circuits (VLSI Circuits), T16–T17.

56 Nakaharai, S., Tezuka, T., Sugiyama, N. et al. (2003). Characterization of 7-nm-thick strained Ge-on-insulator layer fabricated by Ge-condensation technique. *Appl. Phys. Lett.* 83 (17): 3516–3518.

57 Letertre, F., Deguet, C., Richtarch, C. et al. (2004). Germanium-on-insulator (GeOI) structure realized by the Smart Cut™ technology. *MRS Proc.* 809: B4.4.

58 Chao, Y.-L., Scholz, R., Reiche, M. et al. (2006). Characteristics of Germanium-on-insulators fabricated by wafer bonding and hydrogen-induced layer splitting. *Jpn. J. Appl. Phys.* 45 (11): 8565–8570.

59 Yu, D.S., Chin, A., Liao, C.C. et al. (2005). Three-dimensional metal gate-high-k-GOI CMOSFETs on 1-poly-6-metal 0.18-μm Si devices. *IEEE Electron Device Lett.* 26 (2): 118–120.

60 Chui, K.-J., Ang, K.-W., Madan, A. et al. (2005). Source/drain germanium condensation for p-channel strained ultra-thin body transistors. IEEE International Electron Devices Meeting, 2005. IEDM Technical Digest, 493–496.

61 Akatsu, T., Deguet, C., Sanchez, L. et al. (2005). 200mm Germanium-On-Insulator (GeOI) by Smart Cut™ technology and recent GeOI pMOSFETs achievements. 2005 IEEE International SOI Conference Proceedings, 137–138.

62 Wu, H., Conrad, N., Luo, W., and Ye, P.D. (2014). First experimental demonstration of Ge CMOS circuits. 2014 IEEE International Electron Devices Meeting, 9.3.1–9.3.4.

63 Lee, C.H., Nishimura, T., Tabata, T. et al. (2011). Experimental study of carrier transport in ultra-thin body GeOI MOSFETs. IEEE 2011 International SOI Conference, 1–2.

64 Low, T., Li, M.-F., Samudra, G. et al. (2005). Modeling study of the impact of surface roughness on silicon and Germanium UTB MOSFETs. *IEEE Trans. Electron Devices* 52 (11): 2430–2439.

65 Wu, Y.-S., Hsieh, H.-Y., Hu, V.P.-H., and Su, P. (2011). Impact of quantum confinement on short-channel effects for ultrathin-body Germanium-on-insulator MOSFETs. *IEEE Electron Device Lett.* 32 (1): 18–20.

66 Pop, E., Chui, C.O., Sinhaf, S. et al. (2004). Electro-thermal comparison and performance optimization of thin-body SOI and GOI MOSFETs. IEDM Technical Digest. IEEE International Electron Devices Meeting, 411–414.

67 Wu, H., Wu, W., Si, M., and Ye, P.D. (2016). Demonstration of Ge nanowire CMOS devices and circuits for ultimate scaling. *IEEE Trans. Electron Devices* 1–9.

68 Wu, H., Conrad, N., Si, M., and Ye, P.D. (2015). Demonstration of Ge CMOS inverter and ring oscillator with 10 nm ultra-thin channel. 2015 73rd Annual Device Research Conference (DRC), 281–282.

69 Khan, A.I., Chatterjee, K., Duarte, J.P. et al. (2016). Negative capacitance in short-channel FinFETs externally connected to an epitaxial ferroelectric capacitor. *IEEE Electron Device Lett.* 37 (1): 111–114.

70 Salahuddin, S. and Datta, S. (2008). Can the subthreshold swing in a classical FET be lowered below 60 mV/decade? 2008 IEEE International Electron Devices Meeting, 1, 1–4.

71 Si, M., Su, C.-J., Jiang, C. et al. (2017). Steep slope MoS_2 2D transistors: negative capacitance and negative differential resistance. *Nature Nanotechnology.*

72 Su, C.-J., Tang, Y.-T., Tsou, Y.-C. et al. (2017). Nano-scaled Ge FinFETs with low temperature ferroelectric HfZrOx on specific interfacial layers exhibiting 65% S.S. Reduction and improved ION. 2017 Symposium on VLSI Technology (VLSI Technology), 152–153.

73 Zhou, J., Han, G., Li, Q. et al. (2016). Ferroelectric HfZrOx Ge and GeSn PMOSFETs with Sub-60 mV/decade subthreshold swing, negligible hysteresis, and improved Ids. 2016 IEEE International Electron Devices Meeting (IEDM), 12.2.1–12.2.4.
74 Li, K.S., Chen, P.G., Lai, T.Y. et al. (2016). Sub-60 mV-swing negative-capacitance FinFET without hysteresis. Technical Digest – International Electron Devices Meeting IEDM, vol. 2016–February, 22.6.1–22.6.4.
75 Salahuddin, S. and Datta, S. (2008). Use of negative capacitance to provide voltage amplification for low power nanoscale devices. *Nano Lett.* 8 (2): 405–410.
76 Muller, J., Boscke, T.S., Schroder, U. et al. (2012). Ferroelectricity in simple binary ZrO_2 and HfO_2. *Nano Lett.* 12 (8): 4318–4323.
77 Lee, M.H., Fan, S.-T., Tang, C.-H. et al. (2016). Physical thickness 1.x nm ferroelectric HfZrOx negative capacitance FETs. 2016 IEEE International Electron Devices Meeting (IEDM), 12.1.1–12.1.4.
78 Chung, W., Si, M., and Ye, P.D. (2017). Hysteresis-free negative capacitance germanium CMOS FinFETs with Bi-directional Sub-60 mV/dec. 2017 IEEE International Electron Devices Meeting (IEDM), 15.3.1–15.3.4.
79 Jain, A. and Alam, M.A. (2013). Prospects of hysteresis-free abrupt switching (0 mV/decade) in Landau switches. *IEEE Trans. Electron Devices* 60 (12): 4269–4276.
80 Si, M., Jiang, C., Su, C.-J. et al. (2017). Sub-60 mV/dec Ferroelectric HZO MoS_2 Negative Capacitance Field-effect Transistor with Internal Metal Gate: the Role of Parasitic Capacitance. 2017 IEDM, 573–576.
81 Duarte, J.P., Khandelwal, S., Khan, A.I. et al. (2016). Compact models of negative-capacitance FinFETs: Lumped and distributed charge models. 2016 IEEE International Electron Devices Meeting (IEDM), 5(2), 30.5.1–30.5.4.
82 Chung, W., Si, M., Shrestha, P.R. et al. (2018). First Direct Experimental Studies of $Hf_{0.5}Zr_{0.5}O_2$ Ferroelectric Polarization Switching Down to 100-picosecond in Sub-60mV/dec Germanium Ferroelectric Nanowire FETs, vol. 7, 89–90.
83 Muller, J., Boscke, T.S., Schroder, U. et al. (2012). Nanosecond Polarization Switching and Long Retention in a Novel MFIS-FET Based on Ferroelectric HfO_2. *IEEE Electron Device Lett.* 33 (2), 185–187.
84 Krivokapic, Z., Rana, U., Galatage, R. et al. (2017). 14nm Ferroelectric FinFET technology with steep subthreshold slope for ultra low power applications. 2017 IEEE International Electron Devices Meeting (IEDM), 15.1.1–15.1.4.

5

Carbon Nanotube Logic Technology

Jianshi Tang and Shu-Jen Han

IBM, T. J. Watson Research Center, 1101 Kitchawan Road, Yorktown Heights, NY 10598, USA

5.1 Introduction – Silicon CMOS Scaling and the Challenges

Postulating the number of transistors per chip at a constant cost doubles roughly every 18–24 months, the Moore's law scaling has been driving the semiconductor industry for more than four decades. Historically, transistors have been continuously involving: from the very first point-contact Ge transistor (Bell Labs. 1947) to planar Si metal-oxide-semiconductor field-effect transistor (MOSFET), then to strained SiGe source/drain (e.g. 90- and 65-nm technology nodes) and high-κ/metal gate stack (e.g. 45- and 32-nm nodes), and finally to three-dimensional (3D) tri-gate structure (e.g. 22- and 16/14-nm nodes) nowadays. In an ultrascaled transistor, the parasitic resistance will reduce the transistor's drive current, and degrade the circuit speed. Figure 5.1a shows some major resistance components in today's Si CMOS devices, where the gate region is separated from the source and drain (S/D) contacts by spacers. These resistance components consist of the contact resistance (R_{CONT}) including metal itself, a silicide layer and perhaps an epitaxial layer (for the raised S/D), the resistance of S/D region (R_{SD}), the associated resistance of S/D extension (R_{EXT}), and the gate resistance (R_G) that includes the resistance associated with various gate layers (for work function adjustment) and the metal itself. The device design that optimizes the parasitic resistance does not necessarily provide the best performance, and we have to also consider the impact of parasitic capacitance. Large parasitic capacitance not only increases circuit delay but also consumes more active power, which is proportional to CV^2f, where V is the operating voltage and f is the frequency. The major parasitic capacitance components are shown in Figure 5.1b for a better illustration of resistance-capacitance trade-offs. Starting from the S/D contacts, there are junction capacitance (C_J, for non-silicon-on-insulator (SOI) devices), S/D extension and gate overlap capacitance (C_{OV}), inner fringe (C_{IF}) and outer fringe (C_{OF}) capacitances, and S/D contacts to gate electrode capacitance (C_{CG}). For a fixed contacted gate pitch, a thinner spacer can help increase S/D region (lower R_{CONT} and R_{SD}),

Advanced Nanoelectronics: Post-Silicon Materials and Devices,
First Edition. Edited by Muhammad Mustafa Hussain.

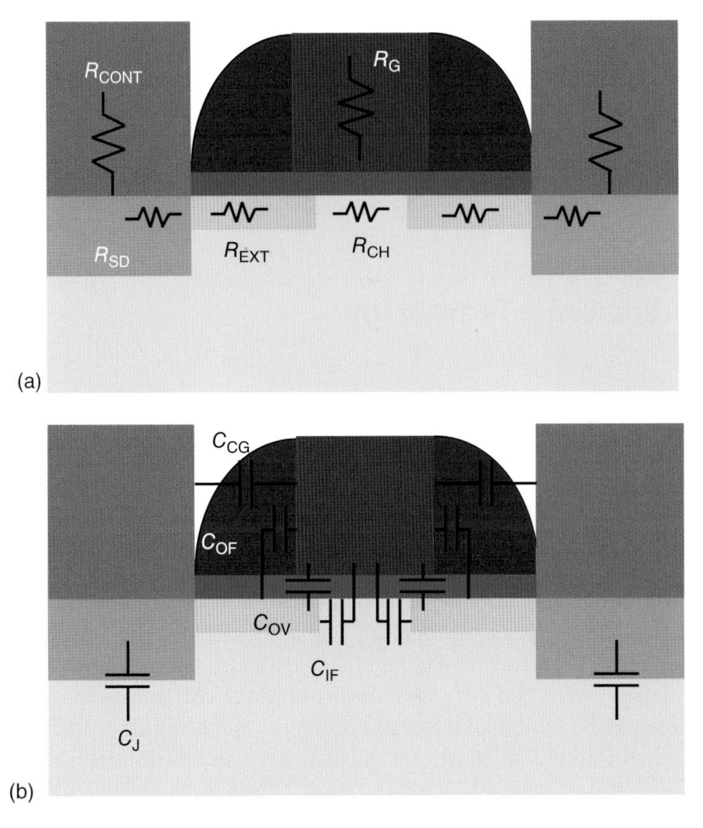

Figure 5.1 (a) Illustrations of resistance components in a typical Si MOSFET, which includes contact resistance (R_{CONT}), S/D resistance (R_{SD}), S/D extension resistance (R_{EXT}), channel resistance (R_{CH}), and gate resistance (R_G). (b) Capacitance components that include junction capacitance (C_J), S/D extension and gate overlap capacitance (C_{OV}), inner fringe (C_{IF}) and outer fringe (C_{OF}) capacitances, and S/D contacts to gate electrode capacitance (C_{CG}).

reduce extension area (lower R_{EXT}), and improve the gate fill (lower R_G), but at the cost of a significantly higher C_{CG}.

Besides the pronounced performance penalty from parasitic components, the degraded electrostatics perhaps has an even larger impact of determining the fate of Moore's law. The so-called "short-channel effects" (SCEs) result from the depletion regions of the S/D junctions that extend into the channel region and reduce the control of the channel charges from the gate. This loss of charge control by the gate causes two undesirable effects in MOSFETs: (i) the threshold voltage (V_{th}) decreases when the drain voltage increases. This drain-induced barrier lowering (DIBL) effect is shown in Figure 5.2a. (ii) The degradation of the subthreshold swing ($\mathrm{SS} = \frac{\mathrm{d}V_G}{\mathrm{d}[\log(I_D)]}$), as shown in Figure 5.2b. A simple method to estimate the improvement of the short channel control is to use the nature length, λ_N, which represents the extension of the electric field lines from the S/D into the channel [1]:

$$\lambda_N = \sqrt{\frac{\varepsilon_S}{N\varepsilon_{ox}} t_{ox} t_S} = \sqrt{\frac{\varepsilon_S}{N\varepsilon_{siO_2}} \mathrm{EOT} \cdot t_S} \tag{5.1}$$

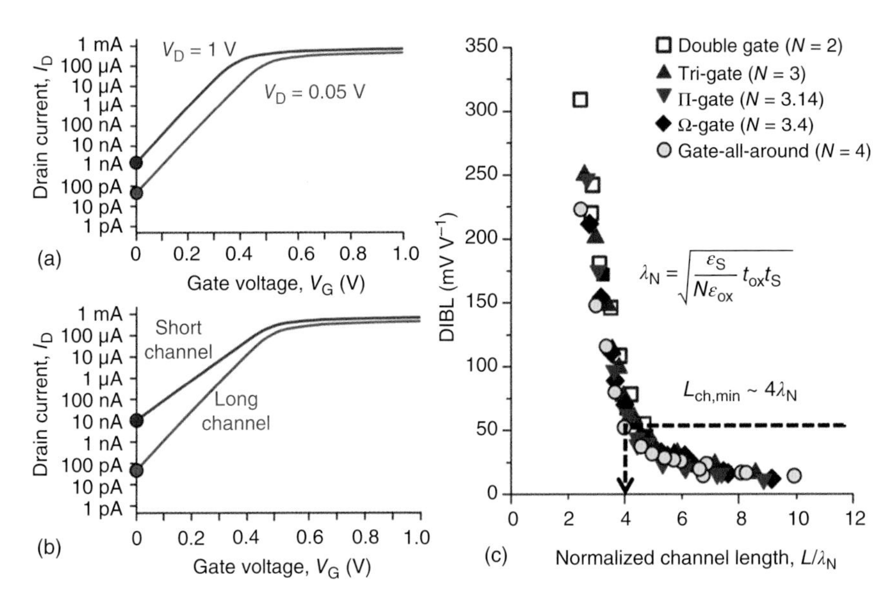

Figure 5.2 (a) Illustration of drain-induced barrier lowering (DIBL) effect. (b) Short-channel effect (SCE). (c) The dependence of DIBL effect on the transistor channel length normalized by the natural length λ_N for various gate structures. Source: Ferain et al. 2011 [1]. Reproduced with permission of Springer Nature.

where $\varepsilon_S(t_S)$ and $\varepsilon_{ox}(t_{ox})$ are the relative dielectric constant (thickness) of the semiconductor channel and the gate oxide, respectively, $\text{EOT} = \frac{\varepsilon_{SiO_2}}{\varepsilon_{ox}} t_{ox}$ is the equivalent gate oxide thickness, and N is the effective number of gates. Figure 5.2c plots the variation of the DIBL effect with the channel length L_{ch} normalized by λ_N for various gate structures. Reducing λ_N enables the scaling of minimum $L_{ch,min} \approx 4\lambda_N$, while ensuring gate electrostatic control over the channel with tolerable SCE (e.g. DIBL $\leq 50\,\text{mV V}^{-1}$, SS $\leq 75\,\text{mV/decade}$). It should be noted that Eq. (5.1) is purely based on electrostatic analysis, and more rigorous analysis considering quantum mechanics effects like source/drain direct tunneling would impose more stringent limits on $L_{ch,min}$.

Now it is clear that the ideal channel materials for the future transistors should possess two properties in order to achieve low power consumption: it should have very *high carrier saturation velocity* for achieving the desired performance with lower biases in order to reduce the dynamic power; and it should also have an *intrinsic ultrathin body* to achieve the best electrostatic control in order to reduce leakage power. From this perspective, low-dimensional nanomaterials, such as one-dimensional (1D) carbon nanotubes (CNTs) and nanowires as well as the expanding library of two-dimensional (2D) van der Waals materials, may provide certain advantages over Si bulk/fin for MOSFET scaling. For example, in the case of single-walled CNT ($t_S \approx 1$ nm), the estimated channel length scaling limit would be reduced to about 2 nm if applying the same structural parameters as Si ($N = 4$, EOT ≈ 0.5 nm) [2]. In addition, the adoption of high-mobility channel materials like CNT allows more aggressive scaling of supply voltage (e.g. V_{dd} as low as 0.2–0.4 V for CNT compared to 0.6–0.8 V for Si [3]).

Despite these many intrinsic advantages of using CNTs in logic transistors, the realization of such revolutionized technology is not as straightforward as it sounds, and is certainly much more difficult and complicated in terms of wafer-scale integration. There are still many challenges to be addressed to build practical nanomaterial-based logic technology with competitive performance, not to mention the manufacturing compatibility to the existing Si foundries. In this chapter, we focus on the discussion of logic transistor technology based on CNT. We first briefly review the electrical properties of CNT and highlight the merits that make it an ideal candidate to replace Si in ultrascaled transistors. We then review recent advances in the development of individual CNT field-effect transistors (CNTFETs). This section covers the research progress in both channel length scaling and contact length scaling as well as n-type field-effect transistor (NFET) solutions for complementary logic technology. After that, we discuss the manufacturability of CNTFETs for potential technology transfer to wafer-scale production. Several key issues, including CNT material preparation, placement and variability, are discussed in this section. Finally, we provide an outlook for future research and development in implementing CNT logic technology.

5.2 Fundamentals of Carbon Nanotube

The discoveries of three sp^2 carbon allotropes (Figure 5.3a), namely, fullerene (1985) [6], CNT (1991) [7], and graphene (2004) [8], are all groundbreaking milestones in the research of nanomaterials. They represent prototype zero-dimensional (0D), 1D, and 2D material systems with peculiar physical properties, and stimulate exploding research interest on other nanomaterials in the same class. As shown in Figure 5.3b, single-walled CNT can be viewed as a graphene sheet rolling up along a certain direction $\vec{C_h}$ to form a hollow cylindrical tube. Here, the chiral vector can be written as $\vec{C_h} = m\vec{a_1} + n\vec{a_2}$, where the integers n and m denote the number of unit vectors along two directions in the honeycomb lattice of graphene. Structurally, if $n = m$, CNTs are called *armchair* nanotubes; if $m = 0$, then they are called *zigzag* nanotubes; otherwise, they are *chiral*. The band structure and hence electronic properties of CNT are largely determined by the chirality represented by a pair of indices (n, m): if $n = m$, CNT is *metallic*; if $|n - m|$ is a multiple of 3, then it is *semiconducting* with a tiny band gap (on the order of milli-electron volts); otherwise, it is *semiconducting* with a moderate band gap that roughly scales with CNT diameter: $E_g \approx 0.8\ \mathrm{eV}/d$ [9]. Here, d is the CNT diameter in the unit of nanometer, and it is given by [10]

$$d = \frac{a}{\pi}\sqrt{n^2 + mn + m^2} \tag{5.2}$$

where $a = 0.246$ nm is the lattice constant. For metallic nanotubes, they can carry extremely high current density, nearly three orders of magnitude larger than typical metals like copper or aluminum [11]. For semiconducting tubes,

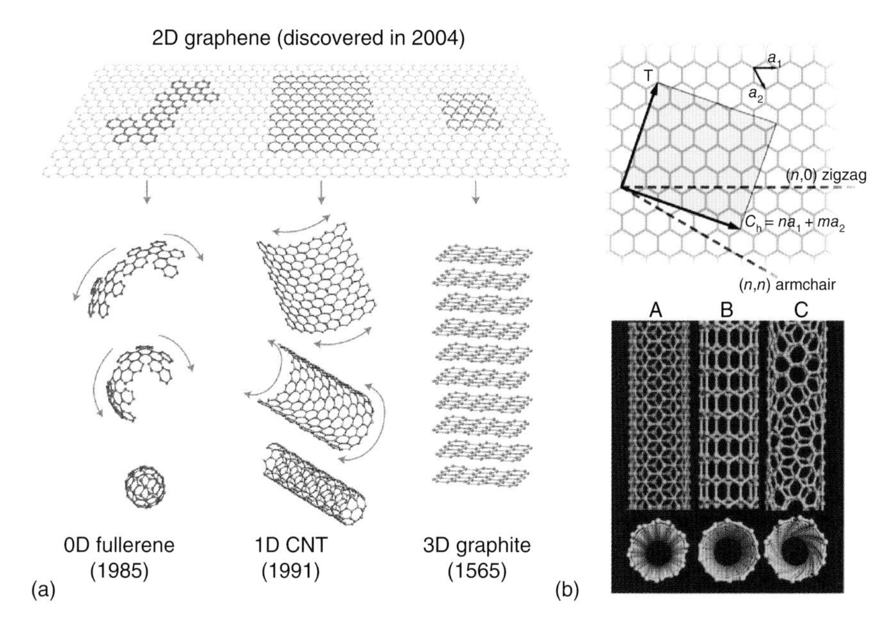

Figure 5.3 (a) Allotropes of sp^2 carbon include 0D buckyball fullerene, 1D CNT, 2D graphene, and also 3D graphite. Source: Reproduced with permission from Geim and Novoselov 2007 [4]. (b) Schematics of CNT lattice structure. The electronic properties of CNT are largely determined by its chirality represented by a pair of indices (*n*, *m*) in the denotation of chiral vector: (A) armchair ($n = m$, all are metallic), (B) armchair ($m = 0$, metallic or semiconducting), (C) chiral (otherwise, metallic or semiconducting). Source: Baughman et al. 2002 [5]. Reproduced with permission of AAAS.

the sizable energy bandgap (typical $E_g \sim 0.6$ eV, much larger than the thermal energy ~ 0.026 eV at room temperature) makes it a much better building block than graphene (Dirac semimetal with a zero bandgap) for digital logic technology.

Extensive experimental studies have revealed that CNT is an extreme 1D channel with superb electrical properties for high-performance electronics. It has high field-effect mobility: typically $\mu_{FE} > 5000\,\text{cm}^2\ \text{V}^{-1}\text{s}^{-1}$ for CNTs grown by chemical vapor deposition (CVD) [12], and $\mu_{FE} \sim 300\,\text{cm}^2\ \text{V}^{-1}\text{s}^{-1}$ for solution-processed tubes [13]. It also has a relatively long mean free path $L_{mfp} \sim 200$ nm, making it readily to achieve ballistic transport with sub-100-nm channel length [14, 15]. In addition, CNT has superior mechanical properties (light weight but with very high strength, modulus, and resiliency) [16, 17]. Therefore, many interesting applications have been proposed for CNTs, such as transistors for electronic switch [18, 19], interconnects for integrated circuits [20, 21], field emitters for flat-panel display [22, 23], nanoprobes for scanning probe microscope and lithography [24–26], an electrochemical material for energy storage [27], artificial scaffolds or transporters for biological applications [28, 29], and many others [30, 31]. In the following, we focus our discussion on potentially the most lucrative application – a logic switch that can extend Moore's law. The high mobility and the immunity of the SCE make CNT an ideal candidate to replace single-crystalline Si in logic transistor scaling [32].

5.3 Complementary Logic and Device Scalability Demonstrations

The first CNT FETs were reported in 1998 (Figure 5.4a), made by randomly depositing CNTs onto prefabricated metal electrodes [18, 19]. Such a back-gated transistor structure was not optimized as the device current was quite low, mainly because of the poor electrical contact between CNT and bottom electrodes. Since then, extensive studies have been made to improve CNTFET performance, such as improving source/drain contact quality and optimizing gate structures. For example, an improved version of CNT back-gated transistor was demonstrated with local bottom gate and top metal contact [33] (Figure 5.4b). Another form of back-gated transistors with a suspended CNT channel was also fabricated [38], where air or vacuum served as the gate dielectric. A suspended transistor was shown to have better noise performance by decoupling from the substrate (trap charges in the gate oxide) [39]. Despite the simple fabrication process, back-gated transistors usually suffer from large hysteresis and variability since the unpassivated CNTs are easily affected by water/oxygen absorption from ambient or charges on the substrate [40]. Therefore, top-gated CNTFETs are generally preferred for practical logic applications [34, 35, 41], and they could also provide better gate control over the channel (Figures 5.4c,d). For ultimate electrostatic control, wrap-around gated (gate-all-around geometry) CNTFET has also been developed [36] (Figure 5.4e), and the structure was further improved using a self-aligned gate to reduce source/drain parasitic resistance [42] (Figure 5.4f). In the atomic layer deposition (ALD) of thin gate dielectric, the CNT surface has no dangling bonds and is chemically inert to precursor molecules; therefore, it is critical to functionalize the CNT surface to facilitate the gate oxide growth [43]. Besides, the presence of oxygen vacancies in ALD oxides or the formation of dipole layer at oxide interfaces could effectively dope CNT and significantly change the transistor type [37]. All these factors need be taken into consideration in the design of CNT top-gated and wrap-around gated transistors. Beyond these standard transistor structures, more complicated CNTFETs involving gate structure engineering (such as asymmetric gate or feedback gate) have also been proposed [44, 45]. Although they may be of interest for the purpose of suppressing off-state leakage current, it is still challenging to implement them in practical logic technology.

5.3.1 CNT NFET and Contact Engineering for CMOS Logic

Because of the ultrathin body of CNT, it is very difficult to introduce substitutional doping as in Si. Although several attempts of chemical doping have been made, the stability of these chemical doping methods is poor [46–50]. As a result, without intentional doping in the source/drain contact regions, CNTFET is essentially a Schottky barrier transistor (Figure 5.5A), whose transistor type is largely determined by the work function of the metal contacts [51–53]. Tuning the metal work function allows electrons and holes to be simultaneously injected into CNT [54], and such ambipolar transport has been exploited in optoelectronics applications such as light emission [55, 56]. Unipolar PFETs and

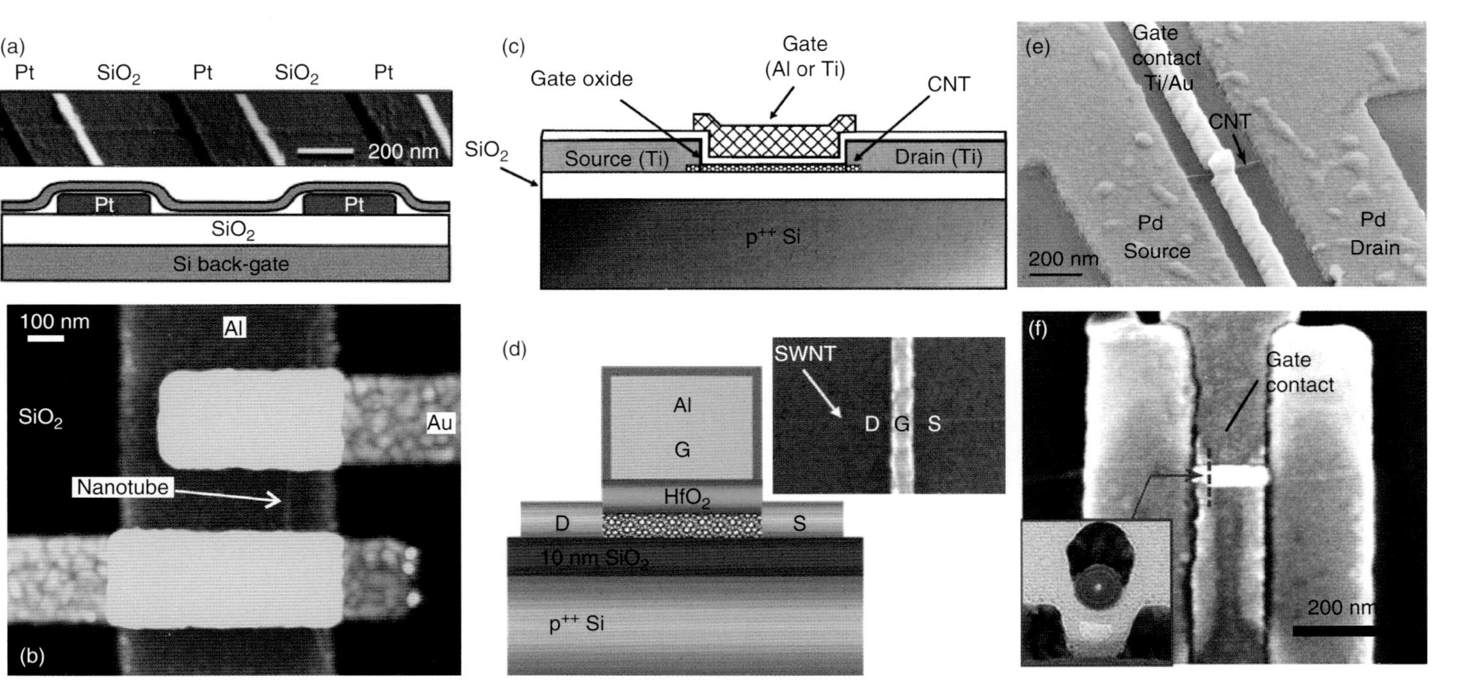

Figure 5.4 Collection of CNTFETs with various gate structures. (a) The first report of CNTFET with back-gate and metal-bottom contact. Souce: Dekker et al. 1998 [18]. Reproduced with permission of Springer Nature. (b) Improved version of CNT back-gated transistor with local bottom-gate and metal-top contact. Source: Bachtold et al. 2001 [33]. Reproduced with permission of AAAS. (c) Top-gated CNTFET. Source: Wind et al. 2002 [34]. Reproduced with permission of AIP Publishing. (d) Self-aligned top-gated CNTFET. Source: Javey et al. 2004 [35]. Reproduced with permission of ACS. (e) Gate-all-around CNTFET. Source: Chen et al. 2008 [36]. Reproduced with of IEEE. (f) Self-aligned gate-all-around CNTFET. Source: Franklin et al. 2013 [37]. Reproduced with permission of ACS.

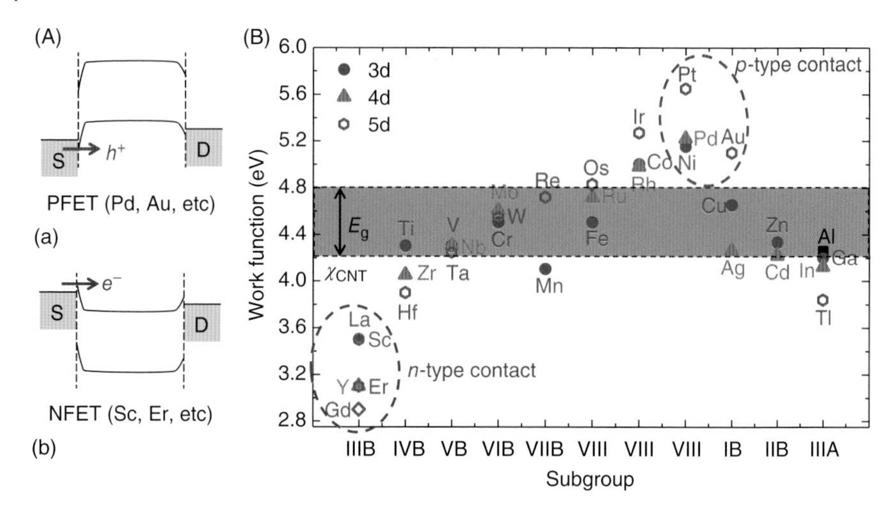

Figure 5.5 (A) Band structure for CNT PFET (a) and NFET (b). The source/drain metal work function determines the Schottky barrier height for electron and hole transport, and hence determines the transistor type. (B) Collection of metal work functions and their band alignment to CNT. Typical *p*-type metal contacts for CNT, including Pd, Au, and Pt, have high work function, while typical *n*-type metal contacts, including Sc, La, Er, Y, and Gd, have very low work function.

NFETs are usually made with high- and low-work-function metals, respectively, to selectively inject either holes or electrons into the intrinsic CNT channel (Figure 5.5B). While *p*-type metal contacts (e.g. Pd [57], Au [18], and Pt [57]) have been extensively studied and proved to be air-stable with good wetting on the CNT surface, *n*-type metal contacts (e.g. Sc [58], Er [59], Y [60], Ca [61], and La [62]) are usually prone to react with oxygen and hence suffer from severe degradation by oxidation (Figure 5.6). This oxidation problem, along with the poor wetting on the CNT surface, lead to unstable NFETs with low-yield, and also result in a large device-to-device variation [62].

In earlier reports of CNT NFETs [58–62], the *n*-type metal contacts are usually large (typically >100 nm in width), and capped with another metal (e.g. Au or Al) to prevent oxidation from the top. However, oxidation from sidewalls could still occur in air (Figure 5.6a), and gradually degrade the contact and device performance [62]. This becomes a serious issue when shrinking the contact size for scaling, and electrical measurements in vacuum are usually needed to preserve the quality of such small contacts [58, 60]. Recently, a much-improved three-step passivation scheme for *n*-type metal contact was reported: (i) sidewall protection by pre-patterned hydrogen silsesquioxane (HSQ) trenches, (ii) Au *in situ* capping after Sc deposition, and (iii) ALD Al_2O_3 full passivation [63]. Through a control experiment (Figure 5.7), it is shown that the HSQ pattern can effectively protect the sidewalls of Sc contacts, and hence significantly improve the NFET device yield and stability (no degradation over eight months). The effect is more pronounced for a small contact size of $L_c = 40$ nm with more than 2× improvement in yield. Furthermore, it is also found that the insertion of an ultrathin Ti wetting layer (nominally 0.2 nm) before Sc deposition dramatically

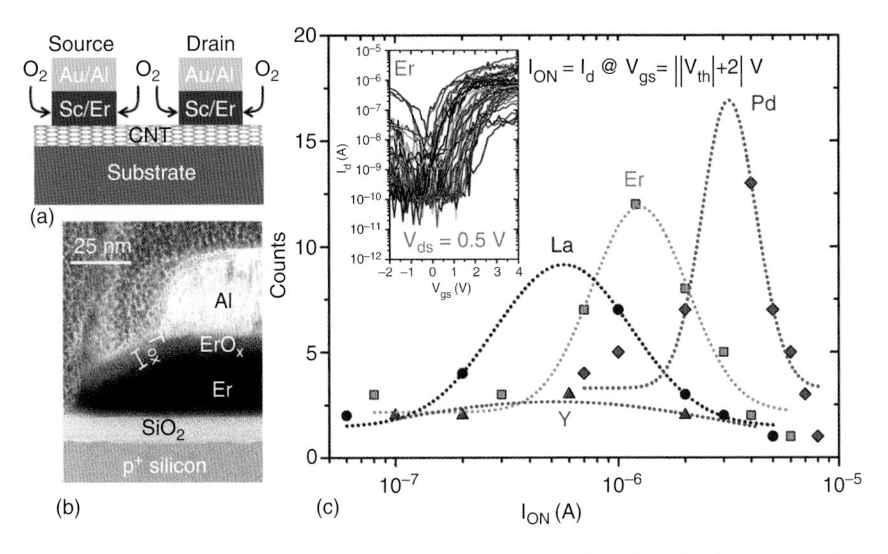

Figure 5.6 CNT NFET issue. (a) Even with top metal capping, the sidewalls of n-type metal contacts are still exposed to oxidation in air. (b) Cross-sectional TEM image of Er electrode capped with Al shows the oxidation of Er even in ultrahigh vacuum condition, and the progression of oxidation continues over time. (c) Comparison of on-state current I_{on} for devices with Pd, Er, La, and Y contacts, showing that NFETs have much larger device variations than PFETs. The inset shows a set of CNT NFETs with Er contacts. Source: Shahrjerdi et al. 2011 [62].Reproduced with permission of IEEE.

improves NFET yield and performance, showing over 2× improvement in yield and on-state current (Figure 5.7e). All these advances in *n*-type contact engineering pave the road to realize high-performance CNT CMOS ring oscillators (ROs) using dual-work-function contacts [64].

5.3.2 Channel Length Scaling in CNTFET

Earlier discussions show that CNT is an excellent candidate for ultimate scaled logic technology. It is then important to study the scaling behavior of CNTFET and benchmark the performance with latest Si technology. In practice, the scaling of transistor size involves multiple pieces: thinning down the gate dielectric, shrinking the channel length, and also reducing the contact size. There have been very limited studies on the first topic, and as expected, scaling down the gate oxide thickness helps reduce variability and suppress hysteresis [65, 66]. Experimentally, CNT is extremely sensitive to charges that are in close proximity, so those at the CNT/oxide interface induce a large device variability and hysteresis [67], which could potentially overwhelm the effect of gate oxide thickness on SS or V_{th} [40]. Nevertheless, more systematic studies on this topic are needed for CNTFET scaling, and achieving clean gate oxide/CNT and CNT/substrate interfaces is one of the key bottlenecks that remains to be solved.

As to the CNT channel length scaling, there are several short channel (from 10s nm down to 5 nm) studies on both back-gated transistors [15, 68, 69], and top-gated transistors [3, 35, 37, 41]. In the case of back-gated transistors with

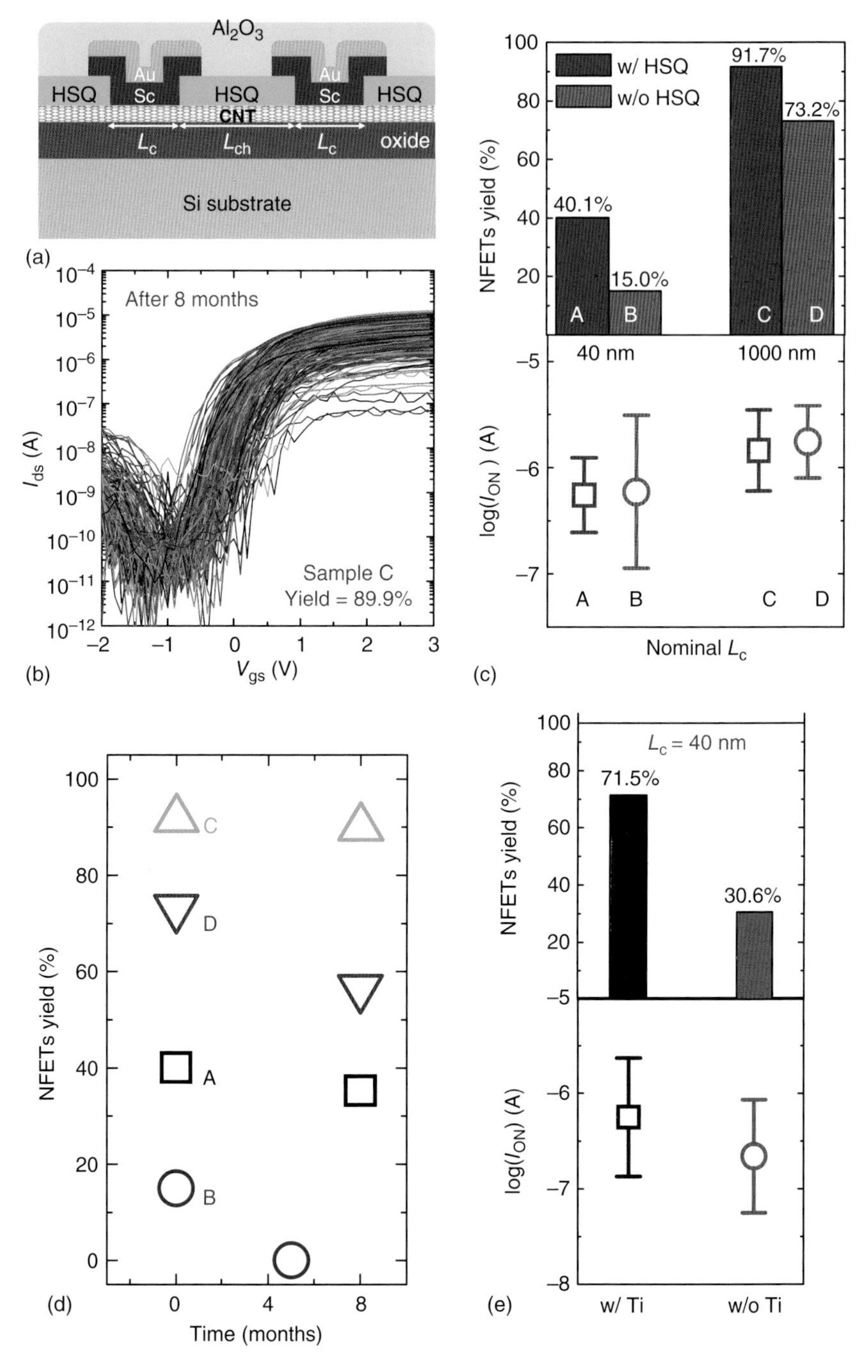

Figure 5.7 (a) All-around protection of *n*-type metal contact with Au in situ capping, HSQ sidewall protection, and ALD Al_2O_3 full passivation. (b) I_{ds}–V_{gs} curves at $V_{ds} = 0.5$ V measured from 207 NFETs with HSQ sidewall protection (nominal $L_c = 1000$ nm) after eight months, showing little degradation and excellent stability. (c) Statistical analysis of NFETs yield (top) and I_{on} (bottom, defined at gate overdrive of $|V_g - V_{th}| = 1$ V) from four sets of samples A to D. (d) Device yield of samples A–D after several months. (e) Statistical analysis of NFET yield (top) and I_{on} (bottom) from samples with and without 0.2 nm Ti wetting layer. Source: Tang et al. 2017 [63]. Reproduced with permission of IEEE.

top contacts, it should be noted that the back gate not only modulates the CNT channel but also the segments underneath source/drain contacts. This could complicate the estimation of effective channel length, especially when L_c is comparable to L_{ch} or even larger. Therefore, CNTFETs with end-bonded contacts (to be discussed) or top/wrap-around gate are favored for systematic studies on the channel length scaling, which, however, is still lacking. A relevant study on the length scaling of back-gated CNTFETs has been carried out using a long CVD tube to minimize variations [15], and some key results are summarized in Figure 5.8. Inspiringly, when scaling L_{ch} over two orders of magnitude (from 3 μm down to 15 nm), CNTFETs showed similar SS and V_{th}, and no SCE was observed even with the smallest $L_{ch} \sim 15$ nm. These ultrascaled CNTFETs exhibited a room-temperature, low-field resistance of 9–11 kΩ, which is very close to the quantum limit ($R_Q \approx 6.5$ kΩ) in the ballistic transport regime. The saturated on-current was up to 25 μA with a transconductance $g_m \approx 40$ μS for a single CNT at $V_{ds} = -0.4$ V, equivalent to an extremely high current density of over 3 mA/μm if assuming a CNT–CNT pitch of 125 tubes μm^{-1}.

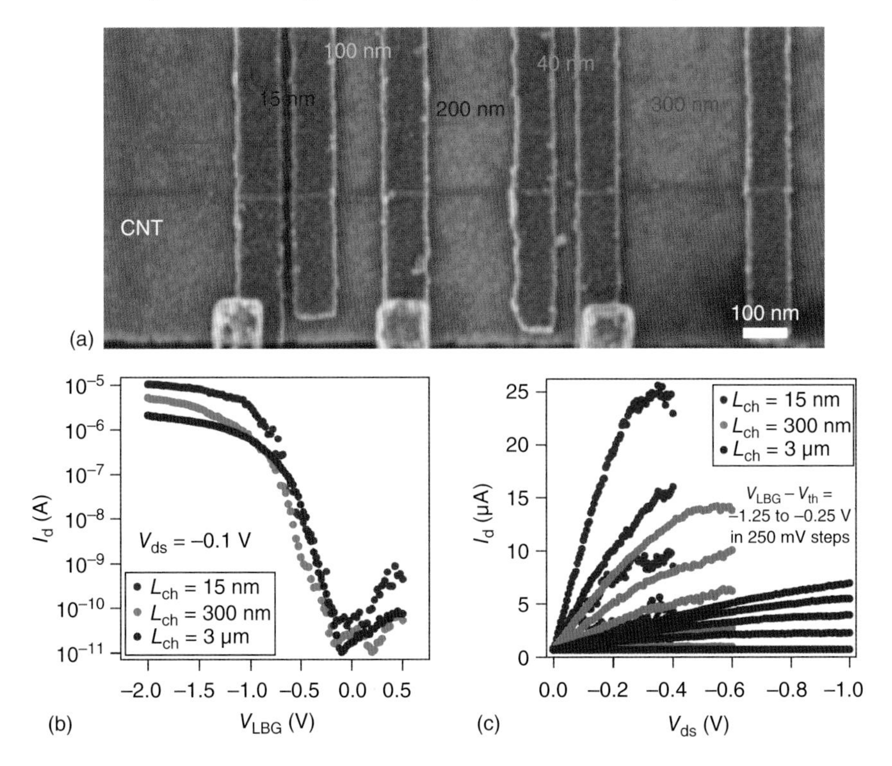

Figure 5.8 Channel length scaling of CNTFET. (a) SEM image of a set of CNTFETs with local bottom gate fabricated on the same CNT with L_{ch} ranging from 15 to 300 nm. The contact size is $L_c = 100$ nm, and the gate dielectric is 10 nm HfO_2. (b) I_{ds}–V_{gs} transfer curves from a set of devices with $L_{ch} \approx 15$ nm, 300 nm, and 3 μm, showing similar SS and V_{th}. (c) I_{ds}–V_{ds} output characteristics from the same set of devices. (d) Transfer length model plot of transistor resistance versus L_{ch} for both semiconducting and metallic CNTs. The linear fitting curves intersect at the quantum limit of $R_Q = 6.5$ kΩ, indicating ballistic transport in ultrasmall L_{ch}. Source: Franklin and Chen 2010 [15]. Reproduced with permission of Springer Nature.

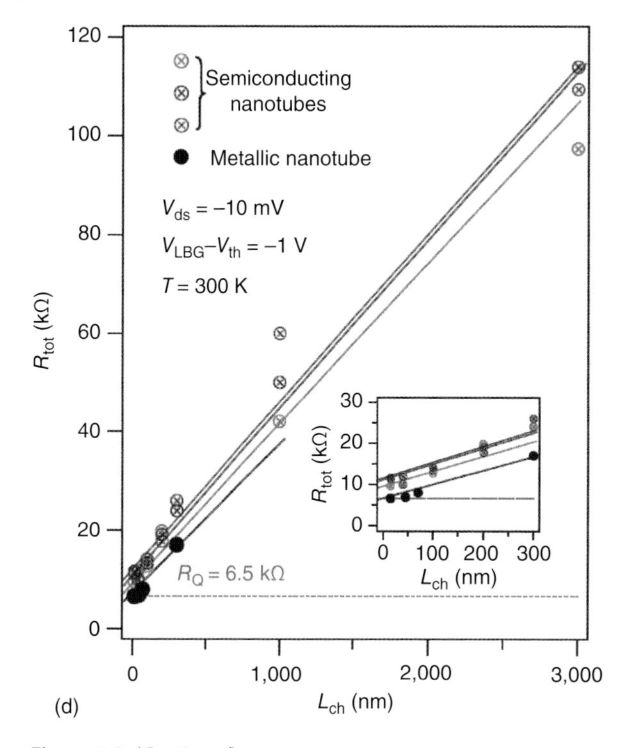

Figure 5.8 *(Continued)*

For further L_{ch} scaling, Figure 5.9 highlights a collection of CNTFETs fabricated with ultrasmall channel lengths (L_{ch} < 10 nm). Figure 5.9a,b shows a back-gated CNTFET with $L_{ch} \sim 9$ nm, and I_{ds}–V_{gs} curve at $V_{ds} = -0.4$ V exhibits a large on-state current of ~10 μA and relatively small SS of ~94 mV/decade [69]. Furthermore, Figure 5.9c,d shows a graphene-contacted top-gated CNTFET with even smaller L_{ch} of ~5 nm, and I_{ds}–V_{gs} curve at $V_{ds} = -0.1$ V exhibits very small SS of ~73 mV/decade in the best device measured, despite clear device-to-device variation observed [3]. In the benchmarking against silicon technology, ultrascaled CNTFETs consistently show superior SS with no sign of SCE down to a 5-nm length scale. As for the drive current I_{on}, it is usually normalized by either diameter or tight CNT–CNT pitch (~125 tubes μm^{-1}, which remains to be demonstrated experimentally (to be discussed later)). Also, the supply voltage V_{dd} can be scaled much more aggressively than Si technology, as simple logic circuits operating at V_{dd} as low as 0.2–0.4 V (compared to 0.6–0.8 V for Si) has been demonstrated (Figure 5.9e,f) [3]. These would offer great benefits in terms of power and performance for CNT logic technology. Besides experimental studies, extensive numerical simulations have also been carried out to investigate the scaling behavior of CNTFETs [70–75], and they project that CNTFETs perform well at the sub-10-nm dimension.

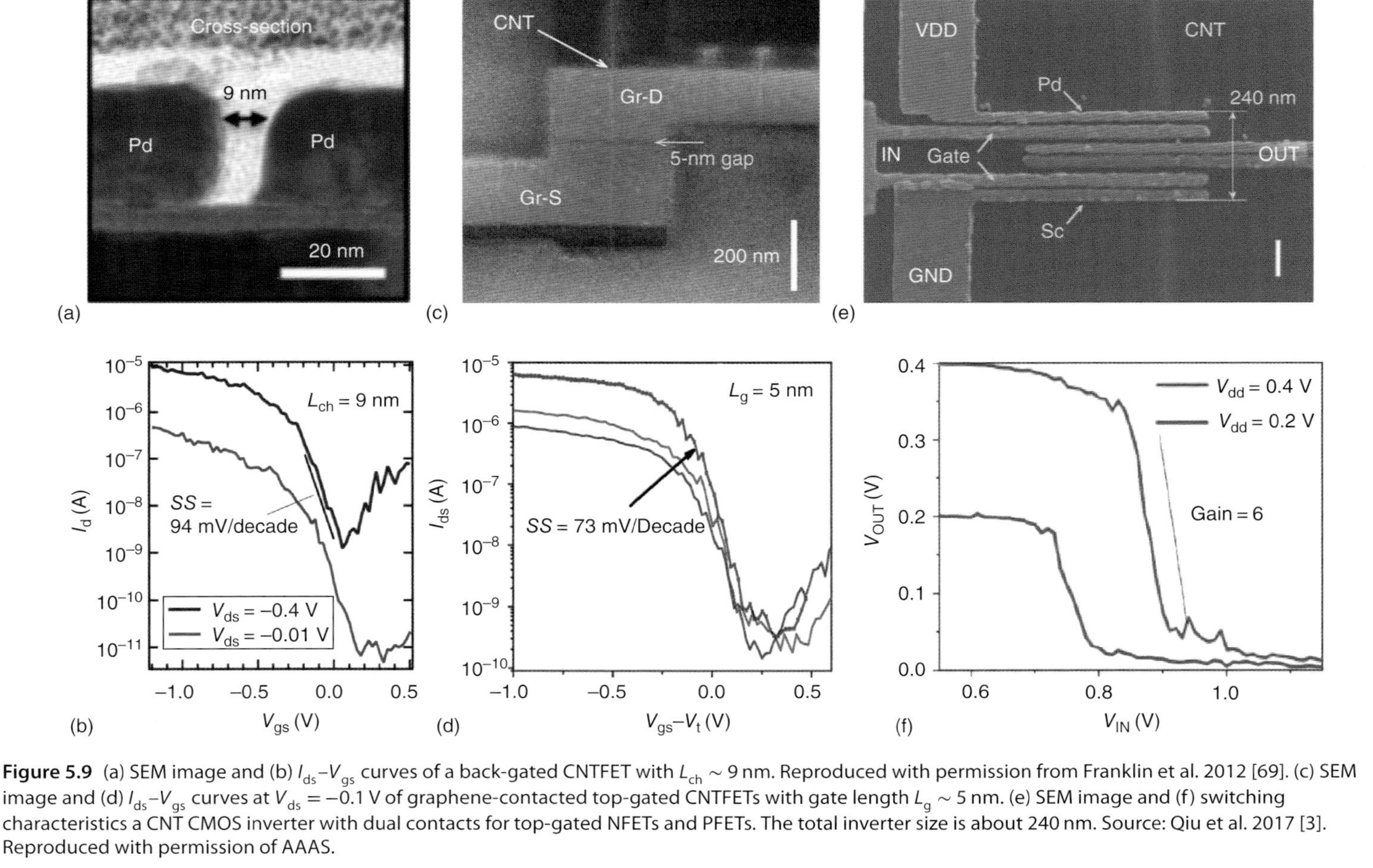

Figure 5.9 (a) SEM image and (b) I_{ds}–V_{gs} curves of a back-gated CNTFET with $L_{ch} \sim 9$ nm. Reproduced with permission from Franklin et al. 2012 [69]. (c) SEM image and (d) I_{ds}–V_{gs} curves at $V_{ds} = -0.1$ V of graphene-contacted top-gated CNTFETs with gate length $L_g \sim 5$ nm. (e) SEM image and (f) switching characteristics a CNT CMOS inverter with dual contacts for top-gated NFETs and PFETs. The total inverter size is about 240 nm. Source: Qiu et al. 2017 [3]. Reproduced with permission of AAAS.

5.3.3 Contact Length Scaling in CNTFET

The miniaturization of MOSFET requires the reduction of both channel length L_{ch} and contact length L_c simultaneously, which reduces channel resistance R_{ch} but increases contact resistance R_c. Such opposite scaling trends make R_c comparable to R_{ch} because of the 32-nm technology node, and R_c becomes a performance limiting factor nowadays [76]. In Si technology, source/drain regions are usually heavily doped to reduce R_c of silicides [77]. For nanomaterials like CNTs, however, because of their ultrathin body nature, it is difficult to adopt the same strategy of ion implantation, which leads to a severe contact resistance issue [78, 79].

Considering CNT geometry, there are three possible contact schemes, as illustrated in Figure 5.10, which have different scaling characteristics. Conventional electrical contacts to CNTs are usually made by directly depositing metals on top of CNTs (i.e. side-bonded contacts), as illustrated in Figure 5.10a. The metal work function, together with the metal–CNT interaction strength (e.g. binding energy and wetting property) [81] largely determine R_c. Perceptibly, the contact area for side-bonded contacts is proportional to L_c, and hence R_c is inversely proportional to L_c:

$$R_c = \frac{2\rho_c}{d_{CNT}L_c} \tag{5.3}$$

where ρ_c is the specific contact resistivity and d_{CNT} is the CNT diameter. It should be noted that Eq. (5.3) is valid only if L_c is less than the charge-transfer length L_T, which is contact dependent and could be less than 200 nm for CNT Pd contact [15]. Here, L_T is defined as the length at which the applied potential drops to $1/e$ of its value, and it describes the length over which the majority of the carriers are injected from the metal contact into CNT [82].

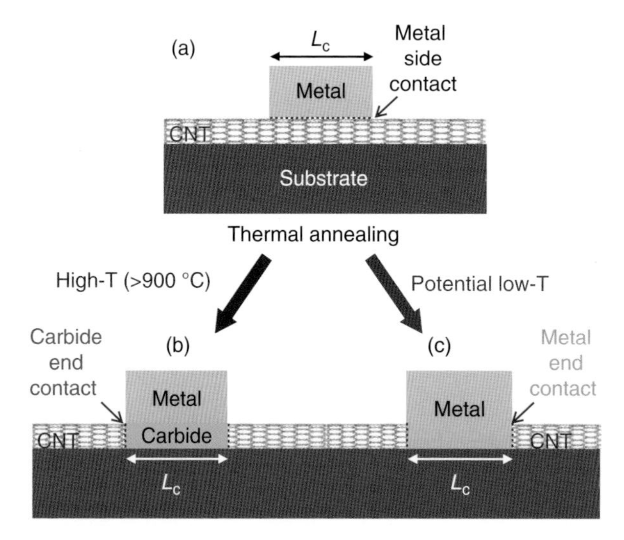

Figure 5.10 Schematic illustrations of three possible contact schemes for CNT: (a) conventional metal side-bonded contact; (b) metal carbide end-bonded contact; (c) direct metal end-bonded contact. Source: Tang et al. 2017 [80]. Reproduced with permission of AAAS.

Experimentally, R_c can be extracted via multiple-point measurements or the transfer length method. The resistance of a CNT transistor can be written as

$$R_{total} = R_{ch} + R_Q + R_c = \frac{h}{4e^2} \cdot \frac{L_{ch}}{L_{mfp}} + \frac{h}{4e^2} + \frac{2\rho_c}{d_{CNT}L_c} \tag{5.4}$$

where h is the Planck's constant, e is the electron charge, $R_Q = 1/G_0 = h/4e^2 \approx 6.5\ k\Omega$ is the quantum resistance of the CNT channel (R_Q is sometimes considered as a part of R_c in literature [3, 15, 78]), L_{mfp} is the mean free path (typically ~200 nm for a CVD tube [15]). For CNT transistors with quasi-ballistic channels ($L_{ch} \ll L_{mfp}$), R_{ch} is negligible, so R_c can be extracted from the low-bias resistance $R_{total} = V_{ds}/I_{ds}$ (after subtracting the metal lead resistance). Using this method, the contact length scaling characteristics of several p-type metal contacts, including Pd [15], Rh, Pt, and Au [78], are shown in Figure 5.11a–c. In general, Pd shows the lowest R_c down to 20 nm size (attributed to its good wetting on the CNT surface), while Rh shows more favorable scaling behavior (inferred as Rh has a shorter L_T). However, even in the Pd case, R_c still increases dramatically from 5 kΩ for large contacts ($L_c \sim 200$ nm) to extrapolated ~65 kΩ for $L_c \sim 9$ nm, and continuously increases for even smaller contacts [15]. For n-type side-bonded contacts, the length scaling study of Sc contacts has shown an even larger R_c (nearly eight times larger than Pd) [63], possibly due to oxidation and poor surface wetting issues of these low-work-function metals (Figure 5.11d). Another recent study showed smaller R_c for Sc via measurements in vacuum to prevent oxidation [3], and R_c was further reduced by using CNT with a larger diameter as suggested by Eq. (5.3). In addition, ultrasonic nanowielding technique can be used to embed CNT into metal electrodes in order to increase the contact area and further lower R_c [83, 84]. Nevertheless, the undesired scaling characteristic of side-bonded contacts still imposes a big challenge for CNTFET scaling to meet the R_c requirement for advanced logic technology nodes [85].

To address the scaling issue of side-bonded metal contacts, various end-bonded contact schemes have been demonstrated. As illustrated in Figure 5.10b, one useful approach is to form metal carbides through thermal annealing, similar to the silicidation process used in Si technology and also Si/Ge nanowire devices [86, 87]. In this end-bonded contact scheme, the contact area, nominally the cross-section of CNT, is independent of L_c, leading to contact-size-independent R_c. Early *in situ* transmission electron microscopy (TEM) studies have observed the formation of various metal carbides via thermal annealing, such as TiC [88], NbC [88], WC [89], etc. For CNTFET, Mo/Mo_2C end-bonded contacts, formed by the solid-state reaction between Mo contacts and carbon atoms through annealing at 850 °C or higher (Figure 5.12), exhibit a size-independent R_c of ~30 kΩ with L_c down to 9 nm [90]. This is so far the smallest contact size demonstrated in CNTFET (Figure 5.12d). Assuming a CNT–CNT pitch of 125 tubes μm^{-1}, such Mo/Mo_2C end-bonded contacts would provide an effective R_c of ~240 Ω-μm, which meets even the most stringent contact requirement ($R_c \sim 256$ Ω μm) for technology nodes down to 1.3 nm by the International Technology Roadmap for Semiconductors (ITRS) [85]. Therefore, such an end-bonded contact scheme is highly favorable for future scaled CNT logic technology.

Notably, the formation of metal carbides in CNT usually requires very high temperature annealing (typically above 850 °C) due to the strong C—C bond.

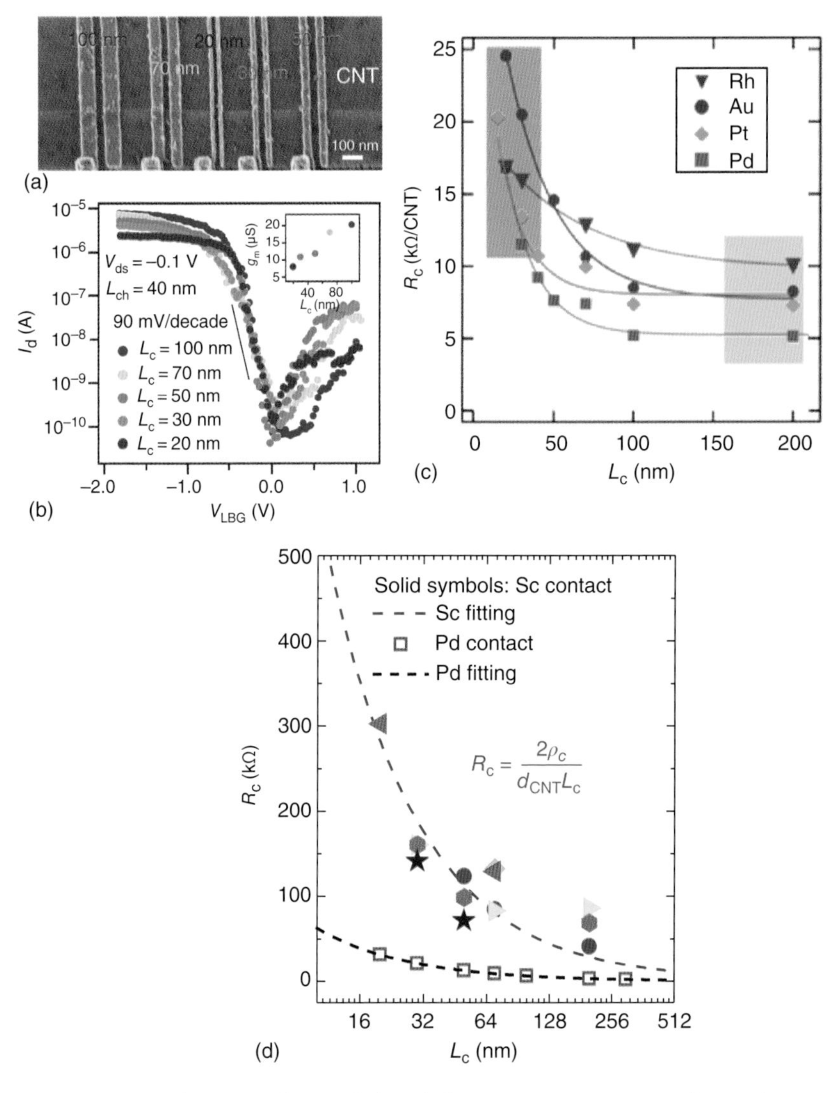

Figure 5.11 Length scaling of CNT side-bonded contacts. (a) SEM image of a set of transistors fabricated on the same CNT with $L_c \approx 20$ nm, 30 nm, 50 nm, 70 nm, and 100 nm. $L_{ch} \approx 40$ nm for all devices. (b) I_{ds}–V_{gs} curves from the same set of CNTFETs, showing a decrease in on-state current with reducing L_c. The inset shows the dependence of transconductance g_m on L_c. Source: Franklin and Chen 2010 [15]. Reproduced with permission of Springer Nature. (c) Estimated length-dependent R_c (including R_Q) for four best *p*-type metal contacts including Pd, Rh, Pt, and Au. Source: Franklin et al. 2014 [78]. Reproduced with permission of ACS. (d) Estimated length-dependent R_c for *n*-type contact Sc, showing much larger values in comparison with Pd. The dotted lines are fitting curves using Eq. (5.3). Source: Tang et al. 2017 [63]. Reproduced with permission of IEEE.

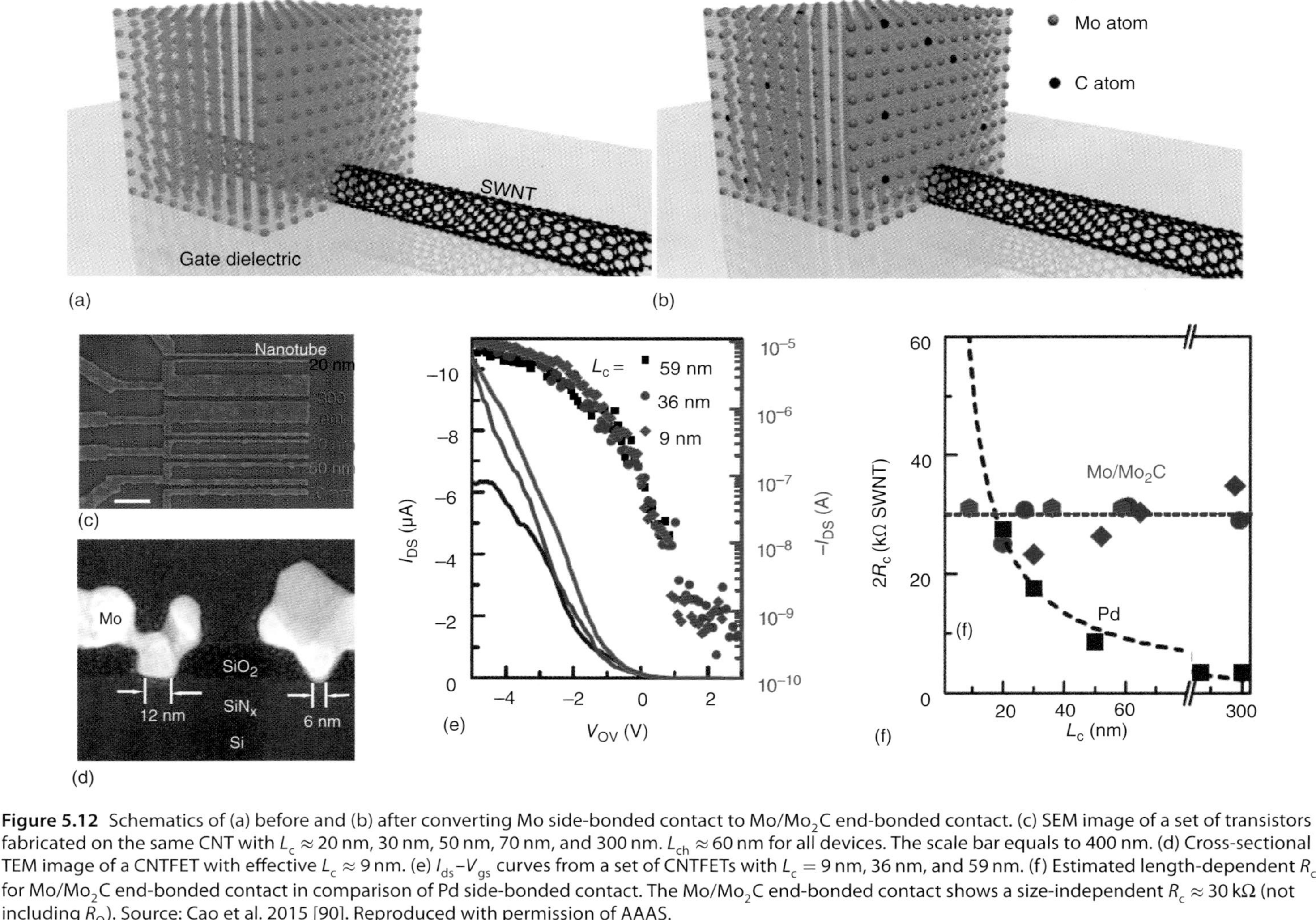

Figure 5.12 Schematics of (a) before and (b) after converting Mo side-bonded contact to Mo/Mo_2C end-bonded contact. (c) SEM image of a set of transistors fabricated on the same CNT with $L_c \approx 20$ nm, 30 nm, 50 nm, 70 nm, and 300 nm. $L_{ch} \approx 60$ nm for all devices. The scale bar equals to 400 nm. (d) Cross-sectional TEM image of a CNTFET with effective $L_c \approx 9$ nm. (e) I_{ds}–V_{gs} curves from a set of CNTFETs with $L_c = 9$ nm, 36 nm, and 59 nm. (f) Estimated length-dependent R_c for Mo/Mo_2C end-bonded contact in comparison of Pd side-bonded contact. The Mo/Mo_2C end-bonded contact shows a size-independent $R_c \approx 30$ kΩ (not including R_Q). Source: Cao et al. 2015 [90]. Reproduced with permission of AAAS.

Such a high temperature process could induce metal crystallization and large roughness, causing reliability as well as manufacturability issues. Also, transistors with metal carbides so far are all p-type, and no existing solution is available for n-type carbide contacts. An alternative form of end-bonded contacts to CNTs has been demonstrated at significantly lower annealing temperatures (Figure 5.10c). This is achieved using metals with high carbon solubility (e.g. Ni, Co) [91]; therefore, carbon atoms dissolve into metal contacts upon annealing, naturally forming end-bonded metal contacts [92]. The dissolution of carbon atoms into Ni contacts upon annealing was verified by scanning electron microscopy (SEM) and Raman spectrum [80]. Using this approach, high-performance CNTFETs with Ni end-bonded contacts have been demonstrated via annealing at moderate temperatures of 400–600 °C. The conversion from weak side-bonded to strong end-bonded contacts led to a device transition from ambipolar FETs to unipolar PFETs (Figure 5.13a,b). The formed Ni end-bonded contacts were proved to be highly reliable, despite all carrier injection occurring over a tiny contact area of 2 nm^2 [90]. Furthermore, a contact length scaling study on Ni end-bonded contacts revealed a size-independent R_c of ~30 kΩ, comparable to R_c of Mo/Mo_2C end-bonded contacts, affirming the signature characteristic of end-bonded contacts (Figure 5.13c).

To further realize CNT-based scalable complementary logic, both PFETs and NFETs constructed with end-bonded contacts are highly desired. Typical n-type metal contacts (e.g. Sc, Er, Y, Ca, and La) do not have high enough carbon solubility to form end-bonded contacts through annealing as Ni does. Alternatively, physicochemical doping through a dielectric layer (i.e. "field-effect" doping by fixed charges or interface dipole) is possible to tune the carrier type in CNTs, so that CNT PFETs can be converted to NFETs with the same end-bonded contacts. Such solid-state doping strategy provides better stability than chemical charge-transfer doping [46–49]. In literature, there have been several reports of air-stable CNT NFETs using SiN_x, HfO_2, or MgO as an n-type doping dielectric [37, 93–96], where robust high-work-function metals (e.g. Pd, Au) can be used as source/drain contacts. Figure 5.14a,b shows a CNT

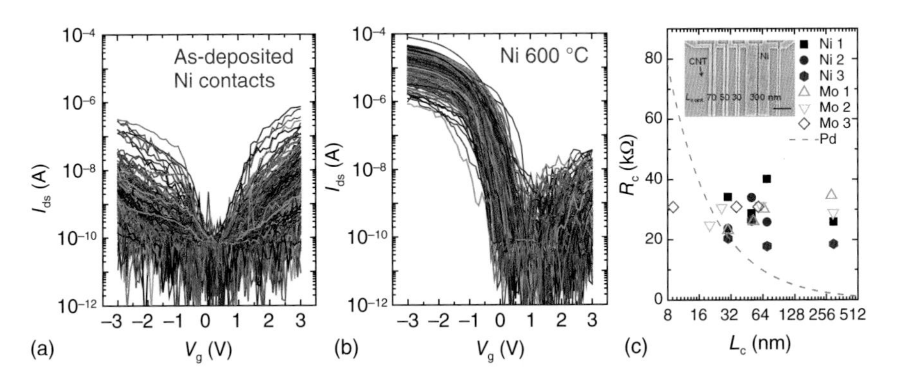

Figure 5.13 $I_{ds}–V_{gs}$ curves at V_{ds} = −0.5 V of 207 CNTFETs with (a) before and (b) after annealing at 600 °C for five minutes to form Ni end-bonded contacts. The nominal L_{ch} = 150 nm. (c) The length scaling study on Ni end-bonded contacts shows a roughly size-independent $R_c \approx 30$ kΩ (not including R_Q). The data are plotted with solid symbols, in comparison with Mo/Mo_2C end-bonded contacts (open symbols) and Pd side-bonded contacts (dashed line). Source: Tang et al. 2017 [80]. Reproduced with permission of IEEE.

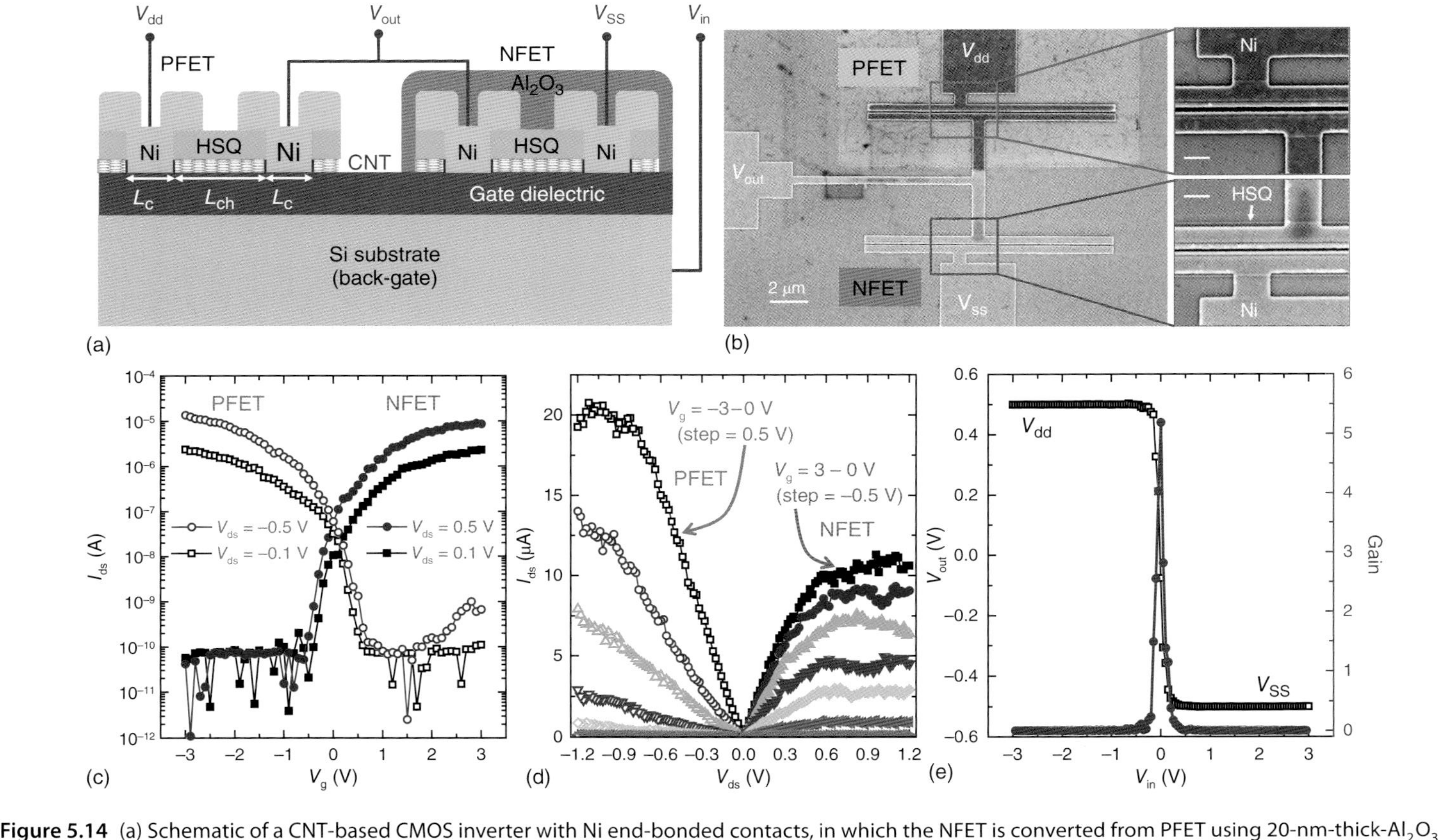

Figure 5.14 (a) Schematic of a CNT-based CMOS inverter with Ni end-bonded contacts, in which the NFET is converted from PFET using 20-nm-thick-Al_2O_3 passivation as *n*-type doping. (b) SEM image of a CMOS inverter with nominal $L_{ch} = 150$ nm and $L_c = 40$ nm. The right panels show the enlarged SEM images of the PFET and NFET. The scale bar equals to 400 nm. (c) I_{ds}–V_{gs} curves of the PFET and NFET in one CNT inverter. (d) Corresponding I_{ds}–V_{ds} curves for the PFET and NFET. (e) The output voltage (left axis) and inverter gain (right axis) as a function of the input voltage. The inverter gain is above 5. Source: Tang et al. 2017 [80]. Reproduced with permission of IEEE.

CMOS inverter with Ni end-bonded contacts, in which the NFET is converted from PFET using HSQ/Al_2O_3 bilayer passivation as *n*-type doping, which was attributed to fixed charges in the ALD-Al_2O_3 layer [97, 98] and also possible electric dipole formation at the Al_2O_3/SiO_x interface [92]. The resulting NFET showed symmetric performance to the original PFET and excellent stability (Figure 5.14c,d) [92]. Using this method, CNT-based CMOS inverters with high yield have been demonstrated (Figure 5.14e), and more sophisticated logic circuits can be readily built with high-quality scalable end-bonded contacts.

5.4 Perspective of CNT-Based Logic Technology

This review of recent advances on CNTFETs has shown that both L_{ch} and L_c can be scaled down to 10 nm without suffering the detrimental SCE and a serious contact resistance issue. To achieve a viable CNT logic technology, Figure 5.15a illustrates a model CNTFET for sub-5-nm technology node, where the key dimensions of L_{ch}, L_c, gate pitch, and also CNT pitch are highlighted. Different from devices built on a single nanotube, high density (equivalently small pitch) of parallel CNTs spanning the source/drain contacts is needed in order to deliver enough current density (on-state current per device width, $\mu A\,\mu m^{-1}$). On the basis of this scaled device structure, simulation using system-level optimizer projects a more than twofold improvement in both performance and power for CNTFET compared to standard Si Fin field-effect transistor (FinFET) in the 5-nm and 10-nm nodes (Figure 5.15b,c). As discussed earlier, the superior electronic property of CNT allows for the reduction of V_{dd} by nearly 50% compared to Si while maintaining the same performance. Most values in Figure 5.15c given by the optimizer can be routinely demonstrated in scaled CNTFETs. In the following, we discuss important challenges regarding the manufacturability of CNT-based logic technology.

5.4.1 CVD-Grown CNT versus Solution-Processed CNT

There are two main approaches to prepare large-scale aligned CNTs: grown-and-transferred method or purified-and-placed method. In the first method, arrays of nanotubes are firstly grown on an ST-cut quartz by the CVD method, followed by a process to transfer them to the final silicon wafer. In literature, CVD-grown CNTs are usually favored for electrical transport studies, mainly because their advantages in length (could span over 100 μm) and structural integrity (mobility in CVD tubes could be one order higher than solution tubes) [13]. However, there are a number of challenges to implement CVD tubes in practical logic technologies. First of all, the CNT density cannot be precisely controlled over a large area, and the reported density by CVD growth so far (typically 1–10 tubes μm^{-1}) is still too low [100]. Although multiple growth or transfer techniques can improve the density up to about 50 tubes μm^{-1} [101], it is still below the requirement (about 125 tubes μm^{-1}) in order to deliver high-enough current. Secondly, the uncontrollable variability in the CNT pitch separation during CVD growth would

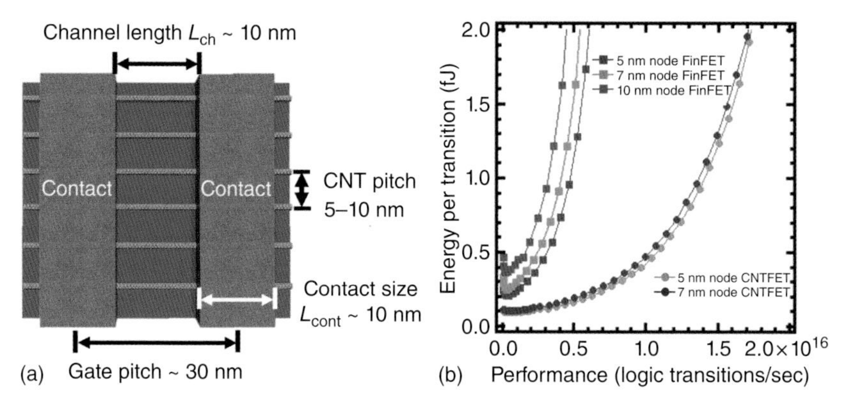

(c)

Parameters	CNT processor
Gate pitch	30 nm
Inter-CNT pitch	8 nm
CNT diameter	1.7 nm
# CNT/transistor	6
Gate length	11.3 nm
On-current	5.93 uA/CNT
Off-current	16.9 nA/CNT
I_{ON}/I_{OFF}	349
Sub-threshold swing	83.2 mV/decade
On-state resistance	20.5 kOhm/CNT
High-K thickness	2 nm

Figure 5.15 (a) Schematic of a model CNTFET for sub-10-nm technology nodes. The gate (not shown for clarity) would be built either on top or wrapping around each CNT. (b) Switching energy versus speed performance for both FinFET and CNTFET technologies. (c) Table of CNTFET parameters determined by the optimizer program (at $V_{dd} \sim 0.3$ V). Source: Tulevski et al. 2014 [99]. Reproduced with permission of ACS.

lead to unacceptable device variations for practical circuit design [102]. Finally, as-synthesized CNTs are usually a mixture of semiconducting (~66%) and metallic (~33%) nanotubes, and it is still very difficult to selectively remove metallic ones without affecting the neighboring semiconducting ones, especially at a tight CNT pitch [103]. Even if such post-growth separation can be achieved, it would induce an even larger variability in CNT pitch and uniformity. On the other hand, in the purified-and-placed method, solution-processed CNTs could avoid the abovementioned intrinsic problems and offer many advantages in terms of manufacturability, such as purity and pitch control. Due to its long mean free path, the lower mobility compared to CVD tubes would not be a limiting factor for device performance in scaled FETs with quasi-ballistic channel lengths [13]. Nevertheless, there are other challenges in order to implement this approach in practical logic technologies, as is discussed later.

5.4.2 Purity and Placement of Solution-Processed CNTs

It has already been demonstrated that high-purity semiconducting CNTs can be readily obtained via several separation schemes in solution [104]. For example, as-synthesized CNTs can be suspended in solution by wrapping them non-covalently with specific surfactants and then they can be separated according to their diameters and electronic types through density-gradient ultracentrifugation (Figure 5.16a), utilizing their contrast in packing density for the complexes formed between different CNTs and surfactants [105]. Another approach is to use column chromatography, and it has been shown that through just three iterations, the purity of semiconducting tubes can be improved to nearly 99.9% (Figure 5.16b,c) [106]. The standard absorption spectroscopy becomes incapable of quantifying the purity of CNT solution beyond 98–99% when the metallic region in the absorption spectra is completely attenuated. Therefore, high-throughput electrical testing of thousands of CNT FETs has been performed to quantify CNT solution purity instead [106]; however, it is impractical to extend this method to millions or even billions of devices. There is no existing technology to precisely quantity the solution purity up to the required level (below 1 ppb) for wafer-scale manufacturing.

Another issue is related to how to assemble solution-processed CNTs into ordered arrays. Several methods have been developed to assemble them into high-density arrays with predefined pitch [107–109]. For example, a widely used method is through chemical recognition between functionalized surface on pre-patterned substrates and CNT/surfactant complexes [107]. For example, the substrate can be patterned into 100-nm-wide HfO_2 trenches with SiO_2 barriers, and the HfO_2 surface is selectively functionalized with 4-(*N*-hydroxycarboxamido)-1-methylpyridinium iodide (NMPI) self-assembled monolayer, while the CNT in solution is wrapped with an anionic surfactant (sodium dodecyl sulphate, SDS). The coulombic attraction between the positively charged NMPI monolayer and the negatively charged SDS surfactant specifically places the CNTs onto HfO_2 trenches with a high degree of alignment (Figure 5.16d,e) [107]. Relatively high device yield (above 90%) has been demonstrated using a trench pitch of 200 nm, equivalent to a CNT density of about 5 tubes μm^{-1} if assuming one tube per trench. However, further scaling down the trench size below 100 nm in order to improve the CNT placement pitch appears to significantly lower the device yield and increase the placement randomness [110]. Further optimization (such as trench dimension, tube length sorting, etc.) and also new patterning techniques (such as directed self-assembly of diblock copolymer, DNA origami, etc.) are needed to meet the required 5- to 10-nm placement pitch. Other challenges include how to minimize the defect formation during placement, such as tube bundling, tube crossing multiple trenches, and so on. Therefore, the development of a robust and scalable method to precisely place CNTs on substrates with stringent pitch remains one of the most critical challenges in the implementation of CNT logic technology.

5.4.3 Variability in CNTFETs

From the chip-level perspective, reducing device variability is as crucial as improving device performance, and it is another key challenge for realizing

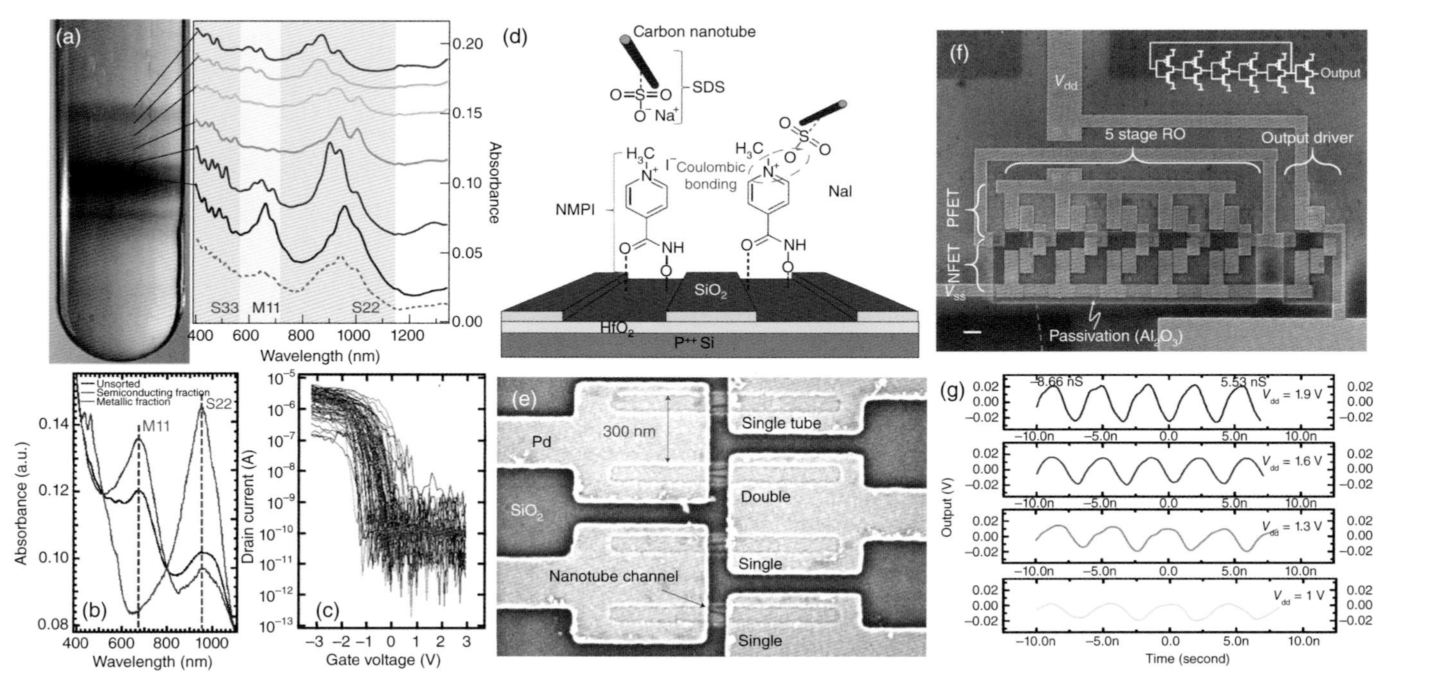

Figure 5.16 (a) Photograph and associated absorbance spectra after CNT separation by its diameter and electronic types using density gradient ultracentrifugation. Source: Arnold et al. 2006 [105]. Reproduced with permission of Springer Nature. (b) Absorption spectra of unsorted (black), metallic enriched (blue), and semiconducting enriched (red) CNTs via column chromatography. (c) I_{ds}–V_{gs} curves of 150 CNTFETs fabricated with semiconducting enriched CNTs. Source: Tulevski et al. 2013Reproduced [106]. Reproduced with permission of ACS. (d) Schematic showing the selective placement of SDS-wrapped CNTs onto NMPI-functionalized HfO_2 trenches by an ion-exchange process. (e) SEM image of CNTFETs fabricated with assembled CNTs on 70 nm-wide, 300 nm-pitch HfO_2 trenches. Source: Park et al. 2012 [107]. Reproduced with permission of Springer Nature. (f) SEM image and (g) output characteristics at different supply voltages of fully integrated five-stage CNT CMOS RO. Source: Han et al. 2017 [64]. Reproduced with permission of Springer Nature.

high-performance CNT logic technology. As discussed in previous sections, there are many factors that could contribute to the variability of CNTFETs, such as CNT diameter, contact resistance (especially for NFETs), CNT placement, purity and doping, and also typical process variations from lithography and etch. A more fundamental factor is the threshold voltage V_{th} variation. In back-gated CNTFETs, it is suggested that the random fluctuation from fixed charges presented on the gate oxide surface near CNT dominates the V_{th} variability [65], and incorporating multiple CNTs per device may help reduce the variability. It is also shown that the variability can be reduced by the scaling of gate dielectric thickness, and it could be further suppressed by employing effective device passivation [111]. Therefore, CNTFETs with a wrap-around gate are expected to show much less variability by fully encapsulating CNT with the gate oxide; however, more systematic studies are still needed to verify this. Again, the growth of high-quality ultrathin ALD oxides on chemically inert tube surface would be the key to reduce the overall variability.

5.4.4 Circuit-Level Integration

Beyond the device-level demonstrations, CNT-based integrated logic circuits have also been exploited, including logic gates [3, 33], ROs [112], and a single-bit processor [113]. However, most circuits built with CNTs so far have been stalled at simple functionality demonstrations, and their operation frequencies are typically in the range of kilohertz to megahertz, which is far behind their Si counterparts with similar device dimensions. Also, some demonstrations rely on unipolar CNT PFETs, while others employ fabrication processes that are not suitable for wafer-scale manufacturing. In order to rationally evaluate the maturity of CNT-based logic technology for high-speed electronic applications, it is therefore important to demonstrate more complexed CMOS circuits through manufacturable process integration. For this purpose, we have demonstrated the first high-performance CNT CMOS ROs using high-purity, solution-processed self-assembled CNT arrays with dual work-function-metal contacts (Figure 5.16f) [64]. Remarkably, this CNT-based true CMOS RO recorded a high-stage switching frequency of 2.82 GHz, the highest value reported from any nanomaterials (Figure 5.16g). To outperform the Si-based counterparts, sub-picosecond stage delay can be potentially achieved for ultrascaled CNT ROs by adopting a self-aligned structure (reducing parasitic capacitance), scaling channel length down to 10 nm, and increasing CNT density to above 100 tubes μm^{-1}. This work represents a significant step forward in realizing CNT logic technology in the not-too-distant future.

5.5 Summary and Outlook

With superior electronic properties, CNT has been identified as one of the top candidates for Si replacement in ultrascaled logic technology. The unique structure with ultrathin body makes CNTFET (with Schottky source/drain) construct and operate differently from conventional MOSFET. This chapter highlights tremendous advances that have been made on the scaling of CNTFETs to provide both performance and power benefits compared to Si counterparts. Recent progresses on the contact engineering, channel length scaling, and also contact length scaling in CNTFET are reviewed. As the Si industry is struggling to extend the scaling roadmap beyond 5 nm, it is imperative to investigate the manufacturing maturity of CNT-based logic technology for high-speed electronic applications. Several key obstacles have been identified, such as further improving and quantifying semiconducting CNT purity, developing robust and scalable placement technique to assemble ordered CNT arrays with very tight pitch, reducing device variability, and integrating circuit- and chip-level demonstrations with manufacturable processes. With extensive work being done on the scaling of the dimension of CNT transistors, more efforts need to be devoted in the developing large-scale integration processes in order to make the technology a viable choice before the end of the Si roadmap.

References

1 Ferain, I., Colinge, C.A., and Colinge, J.-P. (2011). Multigate transistors as the future of classical metal–oxide–semiconductor field-effect transistors. *Nature* 479 (7373): 310–316.
2 Li, Y.-H. and Lue, J.-T. (2007). Dielectric constants of single-wall carbon nanotubes at various frequencies. *J. Nanosci. Nanotechnol.* 7 (9): 3185–3188.
3 Qiu, C., Zhang, Z., Xiao, M. et al. (2017). Scaling carbon nanotube complementary transistors to 5-nm gate lengths. *Science* 355 (6322): 271–276.
4 Geim, A.K. and Novoselov, K.S. (2007). The rise of graphene. *Nat. Mater.* 6 (3): 183–191.
5 Baughman, R.H., Zakhidov, A.A., and de Heer, W.A. (2002). Carbon nanotubes–the route toward applications. *Science* 297 (5582): 787–792.
6 Kroto, H.W., Heath, J.R., O'Brien, S.C. et al. (1985). C60: buckminsterfullerene. *Nature* 318 (6042): 162–163.
7 Iijima, S. (1991). Helical microtubules of graphitic carbon. *Nature* 354: 56–58.
8 Novoselov, K.S., Geim, A.K., Morozov, S.V. et al. (2004). Electric field effect in atomically thin carbon films. *Science* 306 (2004): 666–669.
9 Avouris, P., Chen, Z., and Perebeinos, V. (2007). Carbon-based electronics. *Nat. Nanotechnol.* 2: 605–615.
10 Dresselhaus, M.S., Dresselhaus, G., and Avouris, P. (eds.) (2001). *Carbon Nanotubes*. Berlin, Heidelberg: Springer Berlin Heidelberg.
11 Hong, S. and Myung, S. (2007). Nanotube electronics: a flexible approach to mobility. *Nat. Nanotechnol.* 2 (4): 207–208.
12 Zhou, X., Park, J.-Y., Huang, S. et al. (2005). Band structure, phonon scattering, and the performance limit of single-walled carbon nanotube transistors. *Phys. Rev. Lett.* 95 (14): 146805.
13 Cao, Q. and Han, S. (2013). Single-walled carbon nanotubes for high-performance electronics. *Nanoscale* 5: 8852–8863.

14 Javey, A., Guo, J., Wang, Q. et al. (2003). Ballistic carbon nanotube field-effect transistors. *Nature* 424 (6949): 654–657.
15 Franklin, A.D. and Chen, Z. (2010). Length scaling of carbon nanotube transistors. *Nat. Nanotechnol.* 5 (12): 858–862.
16 Coleman, J.N., Khan, U., Blau, W.J., and Gun'ko, Y.K. (2006). Small but strong: a review of the mechanical properties of carbon nanotube–polymer composites. *Carbon N. Y.* 44 (9): 1624–1652.
17 Yu, M.-F., Lourie, O., Dyer, M.J. et al. (2000). Strength and breaking mechanism of multiwalled carbon nanotubes under tensile load. *Science* 287 (5453): 637–640.
18 Dekker, C., Tans, S.J., and Verschueren, A.R.M. (1998). Room-temperature transistor based on a single carbon nanotube. *Nature* 393 (6680): 49–52.
19 Martel, R., Schmidt, T., Shea, H.R. et al. (1998). Single- and multi-wall carbon nanotube field-effect transistors. *Appl. Phys. Lett.* 73 (17): 2447–2449.
20 Wei, B.Q., Vajtai, R., and Ajayan, P.M. (2001). Reliability and current carrying capacity of carbon nanotubes. *Appl. Phys. Lett.* 79 (8): 1172–1174.
21 Kreupl, F., Graham, A.P., Duesberg, G.S. et al. (2002). Carbon nanotubes in interconnect applications. *Microelectron. Eng.* 64 (1–4): 399–408.
22 de Heer, W.A., Ch telain, A., and Ugarte, D. (1995). A carbon nanotube field-emission electron source. *Science* 270 (5239): 1179–1180.
23 Choi, W.B., Chung, D.S., Kang, J.H. et al. (1999). Fully sealed, high-brightness carbon-nanotube field-emission display. *Appl. Phys. Lett.* 75 (20): 3129–3131.
24 Dai, H., Franklin, N., and Han, J. (1998). Exploiting the properties of carbon nanotubes for lithography. *Appl. Phys. Lett.* 73 (11): 1508–1510.
25 Akita, S., Nishijima, H., Nakayama, Y. et al. (1999). Carbon nanotube tips for a scanning probe microscope: their fabrication and properties. *J. Phys. D. Appl. Phys.* 32 (9).
26 Dai, H., Hafner, J.H., Rinzler, A.G. et al. (1996). Nanotubes as nanoprobes in scanning probe microscopy. *Nature* 384 (6605): 147–150.
27 Che, G., Lakshmi, B.B., Fisher, E.R., and Martin, C.R. (1998). Carbon nanotubule membranes for electrochemical energy storage and production. *Nature* 393 (6683): 346–349.
28 Zanello, L.P., Zhao, B., Hu, H., and Haddon, R.C. (2006). Bone cell proliferation on carbon nanotubes. *Nano Lett.* 6 (3): 562–567.
29 Liu, Z., Winters, M., Holodniy, M., and Dai, H. (2007). siRNA delivery into human T cells and primary cells with carbon-nanotube transporters. *Angew. Chemie - Int. Ed.* 46 (12): 2023–2027.
30 Gong, K., Du, F., Xia, Z. et al. (2009). Nitrogen-doped carbon nanotube arrays with high electrocatalytic activity for oxygen reduction. *Science* 323 (5915): 760–764.
31 Behabtu, N., Young, C.C., Tsentalovich, D.E. et al. (2013). Strong, light, multifunctional fibers of carbon nanotubes with ultrahigh conductivity. *Science* 339 (6116): 182–186.
32 Wong, P. and H.-S. (2002). Beyond the conventional transistor. *IBM J. Res. Dev.* 46 (2/3): 133–168.

33 Bachtold, A., Hadley, P., Nakanishi, T., and Dekker, C. (2001). Logic circuits with carbon nanotube transistors. *Science* 294 (5545): 1317–1320.
34 Wind, S.J., Appenzeller, J., Martel, R. et al. (2002). Vertical scaling of carbon nanotube field-effect transistors using top gate electrodes. *Appl. Phys. Lett.* 80 (20): 3817–3819.
35 Javey, A., Guo, J., Farmer, D.B. et al. (2004). Self-aligned ballistic molecular transistors and electrically parallel nanotube arrays. *Nano Lett.* 4 (7): 1319–1322.
36 Chen, Z., Farmer, D., Xu, S. et al. (2008). Externally assembled gate-all-around carbon nanotube field-effect transistor. *IEEE Electron Device Lett.* 29 (2): 183–185.
37 Franklin, A.D., Koswatta, S.O., Farmer, D.B. et al. (2013). Carbon nanotube complementary wrap-gate transistors. *Nano Lett.* 13: 2490–2495.
38 Cao, J., Wang, Q., and Dai, H. (2005). Electron transport in very clean, as-grown suspended carbon nanotubes. *Nat. Mater.* 4 (10): 745–749.
39 Lin, Y.-M., Tsang, J.C., Freitag, M., and Avouris, P. (2007). Impact of oxide substrate on electrical and optical properties of carbon nanotube devices. *Nanotechnology* 18 (29): 295202.
40 Cao, Q., Han, S.J., Penumatcha, A.V. et al. (2015). Origins and characteristics of the threshold voltage variability of quasiballistic single-walled carbon nanotube field-effect transistors. *ACS Nano* 9 (2): 1936–1944.
41 Choi, S.J., Bennett, P., Takei, K. et al. (2013). Short-channel transistors constructed with solution-processed carbon nanotubes. *ACS Nano* 7 (1): 798–803.
42 Franklin, A.D. (2013). Electronics: the road to carbon nanotube transistors. *Nature* 498: 443–444.
43 Farmer, D.B. and Gordon, R.G. (2006). Atomic layer deposition on suspended single-walled carbon nanotubes via gas-phase noncovalent functionalization. *Nano Lett.* 6 (4): 699–703.
44 Qiu, C., Zhang, Z., Zhong, D. et al. (2015). Carbon nanotube feedback-gate field-effect transistor : suppressing current leakage and increasing On/Off ratio. *ACS Nano* 9 (1): 969–977.
45 Lin, Y.M., Appenzeller, J., and Avouris, P. (2004). Ambipolar-to-unipolar conversion of carbon nanotube transistors by gate structure engineering. *Nano Lett.* 4 (5): 947–950.
46 Derycke, V., Martel, R., Appenzeller, J., and Avouris, P. (2001). Carbon nanotube inter- and intramolecular logic gates. *Nano Lett.* 1 (9): 453–456.
47 Bockrath, M., Hone, J., Zettl, A. et al. (2000). Chemical doping of individual semiconducting carbon-nanotube ropes. *Phys. Rev. B* 61 (16): R10606–R10608.
48 Zhou, C., Kong, J., Yenilmez, E., and Dai, H. (2000). Modulated chemical doping of individual carbon nanotubes. *Science* 290 (5496): 1552–1555.
49 Rao, A.M., Eklund, P.C., Bandow, S. et al. (1997). Evidence for charge transfer in doped carbon nanotube bundles from Raman scattering. *Nature* 388 (6639): 257–259.

50 Javey, A., Tu, R., Farmer, D.B. et al. (2005). High performance n-type carbon nanotube field-effect transistors with chemically doped contacts. *Nano Lett.* 5 (2): 345–348.

51 Guo, J., Datta, S., and Lundstrom, M. (2004). A numerical study of scaling issues for Schottky-barrier carbon nanotube transistors. *IEEE Trans. Electron Devices* 51 (2): 172–177.

52 Heinze, S., Tersoff, J., Martel, R. et al. (2002). Carbon nanotubes as Schottky barrier transistors. *Phys. Rev. Lett.* 89 (10): 106801.

53 Chen, Z., Appenzeller, J., Knoch, J. et al. (2005). The role of metal/nanotube contact in the performance of carbon nanotube field-effect transistors. *Nano Lett.* 5 (7): 1497–1502.

54 Martel, R., Derycke, V., Lavoie, C. et al. (2001). Ambipolar electrical transport in semiconducting single-wall carbon nanotubes. *Phys. Rev. Lett.* 87: 256805.

55 Mueller, T., Kinoshita, M., Steiner, M. et al. (2010). Efficient narrow-band light emission from a single carbon nanotube p–n diode. *Nat. Nanotechnol.* 5 (1): 27–31.

56 Misewich, J.A., Martel, R., Avouris, P. et al. (2003). Electrically induced optical emission from a carbon nanotube FET. *Science* 300 (5620): 783–786.

57 Mann, D., Javey, A., Kong, J. et al. (2003). Ballistic transport in metallic nanotubes with reliable Pd ohmic contacts. *Nano Lett.* 3 (11): 1541–1544.

58 Zhang, Z., Liang, X., Wang, S. et al. (2007). Doping-free fabrication of carbon nanotube based ballistic CMOS devices and circuits. *Nano Lett.* 7 (12): 3603–3607.

59 Han, S.-J., Oida, S., Park, H. et al. (2013). Carbon nanotube complementary logic based on Erbium contacts and self-assembled high purity solution tubes. IEDM Technical Digest, pp. 19.8.1–19.8.4.

60 Ding, L., Wang, S., Zhang, Z. et al. (2009). Y-contacted high-performance n-type single-walled carbon nanotube field-effect transistors: scaling and comparison with Sc-contacted devices. *Nano Lett.* 9 (12): 4209–4214.

61 Nosho, Y., Ohno, Y., Kishimoto, S., and Mizutani, T. (2005). n-Type carbon nanotube field-effect transistors fabricated by using Ca contact electrodes. *Appl. Phys. Lett.* 86 (7): 73105.

62 Shahrjerdi, D., Franklin, A.D., Oida, S. et al. (2011). High device yield carbon nanotube NFETs for high-performance logic applications. IEDM Technical Digest, pp. 23.3.1–23.3.4.

63 Tang, J., Farmer, D.B., Bangsaruntip, S. et al. (2017). Contact engineering and channel doping for robust carbon nanotube NFETs. International Symposium VLSI Technology Systems and Applications, p. T6.1.

64 Han, S.-J., Tang, J., Kumar, B. et al. (2017). High-speed logic integrated circuits with solution-processed self-assembled carbon nanotubes and dual work function contacts. *Nat. Nanotechnol.* 12: 861–865. doi: 10.1038/nnano.2017.115.

65 Cao, Q., Tersoff, J., Han, S.J., and Penumatcha, A.V. (2015). Scaling of device variability and subthreshold swing in ballistic carbon nanotube transistors. *Phys. Rev. Appl.* 4 (2): 1–9.

66 Park, R.S., Hills, G., Sohn, J. et al. (2017). Hysteresis-free carbon nanotube field-effect transistors. *ACS Nano* 11 (5): 4785–4791.
67 Kim, W., Javey, A., Vermesh, O. et al. (2003). Hysteresis caused by water molecules in carbon nanotube field-effect transistors. *Nano Lett.* 3 (2): 193–198.
68 Seidel, R.V., Graham, A.P., Kretz, J. et al. (2005). Sub-20 nm short channel carbon nanotube transistors. *Nano Lett.* 5 (1): 147–150.
69 Franklin, A.D., Luisier, M., Han, S.J. et al. (2012). Sub-10 nm carbon nanotube transistor. *Nano Lett.* 12: 758–762.
70 Cummings, A.W. and Léonard, F. (2012). Enhanced performance of short-channel carbon nanotube field-effect transistors due to gate-modulated electrical contacts. *ACS Nano* 6 (5): 4494–4499.
71 Luo, J., Wei, L., Lee, C.S. et al. (2013). Compact model for carbon nanotube field-effect transistors including nonidealities and calibrated with experimental data down to 9-nm gate length. *IEEE Trans. Electron Devices* 60 (6): 1834–1843.
72 Guo, J., Lundstrom, M., and Datta, S. (2002). Performance projections for ballistic carbon nanotube field-effect transistors. *Appl. Phys. Lett.* 80 (17): 3192–3194.
73 Deng, J., Patil, N., Ryu, K. et al. (2007). Carbon nanotube transistor circuits: circuit-level performance benchmarking and design options for living with imperfections. 2007. *IEEE Int. Solid-State Circuits Conf. Dig. Tech. Pap.* 16: 70–588.
74 Deng, J. and Wong, H.-S.P. (2007). A compact SPICE model for carbon-nanotube field-effect transistors including nonidealities and its application-part I: model of the intrinsic channel region. *IEEE Trans. Electron Devices* 54 (12): 3186–3194.
75 Deng, J. and Wong, H.-S.P. (2007). A compact SPICE model for carbon-nanotube field-effect transistors including nonidealities and its application-part II: full device model and circuit performance benchmarking. *IEEE Trans. Electron Devices* 54 (12): 3195–3205.
76 Haensch, W., Nowak, E.J., Dennard, R.H. et al. (2006). Silicon CMOS devices beyond scaling. *IBM J. Res. Dev.* 50 (4.5): 339–361.
77 Taur, Y. and Ning, T.H. (1998). *Fundamentals of Modern VLSI Devices.* Cambridge: Cambridge University Press.
78 Franklin, A.D., Farmer, D.B., and Haensch, W. (2014). Defining and overcoming the contact resistance challenge in scaled carbon nanotube transistors. *ACS Nano* 8 (7): 7333–7339.
79 Allain, A., Kang, J., Banerjee, K., and Kis, A. (2015). Electrical contacts to two-dimensional semiconductors. *Nat. Mater.* 14 (12): 1195–1205.
80 Tang, J., Cao, Q., Farmer, D.B. et al. (2017). High-performance carbon nanotube complementary logic with end-bonded contacts. *IEEE Trans. Electron Devices* 64 (6): 2744–2750.
81 Maiti, A. and Ricca, A. (2004). Metal-nanotube interactions - binding energies and wetting properties. *Chem. Phys. Lett.* 395 (1–3): 7–11.
82 Schroder, D.K. (2005). *Semiconductor Material and Device Characterization.* Hoboken, NJ, USA: John Wiley & Sons, Inc.

83 Chen, C., Yan, L., Kong, E.S.-W., and Zhang, Y. (2006). Ultrasonic nanowelding of carbon nanotubes to metal electrodes. *Nanotechnology* 17 (9): 2192–2197.

84 Chen, C., Xu, D., Kong, E.S.W., and Zhang, Y. (2006). Multichannel carbon-nanotube FETs and complementary logic gates with nanowelded contacts. *IEEE Electron Device Lett.* 27 (10): 852–855.

85 International Technology Roadmap for Semiconductors (ITRS), 2013 Edition. http://www.itrs2.net/2013-itrs.html.

86 Wu, Y., Xiang, J., Yang, C. et al. (2004). Single-crystal metallic nanowires and metal/semiconductor nanowire heterostructures. *Nature* 430 (1992): 61–65.

87 Tang, J., Wang, C.Y., Xiu, F. et al. (2011). Formation and device application of Ge nanowire heterostructures via rapid thermal annealing. *Adv. Mater. Sci. Eng.* 2011: 316513.

88 Zhang, Y., Ichihashi, T., Landree, E. et al. (1999). Heterostructures of single-walled carbon nanotubes and carbide nanorods. *Science* 285: 1719–1722.

89 Wang, M.-S., Golberg, D., and Bando, Y. (2010). Superstrong low-resistant carbon nanotube-carbide-metal nanocontacts. *Adv. Mater.* 22 (47): 5350–5355.

90 Cao, Q., Han, S.-J., Tersoff, J. et al. (2015). End-bonded contacts for carbon nanotube transistors with low, size-independent resistance. *Science* 350 (6256): 68–72.

91 Mattevi, C., Kim, H., and Chhowalla, M. (2011). A review of chemical vapour deposition of graphene on copper. *J. Mater. Chem.* 21 (10): 3324.

92 Tang, J., Cao, Q., Farmer, D.B. et al. (2016). Carbon nanotube complementary logic with low-temperature processed end-bonded metal contacts. 2016 IEEE International Electron Devices Meeting, pp. 5.1.1–5.1.4.

93 Kojima, A., Shimizu, M., Chan, K. et al. (2005). Air stable n-type top gate carbon nanotube filed effect transistors with silicon nitride insulator deposited by thermal chemical vapor deposition. *Jpn. J. Appl. Phys.* 44 (10): L328–L330.

94 Ha, T., Chen, K., Chuang, S. et al. (2015). Highly uniform and stable n-type carbon nanotube transistors by using positively charged silicon nitride thin films. *Nano Lett.* 15 (1): 392–397.

95 Li, G., Li, Q., Jin, Y. et al. (2015). Fabrication of air-stable n-type carbon nanotube thin-film transistors on flexible substrates using bilayer dielectrics. *Nanoscale* 7 (42): 17693–17701.

96 Moriyama, N., Ohno, Y., Kitamura, T. et al. (2010). Change in carrier type in high- k gate carbon nanotube field-effect transistors by interface fixed charges. *Nanotechnology* 21 (16): 165201.

97 Zhang, J., Wang, C., Fu, Y. et al. (2011). Air-stable conversion of separated carbon nanotube thin-film transistors from p-type to n-type using atomic layer deposition of high-κ oxide and its application in CMOS logic circuits. *ACS Nano* 5 (4): 3284–3292.

98 Kita, K. and Toriumi, A. (2009). Origin of electric dipoles formed at high-k/SiO_2 interface. *Appl. Phys. Lett.* 94 (13): 132902.

99 Tulevski, G.S., Franklin, A.D., Frank, D. et al. (2014). Toward high-performance digital logic technology with carbon nanotubes. *ACS Nano* 8 (9): 8730–8745.
100 Kang, S.J., Kocabas, C., Ozel, T. et al. (2007). High-performance electronics using dense, perfectly aligned arrays of single-walled carbon nanotubes. *Nat. Nanotechnol.* 2 (4): 230–236.
101 Hong, S.W., Banks, T., and Rogers, J.A. (2010). Improved density in aligned arrays of single-walled carbon nanotubes by sequential chemical vapor deposition on quartz. *Adv. Mater.* 22 (16): 1826–1830.
102 Hills, G., Zhang, J., Mackin, C. et al. (2013). Rapid exploration of processing and design guidelines to overcome carbon nanotube variations. *Proc. 50th Annu. Des. Autom. Conf. - DAC '13* 34 (7): 1.
103 Jin, S.H., Dunham, S.N., Song, J. et al. (2013). Using nanoscale thermocapillary flows to create arrays of purely semiconducting single-walled carbon nanotubes. *Nat. Nanotechnol.* 8 (5): 347–355.
104 Hersam, M.C. (2008). Progress towards monodisperse single-walled carbon nanotubes. *Nat. Nanotechnol.* 3 (7): 387–394.
105 Arnold, M.S., Green, A.A., Hulvat, J.F. et al. (2006). Sorting carbon nanotubes by electronic structure using density differentiation. *Nat. Nanotechnol.* 1 (1): 60–65.
106 Tulevski, G.S., Franklin, A.D., and Afzali, A. (2013). High purity isolation and quantification of semiconducting carbon nanotubes via column chromatography. *ACS Nano* 7 (4): 2971–2976.
107 Park, H., Afzali, A., Han, S.-J. et al. (2012). High-density integration of carbon nanotubes via chemical self-assembly. *Nat. Nanotechnol.* 7 (12): 787–791.
108 Lee, M., Im, J., Lee, B.Y. et al. (2006). Linker-free directed assembly of high-performance integrated devices based on nanotubes and nanowires. *Nat. Nanotechnol.* 1 (1): 66–71.
109 Bardecker, J.A., Afzali, A., Tulevski, G.S. et al. (2008). Directed assembly of single-walled carbon nanotubes via drop-casting onto a UV-patterned photosensitive monolayer. *J. Am. Chem. Soc.* 130 (23): 7226–7227.
110 Hu, Z., Comeras, J.M.M.L., Park, H. et al. (2016). Physically unclonable cryptographic primitives using self-assembled carbon nanotubes. *Nat. Nanotechnol.* 11 (6): 559–565.
111 Franklin, A.D., Tulevski, G.S., Han, S.J. et al. (2012). Variability in carbon nanotube transistors: improving device-to-device consistency. *ACS Nano* 6 (2): 1109–1115.
112 Chen, Z., Appenzeller, J., Lin, Y.-M. et al. (2006). An integrated logic circuit assembled on a single carbon nanotube. *Science* 311 (5768): 1735.
113 Shulaker, M.M., Hills, G., Patil, N. et al. (2013). Carbon nanotube computer. *Nature* 501 (7468): 526–530.

6

Tunnel Field-Effect Transistors

Deblina Sarkar

Massachusetts Institute of Technology (MIT), Media Arts and Sciences, 20 Ames Street, Room: E15-416, Cambridge, MA 02139, USA

6.1 Introduction

Designing low-power and energy-efficient integrated circuits constitutes a key area for sustaining the irreversible growth of the global Information Technology industry. Lowering the power in electronic components is of critical importance for energy efficiency, leading to reduction in greenhouse gases, increasing the battery life of portable electronics and implanted medical devices, as well as solving heat dissipation issues in mobile and nonportable systems. The most effective way to control the power density is to scale down the supply voltage (V_{DD}) as the dynamic power displays a quadratic dependence on V_{DD}. Ideally, we would prefer a transistor with very abrupt steplike transfer characteristics (Figure 6.1a) such that with the application of an infinitesimally small gate voltage, the device can be turned ON from the OFF state. However, in conventional field-effect transistors (CFETs), there is a fundamental limitation of the steepness of the turn-ON characteristics (Figure 6.1b) and they need the gate voltage to be changed by at least 60 mV to cause a corresponding change of the drain current by one decade (or 10×). This implies that in order to achieve an ON–OFF ratio of four decades, supply voltage of at least 240 mV (4*60 mV) is required. Hence, in the case of CFETs, it is not possible to reduce the supply voltage and still maintain a reasonable ON–OFF ratio.

The reason behind this fundamental limitation is discussed here. The steepness of the turn-ON behavior of the FET is characterized by a parameter called the subthreshold swing (SS), which is defined as the inverse of the subthreshold slope and is given by $\text{SS} = (\text{d}\log_{10} I_{DS}/\text{d}V_{GS})^{-1}$ where I_{DS}: drain current and V_{GS}: gate-to-source voltage. The SS is dependent on two factors, as shown by the two terms in Eq. (6.1).

$$\text{SS} = \frac{\partial V_{GS}}{\partial \psi} \frac{\partial \psi}{\partial(\log_{10} I_{DS})} \tag{6.1}$$

The first term is determined by the device electrostatics, or, in other words, how well the surface potential of the semiconductor (ψ) at the interface of the

Advanced Nanoelectronics: Post-Silicon Materials and Devices,
First Edition. Edited by Muhammad Mustafa Hussain.

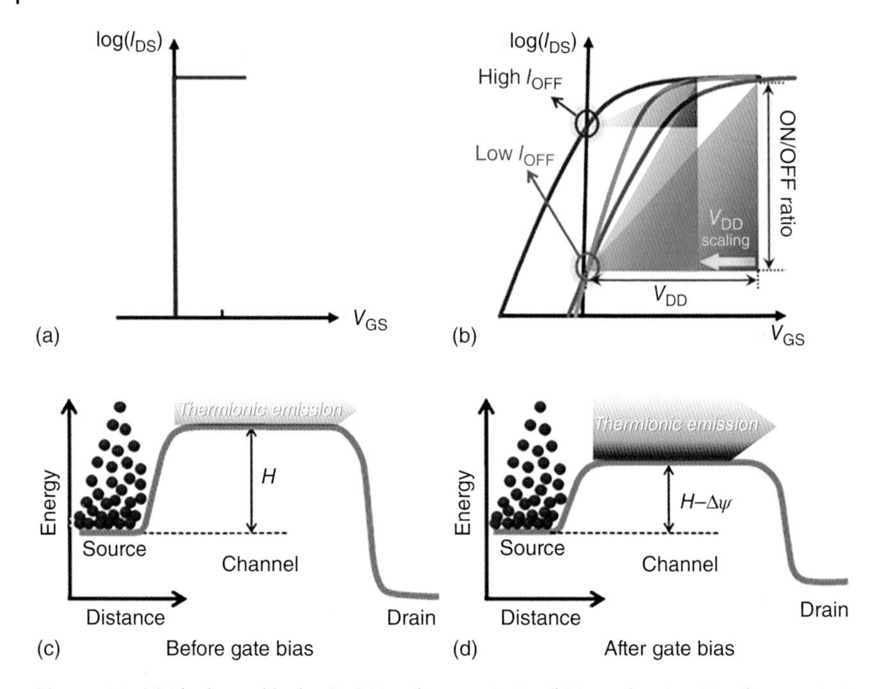

Figure 6.1 (a) Ideal step-like log(I_{DS})-V_{GS} characteristics. (b) In reality, I_{DS}–V_{GS} characteristics have finite slopes and look like the blue curve. If we try to scale down the supply voltage (V_{DD}) and hence make the device turn on with a lower V_{GS}, the characteristics will look like the brown curve leading to significantly higher leakage current and low ON–OFF current ratio. In order to reduce V_{DD}, and at the same time achieve high ON–OFF current ratio, it is necessary to make the slope of the I_{DS}–V_{GS} curve steeper or the inverse of the slope (defined as the subthreshold swing) smaller, as illustrated by the green curve. (c) Band diagrams before and after application of gate potential illustrating the transport mechanism in conventional FETs. Only electrons with energy higher than the barrier height can transport to the drain. H is the height of the barrier before application of gate potential and $\Delta\psi$ is the change in surface potential of the semiconductor channel after application of the gate potential.

semiconductor and gate-dielectric is modulated by the gate voltage and is given by the ratio of the change in V_{GS} to the change in ψ. Even in the case of perfect electrostatics, the lowest possible value of the first factor is 1, which means a particular change in gate voltage (ΔV_{GS}) leads to an almost similar change in surface potential of the semiconductor channel ($\Delta\psi$) such that $\Delta\psi = \Delta V_{GS}$. The second term affecting SS is determined by how efficiently the current is modulated by the change in the surface potential. The lowest possible value of the second factor is 60 mV/decade at room temperature for conventional FETs, as explained later. The transport mechanism (Figure 6.1c) of the conventional FETs is based on thermionic emission over the barrier, which implies that only electrons having energy higher than the barrier height can transport to the drain and contribute to current. Hence, current can be increased (or decreased) by decreasing (or increasing) the barrier height H. Since the electrons in the source are distributed according to the Fermi–Dirac distribution, which can be approximated as Boltzmann distribution for energies much higher than the Fermi level, the occupied density of states, or equivalently the electron density per unit energy

(n) decreases exponentially with increase in energy (E) ($n \propto e^{-E/k_BT}$). Thus, with a decrease in surface potential by $\Delta\psi$, and hence a decrease in barrier by the same amount, current will increase following the trend $I_{DS} \propto e^{\Delta\psi/k_BT}$. Hence, the minimum value for the second factor can be calculated as ($K_BT/q\ln(10)$), which has the value of 60 mV/decade at room temperature. Thus, combining the minimum values for the first and second terms, the limitation of SS in the case of conventional FETs comes to be 60 mV/decade. Note that this minimum value of 60 mV/decade for SS is obtained even in the presence of perfect electrostatics. Degradation of device electrostatics will degrade (increase) the SS further.

Since this limitation in SS arises from the thermionic emission-based transport mechanism in CFETs, it is necessary to develop transistors with a fundamentally different transport mechanism, in order to obtain subthermionic SS. Transistors with steep turn-ON characteristics (i.e. SS below the thermionic limit), which we call the "steep transistors," can lead to the scalability of power supply voltage and hence lower the power dissipation.

Different approaches to achieve steep transistors have been proposed in the literature, such as impact-ionization-based metal-oxide semiconductor (I-MOS) [1], nanoelectro-mechanical switches (NEMS) [2], MOSFETs with ferroelectric layer integrated in the gate stack (Fe-FET) [3] and tunnel-field-effect transistor (TFET) [4]. I-MOS and Fe-FET suffer from reliability and complex material integration issues, respectively. On the other hand, NEMS devices have scalability issues and exhibit significantly small ON currents [5]. Among the new generations of subthermionic transistors, TFETs [4, 6–12] are highly promising as they are less disruptive and offer integrability with CMOS process.

6.2 Tunnel Field-Effect Transistors: The Fundamentals

6.2.1 Working Principle

The schematic diagram of a TFET is shown in Figure 6.2. TFET is also a gated p-i-n diode; but unlike I-MOS, the gate covers the whole of the intrinsic region and the carrier transport mechanism involves band-to-band tunneling (BTBT; i.e. tunneling between valence and conduction bands). The band diagram illustrating the working principle of TFET is shown in Figure 6.3. In the OFF state, electrons below the source valence band cannot transport to the drain, as the width of tunneling barrier (as represented by the length of the magenta arrow) between source and drain is large (Figure 6.3a). Above the source valence band and below the conduction band, i.e. within the bandgap, there are no available density of states and, hence, no carrier transport occurs there as represented by the green arrow with a cross. Above the source conduction band, electrons can flow to the drain as there is no barrier, and this basically constitutes the reverse biased p-n junction leakage. However, the number of electrons available in the conduction band of the source is very low as the source is p-doped with the Fermi level (green dashed line) in the valence band and electron concentration decreases exponentially with the increase in energy above the Fermi level. There can be some channel-to-drain tunneling, as shown by the yellow arrow, but this

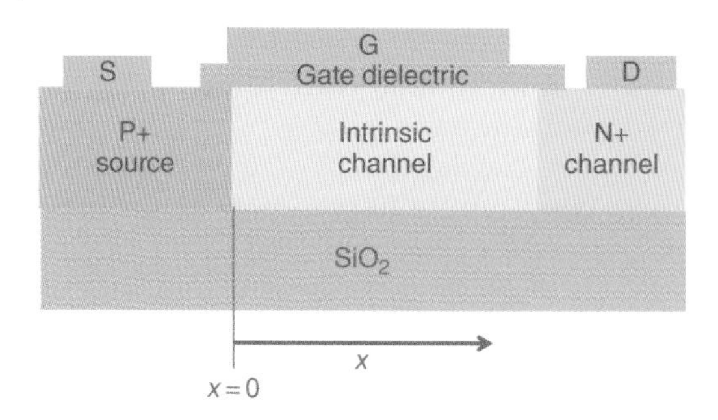

Figure 6.2 Schematic diagram of a TFET.

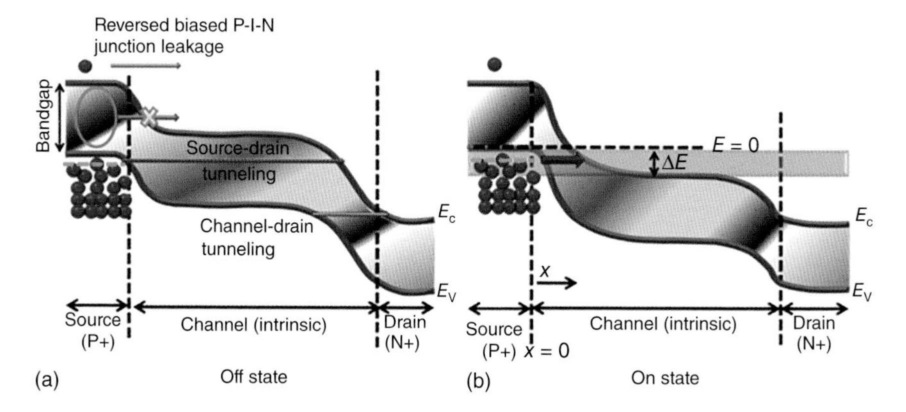

Figure 6.3 Band diagram of a TFET in (a) OFF and (b) ON state. E_C and E_V represent the conduction and valence band, respectively.

can be reduced using a proper drain design such as low doping, higher bandgap near the drain, or underlap gate. In the ON state, after application of positive gate bias, the bands in the channel bend down, and the conduction band of the channel comes below the valence band of the source, lowering the tunneling width (as represented by the length of the red arrow in Figure 6.3b). Hence, electrons from below the source valence band can tunnel to the empty states of the channel conduction band, leading to the increase in the tunneling current (Figure 6.3b). In the case of TFET, particles tunnel through the barrier instead of going over the barrier. However, the forbidden gap has no density of states (DOS) available. Hence, the electrons can no longer follow the Fermi-Dirac distribution within the forbidden gap, and the tail of the Fermi distribution is said to be cut off by the forbidden gap of the semiconductor. As illustrated in the figure, there are no electrons available within the green circle since it lies within the bandgap. This cutting of the Fermi tail can lead to subthermionic SS.

6.2.2 Single-Carrier Tunneling Barrier and Subthreshold Swing

As discussed, in the case of BTBT [13], where carriers tunnel from the valence to the conduction band, it is possible to achieve SS below the thermionic limit. In the introduction section, we discussed that for transport based on thermionic emission over the barrier, the minimum SS achievable is $K_B T/q \ln(10)$. Here,

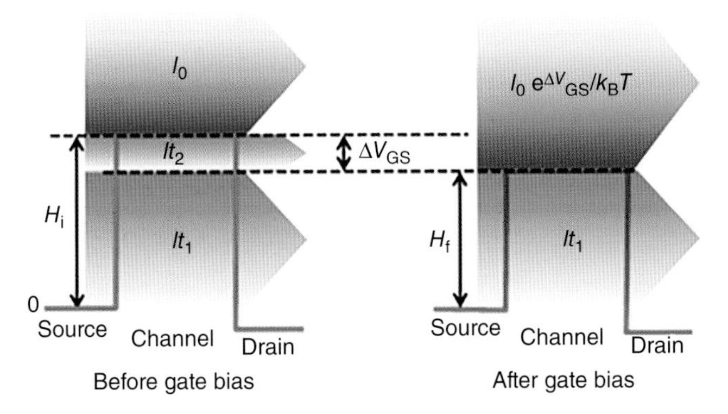

Figure 6.4 Band diagrams of a barrier in which carriers can transport not only through thermionic emission above the barrier but can also tunnel through the barrier, before and after application of gate voltage (V_{GS}).

we show that, for any barrier, which involves a combination of single-carrier tunneling (but not BTBT) and thermionic emission, the SS is also limited by the thermionic limit [14]. By single-carrier tunneling, we mean situations where either electrons or holes are involved in tunneling, but not both. On the other hand, BTBT involves both electrons and holes [15]. A single-carrier tunneling barrier is shown in Figure 6.4, which, for example, can be achieved by vertically stacking a two-dimensional (2D) material (with nonzero bandgap) between two layers of graphene. This barrier can be thought to be similar to that shown in Figure 6.1c. The only difference is that the width of the barrier is small enough, so that apart from transmission over the barrier through thermionic emission, electrons with energy lower than the barrier height can also tunnel directly from source to drain. Initially, before application of gate voltage, let the barrier height be H_i, and after increasing the gate voltage by ΔV_{GS}, the height is reduced to H_f (where $H_f = H_i - \Delta V_{GS}$ in the best case assuming perfect electrostatics). Let us divide the initial current (I_i) into three components: let the current above the barrier through thermionic emission be I_0, tunneling current through the barrier from energy *0* to H_f be It_1, and that from energy H_f to H_i be It_2 (Figure 6.4). After the application of gate voltage, the final current (I_f) can be thought to be comprised of two components, current above the barrier, which is increased when compared to the initial thermionic emission current and is given by $I_0 e^{\Delta V_{GS}/k_B T}$ (as explained in the previous section), and the tunneling current through the barrier given by It_1. For a small change in gate voltage, the SS can be written as

$$\mathrm{SS} = \frac{\Delta V_{GS}}{\log_{10}\left(\dfrac{I_f}{I_i}\right)} = \frac{\ln(10)\Delta V_{GS}}{\ln\left(\dfrac{I_f}{I_i}\right)} \tag{6.2}$$

Now, inserting the current components for I_i and I_f in Eq. (6.2), we get

$$\mathrm{SS} = \frac{\ln(10)\Delta V_{GS}}{\ln\left(\dfrac{I_0 e^{\Delta V_{GS}/k_B T} + I_{t1}}{I_0 + I_{t1} + I_{t2}}\right)} \tag{6.3}$$

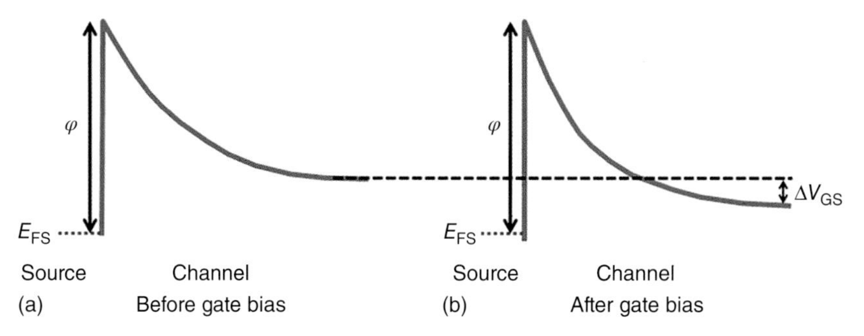

Figure 6.5 Band diagrams of Schottky barrier at the source-channel junction, before and after application of gate potential. φ denotes the Schottky barrier height and E_{FS} denotes the Fermi level of the source.

Now, since

$$\frac{I_0 e^{\Delta V_{GS}/k_B T} + I_{t1}}{I_0 + I_{t1} + I_{t2}} < \frac{I_0 e^{\Delta V_{GS}/k_B T}}{I_0} \tag{6.4}$$

we get

$$\frac{\ln(10)\Delta V_{GS}}{\ln\left(\dfrac{I_0 e^{q\Delta V_{GS}/k_B T} + I_{t1}}{I_0 + I_{t1} + I_{t2}}\right)} > \frac{\ln(10)\Delta V_{GS}}{\ln\left(\dfrac{I_0 e^{q\Delta V_{GS}/k_B T}}{I_0}\right)} = \ln(10)\frac{k_B T}{q} \tag{6.5}$$

This implies that, in the case when single-carrier tunneling is present along with thermionic emission, not only is it fundamentally impossible to achieve SS lower than $K_B T/q$ ln(10) but, in fact, the SS is further degraded from this minimum value.

Even in the case of a Schottky barrier (Figure 6.5), it is not possible to achieve SS lower than the fundamental limit of MOSFET. Solomon has showed using a simple mapping technique that it is not possible to achieve lower SS than $K_B T/q$ ln(10) when single-carrier tunneling is involved. His methodology is valid for both Schottky barriers as well as direct source-drain tunneling. Readers are referred to his paper [16] for further details of the methodology.

6.3 Modeling of TFETs

Although rigorous quantum mechanical treatment of tunneling can be achieved through the *non-equilibrium Green's function* (NEGF) formalism, computationally efficient analytical models are required for fast calculations. Hence, here we develop analytical modeling for BTBT probability and current [17]. First, we develop an analytical model for the potential profile in the device and thereby the BTBT current (I_{BTBT}) in response to a particular gate-dielectric surface potential (ϕ). This surface potential equals the gate voltage (V_{GS}) in the case of a digital FET; while in the case of an FET-based biosensor, it is the potential developed on the surface of the dielectric due to attachment of biomolecules. The modified 1D Poisson equation can be written as

$$\frac{d^2\psi_f(x)}{dx^2} - \frac{\psi_f(x) + \varphi}{\lambda^2} = 0 \tag{6.6}$$

where ψ_f is the potential at the semiconductor–oxide interface, x is the direction along the channel as shown in Figure 6.3b, and λ is defined as the natural length scale [18]. The right-hand side of the equation is set to zero since for TFETs the channel is intrinsic or very lightly doped. For the boundary condition at the left side, the potential at the source-channel junction is set to the potential of the source and hence $\psi_f\,(0)$ is set to 0. This assumption is valid when the depletion region and the potential drop in the source are negligible, which is usually valid due to high doping of the source. Note that $\psi_f(x)$ is defined as the potential energy profile of the valence band. The potential at the channel-drain junction is taken as U_d so that $\psi_f(L_{ch}) = U_d$, where L_{ch} is the length of the channel region of the TFET. Using these boundary conditions, ψ_f can be derived as

$$\psi_f = \frac{e^{\frac{x}{\lambda}}\left(-U_d - \varphi + \varphi e^{-\frac{L_{ch}}{\lambda}}\right)}{e^{-\frac{L_{ch}}{\lambda}} - e^{\frac{L_{ch}}{\lambda}}} - \frac{e^{-\frac{x}{\lambda}}\left(-U_d - \varphi + \varphi e^{\frac{L_{ch}}{\lambda}}\right)}{e^{-\frac{L_{ch}}{\lambda}} - e^{\frac{L_{ch}}{\lambda}}} - \varphi \tag{6.7}$$

For $L_{ch} \gg \lambda$ and considering only the potential profile near the source-channel junction (ψ_{fJ}), we can simplify the above equation as

$$\psi_{fJ} = \varphi(e^{-x/\lambda} - 1) \tag{6.8}$$

This formula is valid when the effect of drain voltage on the potential profile at the source-channel junction is negligible. The electric field/force can be obtained by taking the derivative of the potential profile with respect to distance x and is given by

$$F = \varphi e^{-x/\lambda}/\lambda \tag{6.9}$$

For obtaining an analytical expression for BTBT probability, the maximum value of electric field ($F_{\text{mid-max}}$) in the middle of the bandgap (since the region near the middle of the bandgap has maximum contribution to the tunneling probability [19]) is required to be derived. The effective $F_{\text{mid-max}}$ occurs at energy $E = 0$, as shown in Figure 6.3b, since above $E = 0$, the current is cut off by the bandgap of the semiconductor and as we go below $E = 0$ the electric field decreases. For deriving $F_{\text{mid-max}}$ at $E = 0$, we first find the value of x_{mid}, which is the point at which the intrinsic potential, i.e. the midgap potential falls to $E = 0$. x_{mid} can be found by solving the equation

$$\varphi e^{-x_{\text{mid}}/\lambda} - \varphi + E_G/2 = 0 \tag{6.10}$$

Hence, we derive the x_{mid} as

$$-\lambda \ln((\varphi - E_G/2)/\varphi) \tag{6.11}$$

Substituting this value of x_{mid} in the equation for the electric field given by Eq. (6.9), $F_{\text{mid-max}}$ can be derived as

$$F_{\text{mid-max}} = (2\varphi - E_G)/(2\lambda) \tag{6.12}$$

where E_G is the bandgap of the semiconductor material. Now that the effective force is determined, we next derive the tunneling probability as a function of force (F) using the Wentzel, Kramers, and Brillouin (WKB) approximation and the two-band approximation as follows.

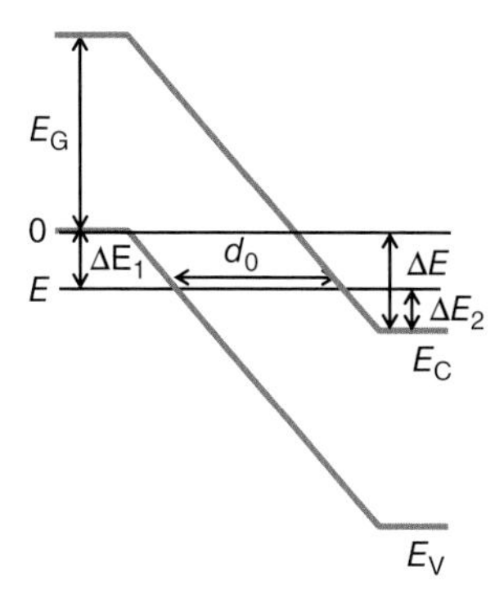

Figure 6.6 Schematic diagram of band bending under a constant electric field.

From the two-band approximation, the relation between energy and wavevector is given by

$$\frac{\hbar^2 k^2}{2m_c} = \frac{(E - E_V)(E - E_V - E_G)}{E_G}\left(1 - \alpha\frac{(E - E_V)}{E_G}\right) \quad \text{where } \alpha = 1 - \sqrt{\frac{m_c}{m_v}} \tag{6.13}$$

where m_c and m_v are the effective masses of the conduction and valence band, respectively, in the transport direction and $\hbar = h/2\pi$, where h is the Planck's constant. Note that while EHD theory is general and can be used for any semiconductor, the two-band approximation is useful for easily deriving the tunneling probability in case of conventional semiconductors with parabolic band structure.

From the WKB equation, the tunneling probability can be written as

$$T = \exp\left\{-2\left(\int_0^{d_0} \kappa \, dx\right)\right\}$$
$$k = i\kappa \tag{6.14}$$

Consider the band bending shown in Figure 6.6. Let F be the electric field representing the slope of the bands. Let us consider an energy level E which cuts the valence band at $x = 0$ and the conduction band at $x = d_0$ where $d_0 = E_G/qF$. We can write $E{-}E_V = qFx$. Using this relation, the tunneling probability can be derived as

$$T = \exp\left(-\frac{\sqrt{2\pi m_{\text{low}}}^{1/2} E_G^{3/2}}{q\hbar F}\frac{(2 - 2r^{1/4} - (1 - \sqrt{r}))}{(1 - \sqrt{r})^2}\right)$$
$$\text{where } r = \frac{m_{\text{low}}}{m_{\text{high}}} \tag{6.15}$$

where m_{low} ($/m_{\text{high}}$) are the lower and higher of m_c and m_v. The tunneling probability can be written in a simplified form as follows:

$$T = \exp\left(-\frac{\sqrt{2\pi m_{\text{low}}}^{1/2} E_G^{3/2}}{4q\hbar F} \times \text{retard}\right) \tag{6.16}$$

$$\text{retard} = \frac{4(2 - 2r^{1/4} - (1 - \sqrt{r}))}{(1 - \sqrt{r})^2}$$

For similar effective masses (m^*) of valence and conduction bands in the tunneling direction, the equation reduces to the well-known Kane's formula given by

$$T = \exp(-\pi m^{*1/2} E_G{}^{3/2} / 2\sqrt{2} q\hbar F) \tag{6.17a}$$

Putting $F = F_{\text{mid-max}}$ in the given equation, the tunneling probability can be written as

$$T = \exp\left(-\pi\sqrt{q} m^{*1/2} E_G{}^{3/2} \lambda / (\sqrt{2}\hbar(2\varphi - E_G))\right) \tag{6.17b}$$

These tunneling probabilities are valid for 1D systems. In 3D systems, the effect due to the presence of momentum in directions (y and z) transverse to the transport direction (x) needs to be considered. The effect of transverse momentum (k_t) can be taken into account by effectively increasing the bandgap in the tunneling probability as given by

$$T = \exp\left(-\frac{\sqrt{2}\pi m_{\text{low}}{}^{1/2} \text{retard}}{4q\hbar F}\left(E_G + \frac{\hbar^2 k_t{}^2}{2m_{ct}} + \frac{\hbar^2 k_t{}^2}{2m_{vt}}\right)^{3/2}\right) \tag{6.18}$$

$$k_t{}^2 = k_y{}^2 + k_z{}^2$$

where m_{ct} and m_{vt} are the effective masses of the conduction and valence band, respectively, in the transverse direction. If the effective masses in the y and z directions are different, then their average values should be considered. Putting reduced transverse effective mass as $1/m_{rt} = 1/m_{ct} + 1/m_{vt}$ and assuming the bandgap is much higher than the transverse energy, we can write

$$E_G{}^{3/2}\left(1 + \frac{\hbar^2 k_t{}^2}{2m_{rt} E_G}\right)^{3/2} = E_G{}^{3/2}\left(1 + \frac{3\hbar^2 k_t{}^2}{4m_{rt} E_G}\right) \tag{6.19}$$

Thus, the tunneling probability can be written as

$$T = \exp\left(-\frac{\sqrt{2}\pi m^{*1/2} E_G{}^{3/2} \text{retard}}{4q\hbar F}\right) \exp\left(-\frac{k_t{}^2}{\text{cnst}}\right) \tag{6.20}$$

$$\text{cnst} = \frac{16qFm_{rt}}{3\sqrt{2}\pi\hbar\sqrt{E_G m_{\text{low}}}\,\text{retard}}$$

Next we derive the energy window ΔE through which the effective tunneling occurs. As is clear from Figure 6.3b, ΔE is defined from the valence band of the source to the point at which the conduction band in the channel becomes flat. By putting the double derivative of the conduction band potential profile with respect to x to zero and assuming $L_{ch} \gg \lambda$, ΔE can be derived as

$$\Delta E = (\phi - E_G) \tag{6.21}$$

Note that here ΔE is derived for relatively small gate voltages such that $\Delta E < U_d$, in which we are interested. For larger gate voltages, ΔE will be given by the difference in energy between the valence band of the source and conduction band of the drain. Finally, using the Launderer's formula, the tunneling current for 1D systems can be written as

$$I_{\mathrm{BTBT}}(\varphi) = \frac{2q^2 \exp\left(-\pi\sqrt{q} m^{*1/2} E_{\mathrm{G}}{}^{3/2} \lambda / (\sqrt{2}\hbar(2\varphi - E_{\mathrm{G}}))\right)}{h} \times \mathrm{Fnc}(\varphi - E_G) \tag{6.22}$$

where the function $\mathrm{Fnc}(\phi - E_{\mathrm{G}})$ is given by

$$\mathrm{Fnc}(\varphi - E_{\mathrm{G}}) = \int_0^{\varphi - E_{\mathrm{G}}} (f_{\mathrm{S}}(E - E_{\mathrm{fS}}) - f_{\mathrm{D}}(E - E_{\mathrm{fD}}))\mathrm{d}E \tag{6.23}$$

Here, $f_{\mathrm{S}}, f_{\mathrm{D}}$ are the Fermi functions and $E_{\mathrm{fS}}, E_{\mathrm{fD}}$ are the Fermi levels at source and drain, respectively. In the energy window ΔE, the source has plenty of available electrons and hence f_{S} can be set to 1, while the drain is devoid of electrons and hence f_{D} can be set to 0. Thus, further simplification can be achieved and $\mathrm{Fnc}(\phi - E_{\mathrm{G}})$ reduces to $(\phi - E_{\mathrm{G}})$.

In case of 3D systems, the current needs to be integrated over the transverse momentum states, and, hence, the limits on the transverse momentum need to be determined. We consider the case where no inelastic scattering is involved during tunneling and the transverse momentum is conserved. From Figure 6.6, we can write

$$\frac{\hbar^2 k_{\mathrm{vx}}{}^2}{2m_{\mathrm{v}}} + \frac{\hbar^2 k_{\mathrm{vt}}{}^2}{2m_{\mathrm{vt}}} = \Delta E_1 \tag{6.24a}$$

$$\frac{\hbar^2 k_{\mathrm{cx}}{}^2}{2m_{\mathrm{c}}} + \frac{\hbar^2 k_{\mathrm{ct}}{}^2}{2m_{\mathrm{ct}}} = \Delta E_2 \tag{6.24b}$$

$$\Delta E_1 + \Delta E_2 = \Delta E \tag{6.24c}$$

As the transverse momentum is conserved $k_{\mathrm{ct}} = k_{\mathrm{vt}}$, we get the following limiting conditions on transverse momentum given by

$$k_{\mathrm{t}}{}^2 \leq \frac{2m_{\mathrm{vt}}\Delta E_1}{\hbar^2} \tag{6.25a}$$

$$k_{\mathrm{t}}{}^2 \leq \frac{2m_{\mathrm{ct}}\Delta E_2}{\hbar^2} \tag{6.25b}$$

Thus, considering an energy level near the valence band (ΔE_1 is small), k_{t} will be limited by Eq. (6.25a); while considering that near to the conduction band (ΔE_2 is small), it will be limited by Eq. (6.25b). At the energy level where the crossover between the two limiting conditions occur, we get

$$\frac{2m_{\mathrm{vt}}\Delta E_{\mathrm{1cross}}}{\hbar^2} = \frac{2m_{\mathrm{ct}}\Delta E_{\mathrm{2cross}}}{\hbar^2} \tag{6.26}$$

Using Eq. (6.24c), we get

$$\Delta E_{\mathrm{1cross}} = \frac{m_{\mathrm{ct}}\Delta E}{m_{\mathrm{ct}} + m_{\mathrm{vt}}} \tag{6.27}$$

Now, considering the energy at the valence band edge at the P-type semiconductor to be zero, and writing $\Delta E_{1\text{cross}}$ as E_{cross}, the current (I_t) can be written using Landauer's formula as

$$I_t = \frac{q}{h\pi} T_{\text{eff}} \left[\int_0^{E_{\text{cross}}} dE \int_0^{\sqrt{2m_{\text{vt}}E/\hbar^2}} k_t e^{-k_t^2/\text{cnst}} + \int_{E_{\text{cross}}}^{\Delta E} dE \int_0^{\sqrt{2m_{\text{ct}}(\Delta E - E)/\hbar^2}} k_t e^{-k_t^2/\text{cnst}} \right] \quad (6.28)$$

where $E_{\text{cross}} = m_{\text{ct}}\Delta E/(m_{\text{ct}} + m_{\text{vt}})$.

Solving the integrals in the equation, we get

$$I_t = \frac{q}{h\pi} T_{\text{eff}} \text{cnst} \left[\Delta E + \frac{C_1 + C_2}{C_1 C_2} (e^{-A\Delta E} - 1) \right] \quad (6.29)$$

where $C_1 = \frac{2m_{\text{vt}}}{\hbar^2 \text{cnst}}$, $C_2 = \frac{2m_{\text{ct}}}{\hbar^2 \text{cnst}}$, and $A = \frac{2\Delta E}{\hbar^2 \text{cnst}} \frac{m_{\text{ct}} m_{\text{vt}}}{m_{\text{ct}} + m_{\text{vt}}}$.

6.4 Design and Fabrication of TFETs

6.4.1 Design Considerations

To design an efficient TFET, several factors must be taken into consideration. As is clear from Eq. (6.15), the tunneling probability is a strong function of the semiconductor bandgap. Hence, to increase the tunneling current, it is desirable to have a lower bandgap. However, low bandgap also increases the leakage of the current by enhancing channel-to-drain tunneling. The solution is to use a heterostructure such that low effective bandgap at the source-channel region and high effective bandgap elsewhere can be achieved. Both staggered and broken-gap heterostructures have been explored for TFETs. While for a broken-gap heterostructure, the barrier to tunneling is ideally zero, phonon-assisted carrier transport can degrade the SS. A diverse way to increase the tunneling current through bandgap engineering is to introduce metallic nanoparticles at the source-channel junction.

The tunneling current can also be increased by decreasing the effective mass along the tunneling direction and increasing the effective masses in the transverse direction, as is evident from Eq. (6.18). Increasing the transverse effective mass also helps in bringing the Fermi level closer to the valence band, which ensures the availability of enough carriers for tunneling in the source region just below the valence band (Figure 6.3). It is necessary to have high carrier density just below the Fermi level in the source, in order to obtain high current when the conduction band in the channel comes just below the valence band of the source. This helps in pushing the steep subthreshold region to higher current values, as the steepest SS occurs when the conduction band in the channel comes just below the valence band of the source. If the Fermi level is much below the valence band, the number of carriers decreases near the valence band of the source and begins to follow the Fermi-tail distribution and thus, the SS degrades.

Doping of the TFET source-channel and drain regions should also be carefully optimized. High source doping can increase the electric field in the source-channel junction, thus increasing tunneling. However, too high doping can push the Fermi level much below the source valence band, which is detrimental, as discussed earlier. A n+ delta doped region at the source-channel junction, has been shown to increase tunneling by increasing the electric field [20]. A moderate doping in the drain region is desirable to decrease channel-drain tunneling, but at the same time not to increase the parasitic resistance.

Improved electrostatics can help increase the electric field in the source-channel junction, resulting in higher tunneling current. This is evident from Eq. (6.17b), as lower natural length scale leads to higher tunneling probability. Tunneling current can also be increased by increasing the tunneling area and making the tunneling direction vertical to the gate (vertical TFET).

The scalability of TFET is limited by direct source-to-drain tunneling when the channel length is scaled down. In a lateral TFET (tunneling direction parallel to gate), the current is almost independent of gate length as the current depends mainly on the electric field in the source-channel junction. In the case of vertical TFET, the current decreases with scaling as the tunneling area is decreased. This can be overcome by using a raised source-channel architecture.

6.4.2 Current Status of Fabricated TFETs

Table 6.1 summarizes the key features of the experimental TFETs in literature, reporting minimum SS of sub-60 mV/decade at room temperature. It is evident from the table that out of all such previously reported TFETs, only those reported by the four groups (highlighted in the second to fifth rows in bold) achieved an average SS below 60 mV/decade (at least over three decades). The International Technology Roadmap for Semiconductors (ITRS) has prescribed attainment of average sub-thermionic SS over four decades, and the TFETs achieving the same are highlighted in blue in the table. It can be observed that there have been two reports, so far, achieving this. Tomioka et al. have reported average sub-thermionic SS over four decades at V_{DS} = 1V. Since the interest in TFETs is for low-power applications, it is necessary to obtain sub-thermionic SS at ultralow voltages. Recently, we developed a TFET with atomically thin channel which achieved sub-thermionic SS over four decades up to a low V_{DS} of 0.1 V.

We built this TFET by engineering the substrate, portions of which are configured as a highly doped semiconductor source and other portions are etched and filled with a dielectric for hosting the drain and gate metal contacts, while ultrathin 2D transition metal dichalcogenide (TMD) forms the channel (Figure 6.7a). This TFET structure offers several unique advantages, as explained later. First, the use of 2D TMD material as the channel attributes excellent electrostatics and, thus, this TFET is promising for simultaneous scaling of device dimensions and supply voltage. Note that using 3D material as the source does not hamper device electrostatics, as the channel region is the one that needs to get modulated by the gate and it is atomically thin in our case. Second, the 2D channel also allows an extremely small tunneling distance (which is determined by the channel thickness), needed for increasing the BTBT current. Third, combining 3D and

Table 6.1 Comparison with experimental TFETs with sub-thermionic minimum-SS.

References	Channel material	Structure (channel dimension) (nm)	$\|V_{DS}\|$ (V)	SS_{min} (mV/decade)	SS_{avg} over 3 decades (mV/decade)	Subthermal SS_{avg} over 4 decades (mV/decade)	I_{on} (μA μm^{-1})	I_{on}/I_{off} ratio
Sarkar et al. [14]	MoS_2	*Planar [1.3]*	*1*	*4.3*	*31*	*34.3*	*1*	1.6×10^{8}
			0.5	*3.9*	*22*	*31*	*0.5*	8.3×10^{7}
			0.1	*3.8*	*33*	*36.5*	*0.11*	1.8×10^{7}
Tomioka et. al. [21] **(Hokkaido Univ./JST)**	**Si/III–V Heterojunction**	**NW [30]**	**1**	**12**	**21**	**21**	**1**	$\mathbf{2\times10^{6}}$
			0.1	**U**	**21**	**N**	**0.001**	$\mathbf{1\times10^{6}}$
Gandhi et. al. [22, 23] **(NUS/IME/UCSB)**	**Si**	**NW [30–40]**	**0.1**	**30**	**50**	**N**	**0.003**	$\mathbf{1\times10^{4}}$
		NW [18]	**0.1**	**30**	**50**	**N**	**0.44**	$\mathbf{1.6\times10^{5}}$
Jeon et. al.[24] **(SEMATECH/ Berkeley/Texas State Univ.)**	**Si**	**SOI [40]**	**1**	**32**	**~ 55**	**N**	**1.2**	$\mathbf{6\times10^{6}}$
Kim et. al.[25] **(UC. Berkeley)**	**Si**	**SOI [70]**	**0.5**	**38**	**~ 50**	**N**	**0.42**	$\mathbf{3.5\times10^{6}}$
Mayer et. al.[26] (CEA-LETI)	Si	SOI [20]	0.8	42	>60	N	0.02	4×10^{4}
			0.1	42	>60	N	4×10^{-4}	4×10^{4}
Choi et. al.[27] (UC. Berkeley)	Si	SOI [70]	1	52.8	>60	N	12.1	2×10^{3}
			0.1	~52.8	>60	N	1	1×10^{4}
Appenzeller et. al.[28] (IBM)	CNT	SOI [U]	0.5	40	>60	N	0.01	1×10^{3}
Knoll et. al.[29] (Peter Grunberg Inst.)	Si	NW [7 × 45]	0.1	30	>60	N	2	2×10^{7}
			0.5	>30	>60	N	10	2×10^{6}
Krishnamohan et. al.[30] (Stanford Univ.)	Ge	DG [10]	0.5	50	>60	N	1	2×10^{6}
Huang et. al.[31] (Peking Univ.)	Si	SOI [100]	0.6	29	70	N	20	1×10^{8}
			0.05	~29	~70	N	1.5	7.5×10^{6}
Villalon et. al.[32] (CEA-LETI)	Ge_xSi_{1-x}	SOI [7–11]	0.1	33	>60	N	1	1×10^{7}
Ganjipour et. al.[33] (Lund Univ.)	InP/GaAs Heterojunction	NW [85–100]	0.75	50	150	N	2.2	1.1×10^{7}
Kim et. al.[34] (U. Tokyo/JST-CREST)	Si	SOI [10–13]	0.05	28	70	N	0.1	1×10^{7}
Dewey et. al.[35] (Intel)	III–V Heterojunction	Bulk	0.05	~58	>60	N	0.3	3×10^{4}

Note: U: unknown, N: did not achieve subthermal SS_{avg} over four decades.
CNT, carbon nanotube; SOI, silicon-on-insulator.

2D materials opens up unprecedented opportunities for designing custom-built heterostructures. We have chosen germanium as the 3D material since it has low electron affinity (EA) and bandgap among the commonly used group IV and III–V semiconductors, while MoS_2 is chosen as the 2D material since it has relatively high EA among the most commonly explored TMDs. Thus, Ge-MoS_2 forms a staggered heterojunction (Figure 6.7b), with a small band overlap at the interface, which is necessary for increasing the BTBT current. Fourth, the heterojunction is formed with van der Waal's bond and thus has strain-free interfaces. Fifth, while methodologies for obtaining stable as well as high doping in TMDs is very challenging and still under investigation, 3D materials already enjoy well-developed doping technologies that have been leveraged in this work, for forming a highly doped source. This enables the creation of an ultrasharp doping profile and hence a high electric field at the source-channel interface, as there is negligible chance of diffusion of dopant atoms across the heterojunction due to the presence of the van der Waal's gap. Sixth, since MoS_2 is placed on top of Ge forming a vertical source-channel junction, BTBT can take place across the entire area of the MoS_2-Ge overlap, which leads to higher ON-current than that in the case of the line overlap obtained in lateral junctions. Last but not the least, we have achieved this TFET on a planar platform, which is easily manufacturable compared to 1D structures such as nanowires and nanotubes.

Figure 6.7c,d demonstrates the operation of the ATLAS-TFET through the band diagrams along vertical dashed line in Figure 6.1a, in OFF and ON (Figure 6.1b) states. In this paper, the N-type transistor is achieved where positive voltage is applied to the drain electrode contacting the MoS_2 layers with respect to the highly p-doped Ge source. Hence, electrons tend to move from the Ge to the MoS_2, and this electron transport can be modulated by the gate to turn the device ON or OFF. In the OFF state, only electrons above the conduction band of Ge can transport to MoS_2 (purple arrow), leading to ultralow current due to the scarcity of available electrons at high energies above the Fermi level. At lower energies, no electrons can flow either due to the nonavailability of DOS in the Ge source (orange arrow; Figure 6.7c) or in MoS_2 channel (black arrow). Hence, the OFF current is very low. As the gate voltage is increased, the conduction band of MoS_2 is lowered below the valence band of the Ge source (ON state), and hence filled DOS in the source gets aligned with empty DOS in the channel, leading to an abrupt increase in electron flow (green arrow; Figure 6.7d) and, hence, current, which can lead to subthermionic SS. The electrons, after tunneling from the Ge source to the MoS_2, are sucked in laterally by the drain contact, as shown by the red arrows in Figure 6.7a. Note that bilayer MoS_2 is used instead of a monolayer. Although the thickness of bilayer MoS_2 is higher by 0.65 nm compared to that of the monolayer, bilayer MoS_2 still offers excellent electrostatics and ultralow tunneling barrier width and, at the same time, has a smaller bandgap [36] and is more robust to surface scattering [37].

Although the current in ATLAS-TFET is at par with other experimental TFETs with steep SS, further improvement in current is required, especially for digital electronic applications. The current can be improved by removing the interfacial germanium oxide layer, which adds extra tunneling resistance in series in the tunneling path from the Ge to the MoS_2. The current levels in all experimental TFETs

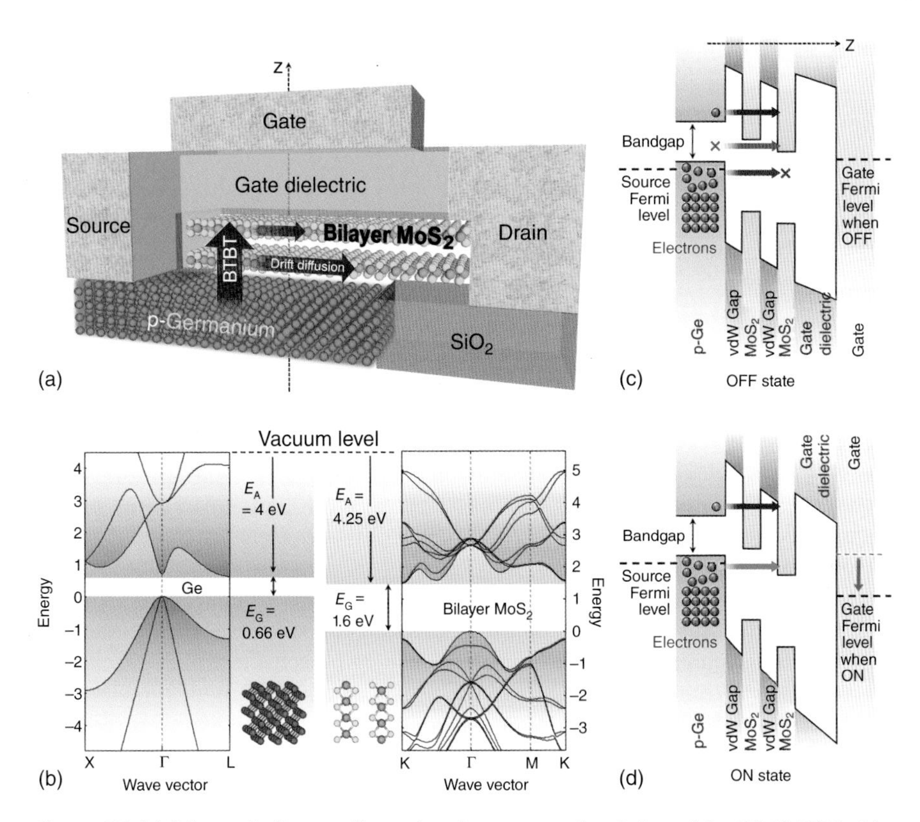

Figure 6.7 (a) Schematic diagram illustrating the cross-sectional view of the ATLAS-TFET with ultrathin bilayer MoS_2 (1.3 nm) as channel and degenerately doped *P*-type Ge as the source. The path for electron transport is shown by the red arrows which run vertically from the Ge source to the MoS_2 and then laterally through the MoS_2 layers to the drain. As the Ge is highly doped, the tunneling barrier height is mainly determined by the effective bandgap of MoS_2 (including the van der Waals gap), while the tunneling width is determined by the MoS_2 thickness. (b) Band alignment of Ge and bilayer MoS_2 showing their electron affinities (E_A) and bandgaps (E_G) and thus illustrating the formation of a staggered vertical heterojunction. The crystal structure of both the materials are shown below, while the bandstructures are shown on both sides. Band diagrams along vertical dashed line in (a), in both (c) OFF and (d) ON states. The white regions represent the forbidden gaps (zero density of states). While the effective bandgap of bilayer MoS_2 has been illustrated in (b), here the bands for the two layers are shown separately with the van der Waal's (vdW) gap between them for better visual interpretation of current flow. Note that the drain contact is located perpendicular to the plane of the figure and is not shown in it. In the OFF state, electrons from the valence band of Ge cannot transport to MoS_2 due to the nonavailability of density of states (DOS) in MoS_2 (black arrow and cross sign). At higher energies, empty DOS is available in MoS_2, but no DOS is available in Ge, again forbidding electron flow (orange arrow and cross mark). With further increase in energy reaching above the conduction band of Ge, DOS is available in both Ge and MoS_2. However, the number of electrons available in the conduction band of the Ge source is negligible due to the exponential decrease in electron concentration with increase in energy above the Fermi level, according to Boltzmann distribution. Thus, very few electrons can flow to the MoS_2 (purple arrow), leading to very low OFF-state current. With the increase in gate voltage (d), when the conduction band of MoS_2 at the dielectric interface is lowered below the valence band of the Ge source, electrons start to flow (green arrow), resulting in an abrupt (sub-thermionic) increase in BTBT current.

with steep SS, demonstrated so far, is much lower than that required specially for digital electronic applications. Both the ON-current as well as the current level where steep SS is exhibited needs three to four orders of magnitude improvement. Apart from the challenge in terms of current, there is also a limit to the minimum SS that can be obtained from TFETs. If the band edges are assumed to be perfectly sharp, then infinitely small SS can be obtained as the channel conduction band comes just below the valence band of the source. However, due to Heisenberg's uncertainty principle, there is a broadening of the energy of the band edges, which poses a fundamental limitation on the SS of TFETs.

6.5 Beyond Low-Power Computation

In the previous sections, we discussed that the TFETs, due to their steep subthermionic turn-ON characteristics, are highly advantageous for low-power electronic computational elements. In this section, we show that, beyond low-power computation, this attribute can revolutionize a diverse arena of gas/biosensing technology [17, 38]. Sensors, especially biosensors, are indispensable to modern society due to their wide applications (Figure 6.8) in public health care, national and homeland security, and forensic industries, as well as in environmental protection. There is a great demand for ultrasensitive biosensors which can detect biomolecules with high reliability and specificity in complex environments such as whole blood. Moreover, detection ability at low biomolecule concentration is necessary for screening many cancers [39], neurological disorders [40, 41] and early-stage infections [42] such as human immunodeficiency virus (HIV).

Currently, enzyme-linked immunosorbent assay (ELISA) based on optical sensing technology is widely used as a medical diagnostic tool as well as a quality-control check in various industries. For ELISA, the labeling of biomolecules is needed, which might alter the target–receptor interaction by conformation change. Moreover, ELISA requires the use of bulky, expensive optical instruments, and hence is not suitable for fast point-of-care clinical applications. On the other hand, the biosensors based on FETs [43–47] are highly attractive as they promise real-time, label-free electrical detection, scalability, inexpensive mass production, and possibility of on-chip integration of both sensor and measurement systems (Figure 6.9). In a conventional FET used for digital computational applications, two electrodes (source and drain) are used to connect a semiconductor material (channel). Current flowing through the channel between the source and the drain is modulated by a third electrode called the gate, which is capacitively coupled through a dielectric layer covering the channel region. In the case of an FET biosensor, the physical gate present in a logic transistor is removed and the dielectric layer is functionalized with specific receptors for selectively capturing the desired target biomolecules. The charged biomolecules when captured produce a *gating effect*, which is transduced into a readable signal in the form of change in electrical characteristics of the FET such as drain-to-source current or channel conductance.

The critical parameter of an FET, which gives an indication of the efficiency of gating effect and thereby the sensitivity of the biosensor, is the SS [17]. SS, as

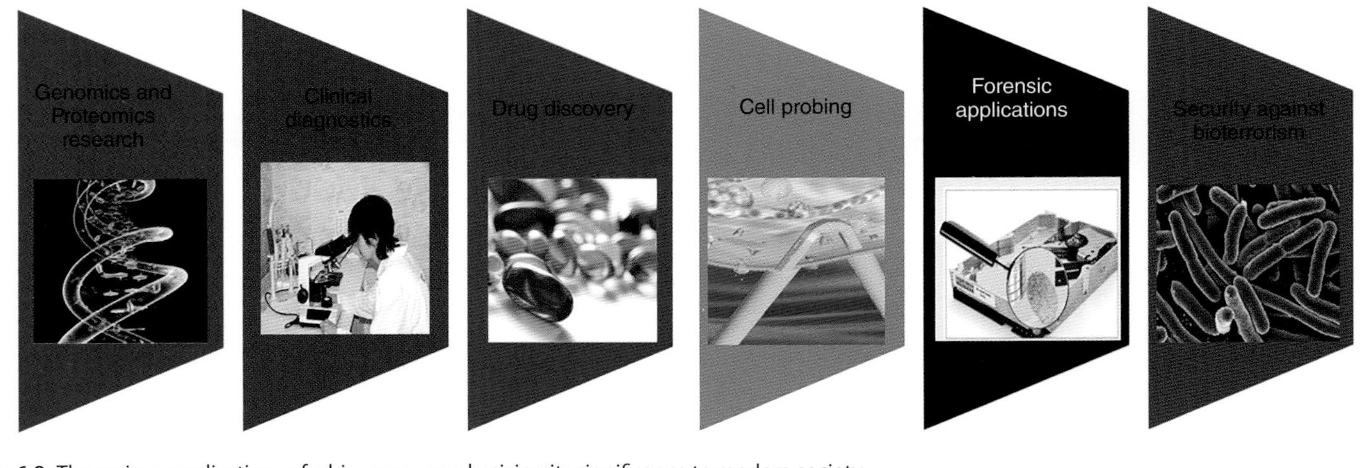

Figure 6.8 The various applications of a biosensor emphasizing its significance to modern society.

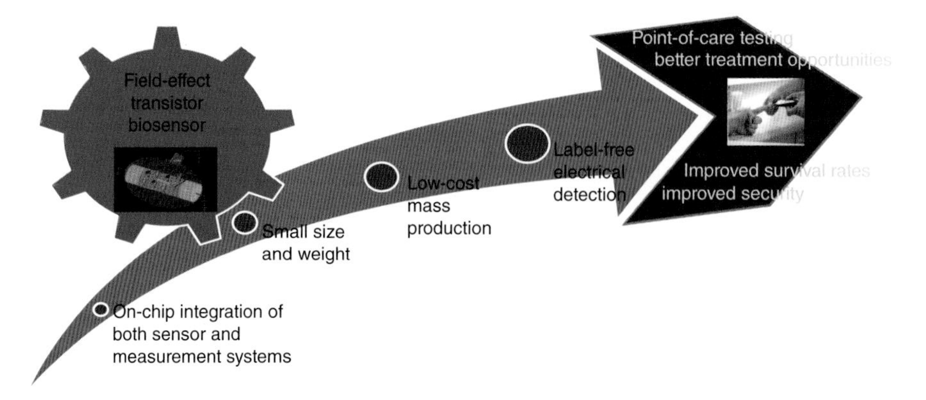

Figure 6.9 Schematic showing the potential of biosensors based on field-effect transistors.

discussed, is given by the inverse of the slope of their $\log_{10}(I_D)-V_{GS}$ curve, where I_D and V_{GS} are the drain-to-source current and gate-to-source voltage, respectively. Therefore, the SS of a device essentially indicates the change in gate voltage required to change the subthreshold current by one decade ($SS = dV_{GS}/d(\log_{10}(I_D))$). Thus, the smaller the SS, the higher will be the change in current for a particular change in the dielectric surface potential due to the gating effect produced by the pH change or attachment of biomolecules and thereby the higher the sensitivity. The dependence of sensitivity (S_n) on SS is given by [17]

$$S_n = 10^{\left[\int_{\varphi_i}^{\varphi_f} \frac{1}{SS} d\varphi\right]} - 1 \tag{6.30}$$

which depicts the exponential dependence of sensitivity on SS where ϕ_i and ϕ_f denote the initial and final surface potential on the gate dielectric before and after attachment of the target biomolecules. In Figure 6.10 we have plotted the sensitivity as a function of SS, which clearly illustrates that by lowering the SS of a device, sensitivity can be increased substantially.

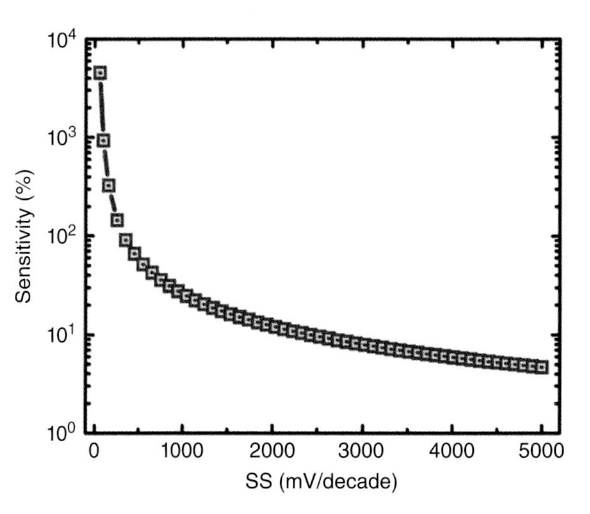

Figure 6.10 Sensitivity as a function of the subthreshold swing (SS) showing that sensitivity increases exponentially with the reduction of SS.

From Eq. (6.30), it is clear that the fundamental limitation on the minimum achievable SS also poses a fundamental limitation on the sensitivity of conventional field-effect transistor (CFET)-based biosensors, irrespective of the channel material [48]. While the relation between sensitivity and SS can be understood easily through intuition, it is not quite intuitive that the thermionic SS also limits the response time of the biosensors; and this effect is discussed here. In the following sections, we theoretically demonstrate that these limitations can be overcome by employing steep transistors with subthermionic SS, such as TFET.

6.5.1 Ultrasensitive Biosensor Based on TFET

The working mechanism of TFET-based biosensor (Figure 6.11) [48] is similar to that of TFETs (Figure 6.3) used for digital applications which were discussed in the previous section. Before the attachment of biomolecules to the sensor surface, the tunneling barrier between source and channel is high (the width of the barrier is depicted by the length of the blue arrow) and hence the current in TFET is low. After biomolecule–receptor conjugation, due to the charges present in the biomolecules (positive charge is assumed here), the bands in the channel bend down, leading to a decrease in the tunneling barrier (the width of the barrier is depicted by the length of the brown arrow), and hence an increase in the tunneling current. Thus, the biomolecules can be detected by monitoring the change in current through the TFET biosensor device.

Here, we establish the supremacy of the TFET biosensor compared to that based on CFETs. We present extensive numerical simulations based on nonequilibrium Green's function formalism for accurate results as well as analytical solutions with the aim of providing easy physical insights. The modeling scheme can be divided into two major parts. The first part deals with the kinetics of biomolecules within the electrolyte, their capture by the receptors, and thereby the development of surface potential (ϕ_{bio}) on the oxide in the presence of electrostatic screening by the ions present in the electrolyte. The second part deals with the electrical response of the TFET or the change in tunneling current with the change in the surface potential developed due to attachment of the biomolecules. Here, we focus on 1D structures for computational efficiency.

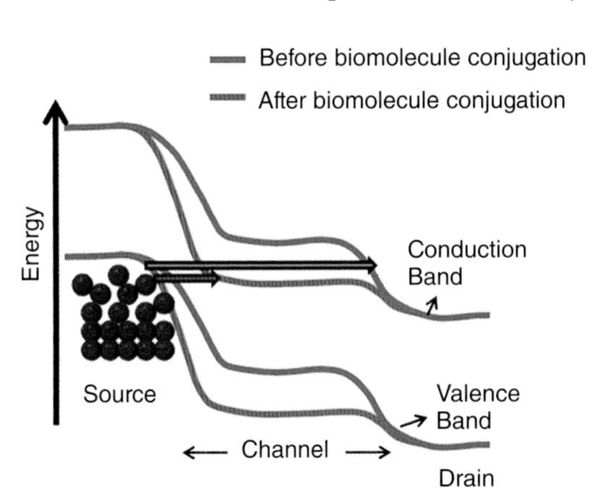

Figure 6.11 Schematic diagram depicting the working mechanism of a TFET-based biosensor.

However, the conclusions derived are general and can also be applied to 2D or 3D structures.

First, we deal with the first part of the modeling scheme, i.e. derivation of surface potential (ϕ_{bio}) due to the binding of charged biomolecules by receptors. The biomolecule–receptor conjugation occurs through two processes [49]. The first process involves the diffusion of the biomolecules to the oxide surface, which has been functionalized with specific receptors and is described by the equation

$$\frac{d\rho}{dt} = D\nabla^2 \rho \tag{6.31}$$

where ρ is the concentration and D is the diffusion coefficient of the biomolecules in the solution. The second process involves the capture of the biomolecules by the receptors and is described by the equation

$$\frac{dN_{bio}}{dt} = k_F(N_0 - N_{bio})\rho_s - k_R N_{bio} \tag{6.32}$$

where N_{bio} is the surface density of conjugated receptors or, in other words, the surface density of the captured biomolecules, N_0 is the surface density of the receptors used to functionalize the surface of the oxide, ρ_s is the concentration of the biomolecules on the surface of the oxide, k_F is the capture constant, and k_R is the dissociation constant. Using these two equations, the surface density of charge due to the attached biomolecules on the sensor surface can be calculated [49].

Now, the surface charge formed on the sensor surface attracts the ions within the electrolyte, which forms a second layer of charge (of opposite polarity). This second layer electrostatically screens the first layer and thereby decreases the effective potential developed on the oxide surface. This double-layer charge density can be calculated using the nonlinear Poisson–Boltzmann equation, which for a 1–1 electrolyte is given by [50]

$$-\nabla^2 \varphi(r) + \frac{K_B T}{{\lambda_{DH}}^2 q} \sinh\left(\frac{q\varphi}{K_B T}\right) = \frac{q}{\varepsilon_w} \sum_i z_i \delta(r - r_i) \tag{6.33}$$

Here, λ_{DH} denotes the Debye–Huckel screening length and is given by

$$\sqrt{\varepsilon_w K_B T / 2q^2 I_0 N_{avo}} \tag{6.34}$$

where, ε_w is the dielectric constant, I_0 is the ion concentration of the electrolyte, N_{avo} denotes the Avagadro number, z_i is the partial charge, and r_i is the location of the atoms within the biomolecule. Finally, ϕ_{bio} can be found by equating the surface charge on the oxide due to the conjugated receptors to the sum of the charge in the electrolyte double layer and the charge developed within the semiconductor 1D structure [49, 51]

Next, we discuss the second part of the modeling scheme, and it is this part that dictates the critical differences between the CFET- and TFET-based biosensors. The analytical formula for BTBT current (I_{BTBT}) was derived in the previous section, and can be used to obtain the sensitivity of the biosensor, which is defined as

$$S_n = \{I_{BTBT}(\varphi_0 + \varphi_{bio}) - I_{BTBT}(\varphi_0)\}/I_{BTBT}(\varphi_0) \tag{6.35}$$

where ϕ_0 denotes the initial surface potential on the oxide before the attachment of biomolecules. In Eq. (6.35), it is implicit that ϕ_0 is tuned such that the current is dominated by the source-channel BTBT and the energy window $\Delta E \geq 0$. The TFETs exhibit ambipolarity and for $\Delta E < 0$ the current is mainly dominated by channel-drain tunneling (Figure 6.3a). Hence, to avoid undesirable ambipolar effects, it is required to tune ϕ_0 such that the operational mode of the biosensor always remains in the regime where source-channel current dominates. Using Eqs. (6.23) and (6.35), the analytical formula for sensitivity can be derived as

$$S_{\text{n}} = \exp\left(\frac{\pi\sqrt{2q}m^{*1/2}E_{\text{G}}^{3/2}\lambda\varphi_{\text{bio}}}{\hbar(2\varphi_0 - E_{\text{G}})(2\varphi_0 + 2\varphi_{\text{bio}} - E_G)}\right)\left(1 + \frac{\varphi_{\text{bio}}}{\varphi_0 - E_{\text{G}}}\right) - 1 \quad (6.36)$$

This analytical formula provides important insights regarding the dependence of sensitivity on the initial surface potential ϕ_0. It can be observed that the sensitivity increases as ϕ_0 is decreased (keeping $\Delta E \geq 0$). This is because, for TFETs, the rate of increase in current with gate voltage is higher for smaller values of ΔE (and thus for smaller values of ϕ_0) giving rise to increased sensitivity at lower values of ϕ_0. Note that small value of ΔE indicates TFET operation in the subthreshold region. Thus, Eq. (6.36) indicates that in order to achieve high sensitivity, the TFET biosensor should be operated in the subthreshold regime. Equation (6.36) also provides direct physical insights regarding the dependence of the sensitivity on the bandgap of the material. As is evident from the equation, sensitivity increases with increase in bandgap. This is because of the decrease in the current before the capture of biomolecules, i.e. IBTBT(ϕ0) with the increase in the bandgap.

For TFETs with relatively large bandgap materials or employing asymmetric design techniques at source and drain to reduce ambipolarity, ϕ0 may be tuned so that the current is mainly dominated by the relatively smaller reverse biased P-I-N junction current (Irev) and the sensitivity will be given by

$$S_{\text{n}} = (I_{\text{BTBT}}(\varphi_0 + \varphi_{\text{bio}}) - I_{\text{rev}}(\varphi_0))/I_{\text{rev}}(\varphi_0) \quad (6.37)$$

In this case, the sensitivity will increase with decreasing bandgap at the source-channel junction due to the exponential increase in $I_{\text{BTBT}}(\phi_0 + \phi_{\text{bio}})$.

The subthreshold regime forms the optimal sensing domain not only for TFET biosensors as discussed earlier but also for the conventional FET biosensors [51]. CFETs suffer from a fundamental limitation on the minimum achievable SS of $[K_{\text{B}}T/q\ln(10)]$ due to the *Boltzmann tyranny* effect, where K_{B} is the Boltzmann constant and T is the temperature. However, the TFETs overcome this limitation due to the *Fermi-tail cutting* by the bandgap of the semiconductor. The charged biomolecules essentially produce a gating effect on the semiconductor channel. Hence, the change in current in TFET biosensors, because of their smaller SS, is substantially higher than that for CFET biosensors in the subthreshold region for the same surface potential developed due to attachment of biomolecules (ϕ_{bio}), as illustrated in Figure 6.12a.

Figure 6.12b shows the current as a function of the drain voltage for both CFET and TFET biosensors before and after the biomolecule conjugation. It is observed that for similar currents in both biosensors before biomolecule conjugation, the

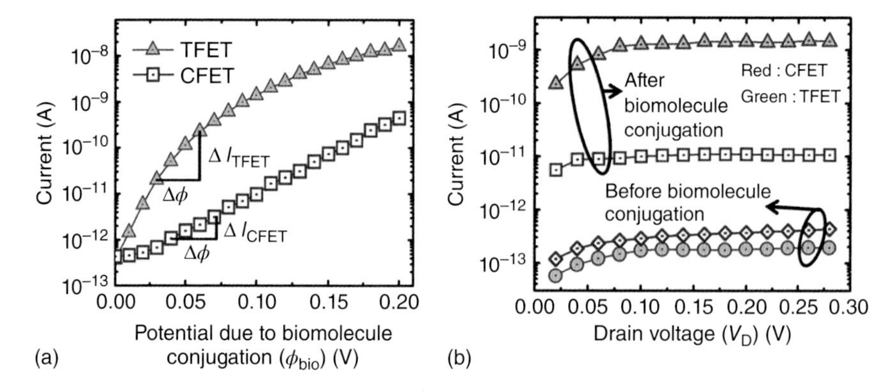

Figure 6.12 (a) Current as a function of surface potential developed due to biomolecule conjugation (ϕ_{bio}) at a drain voltage (V_D) of 0.3 V. Due to the smaller subthreshold swing in TFETs, they can lead to higher change in current compared to CFETs for the same change in surface potential. (b) Current as a function of drain voltage before and after biomolecule conjugation for $\phi_{bio} = 0.1$ eV.

current in TFET biosensors can be more than 2 orders of magnitude higher than that in CFETs after the attachment of the biomolecules, which obviously indicates a significant increase in the sensitivity.

Comparison of the performance of CFET and TFET biosensors, for biomolecule as well as pH sensing, clearly shows that the sensitivity of TFET biosensors can surpass that of CFET biosensors by several orders of magnitude (Figure 6.13a,b). The dependence of Sn on SS is depicted by Eq. (6.30), which depicts the strong relation between the two. Thus, TFETs can harness the benefits of the substantial increase in sensitivity (up to more than 4 orders of magnitude) with decreasing SS and lead to ultrasensitive biosensors, while CFET

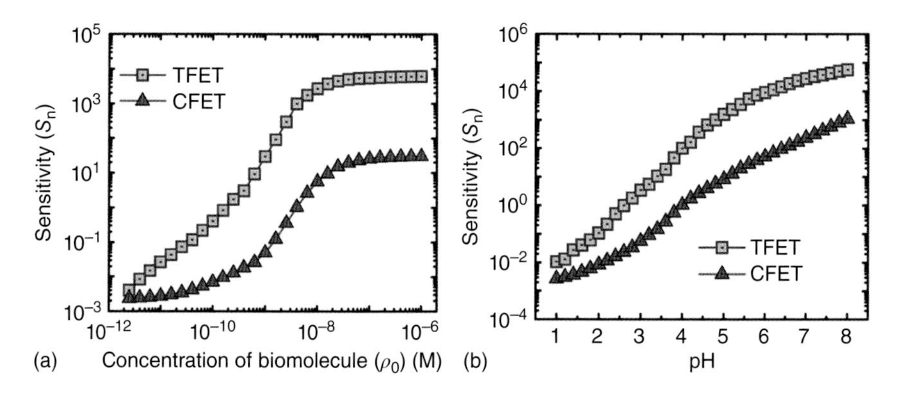

Figure 6.13 (a) Sensitivity for sensing of biomolecules as a function of biomolecule concentration. (b) Sensitivity for pH sensing for different pH values. ϕ_0 is tuned for TFET and CFET so that they operate in the subthreshold regime. The bandgap and the effective masses used in the simulations are 0.4 eV and $0.15m_0$, respectively (where m_0 denotes the mass of a free electron) and the diameter of nanowire is taken as 5 nm.

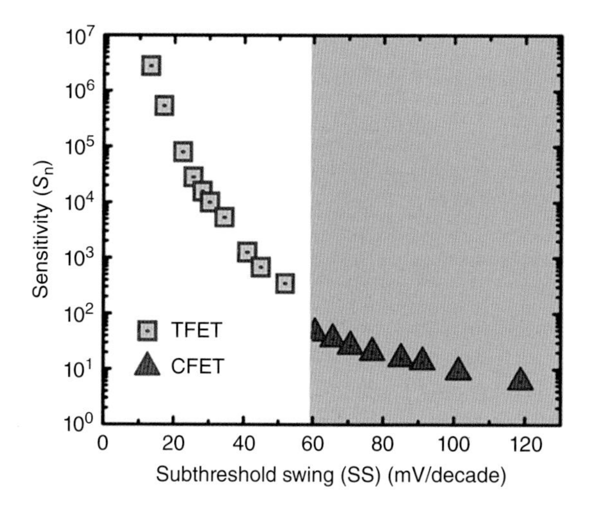

Figure 6.14 Sensitivity as a function of subthreshold swing averaged over 4 orders of magnitude of current for both CFET- and TFET-based biosensors. Surface potential change due to attachment of biomolecules (ϕ_{bio}) is taken to be 0.1 V. Sensitivity increases substantially with the decrease in subthreshold swing. The shaded region shows the sensitivity values for CFET biosensors, indicating that there is a restriction on the maximum achievable sensitivity since the subthreshold swing in CFETs cannot be minimized below 60 mV/decade at room temperature. All simulations in this figure are performed through self-consistent solutions of Poisson's and Schrodinger's equations within the framework of nonequilibrium Green's function formalism.

biosensors are strictly restricted to a higher limit on the maximum achievable sensitivity as highlighted in Figure 6.14.

6.5.2 Improvement in Biosensor Response Time

In the following discussions, we show that TFET biosensors not only lead to a substantial increase in sensitivity but also provide significant improvement in terms of the response time, which is defined as the time required to obtain a desired sensitivity (more specifically, the time needed to capture a certain number of biomolecules in order to achieve a desired change in the electrical signal). First, we derive an analytical formula for the surface density of biomolecules (N_{bio}) that is required to be captured in order to obtain a particular sensitivity. N_{bio} can be related to ϕ_{bio} as $((1/C_{ox}+1/C_{NW})^{-1} + C_{DL})\phi_{bio}$ [51], where C_{ox}, C_{NW} and C_{DL} represent the oxide, quantum, and electrolyte double-layer capacitance, respectively. From Eq. (6.30), we can write ϕ_{bio} as $SS_{avg} \times \log_{10}(S_n + 1)$, where SS_{avg} denotes the average value of SS over the range ϕ_0 to $(\phi_0 + \phi_{bio})$. In the subthreshold region $(1/C_{ox} + 1/C_{NW})^{-1} + C_{DL} \approx \sim C_{DL}$ and hence, N_{bio} can be written as

$$N_{bio} = \frac{\pi\varepsilon_w R_{NW} K_1(R_{NW}/\lambda_{DH}) SS_{avg}}{\lambda_{DH} K_0(R_{NW}/\lambda_{DH})} \log_{10}(S_n + 1) \tag{6.38}$$

In this equation, we have used the expression for C_{DL} as $\pi\varepsilon_w R_{NW}/\lambda_{DH} \times K_1(R_{NW}/\lambda_{DH})/K_0(R_{NW}/\lambda_{DH})$ [51]. Here, λ_{DH} denotes the Debye–Huckel

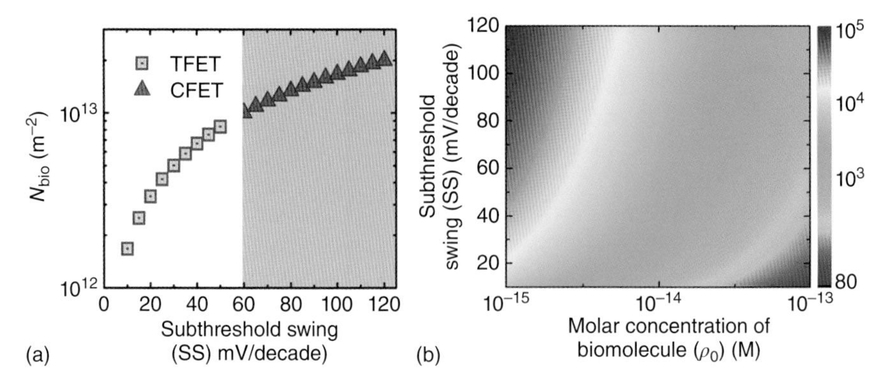

Figure 6.15 (a) Surface density of biomolecules (N_{bio}) required to be attached to the sensor surface for both CFET and TFET biosensors in order to achieve the same sensitivity value in both, as a function of subthreshold swing. It is observed that N_{bio} decreases significantly with decrease in the subthreshold swing. (b) 2D color map showing the response time (in seconds) of the biosensor as a function of the subthreshold swing and the molar concentration of the biomolecules in the solution.

screening length, ε_w is the dielectric constant of the electrolyte, RNW is the radius of the 1D structure, and K_0 and K_1 are the zero- and first-order modified Bessel functions of the second kind. It is clear that N_{bio} decreases with decreasing values of the swing (Figure 6.15a). This can be easily explained by the fact that the better the response of the sensor to the gating effect, the lower would be the required change in oxide surface potential (ϕ_{bio}) and hence in the required N_{bio} for achieving the same sensitivity. The response time (t_r) can be related to N_{bio} [52]. Now, using Eq. (6.38), the dependence of response time to the SS is derived as

$$t_r = \frac{\pi\varepsilon_w R_{NW}{}^2 K_1(R_{NW}/\lambda_{DH})\log_{10}(S_n+1)}{\lambda_{DH}K_0(R_{NW}/\lambda_{DH})DN_{avo}} \times \frac{SS_{avg}}{\rho_0} \tag{6.39}$$

where ρ_0 is the concentration of biomolecules, N_{avo} denotes the Avagadro number, and D is the diffusion coefficient of the biomolecules in the solution. Since the CFETs are plagued by the Boltzmann tyranny effect, there exists fundamental limitations to the minimum response time that can be obtained from biosensors based on them. This lower limit in CFETs can be derived using Eq. (6.39) as

$$t_{r-\min} = \frac{\pi\varepsilon_w R_{NW}{}^2 K_1(R_{NW}/\lambda_{DH})\ln(S_n+1)K_B T}{\lambda_{DH}K_0(R_{NW}/\lambda_{DH})DN_{avo}\rho_0} \tag{6.40}$$

In Figure 6.15b, the response time is plotted as a function of both the SS and the biomolecule concentration in the electrolyte. Since TFET biosensors are not bound by a lower limit on the SS, they can be highly advantageous for reduction of response time (up to more than an order of magnitude) and detection of biomolecules at low concentrations.

It is to be noted that here we have presented the results for n-TFET assuming a positive charge of the biomolecules. In general, the biomolecules such as DNA possess a negative charge. However, this sign change does not affect the general discussion and results presented here.

References

1 Gopalakrishnan, K., Griffin, P.B., and Plummer, J.D. (2002). Novel Semiconductor Device with a Subthreshold Slope lower than kT/q. IEEE International Electron Devices Meeting, pp. 289–292.
2 Fritschi, R., Boucart, K., Casset, F. et al. (2005). Suspended-gate MOSFET: bringing new MEMS functionality into solid-state MOS transistor. *IEEE International Electron Devices Meeting*, pp. 479–481.
3 Salahuddin, S. and Datta, S. (2008). Use of negative capacitance to provide voltage amplification for low power nanoscale devices. *Nano Lett.* 8 (2): 405–410.
4 Baba, T. (1992). Proposal for surface tunnel transistors. *Jpn. J. Appl. Phys.* 31 (2): 455–457.
5 Dadgour, H.F. and Banerjee, K. (2009). Hybrid NEMS–CMOS integrated circuits: a novel strategy for energy-efficient designs. *IET Comput. Digit. Tech.* 3 (6): 593.
6 Quinn, J.J. and Kawamoto, G. (1978). Subband spectroscopy by surface channel tunneling. *Surf. Sci.* 73: 190–196.
7 Bhuwalka, K.K., Sedlmaier, S., Ludsteck, A.K. et al. (2004). Vertical tunnel field-effect transistor. *IEEE Trans. Electron Devices* 51 (2): 279–282.
8 Zhang, Q., Zhao, W., Member, S., and Seabaugh, A. (2006). Low-subthreshold-swing tunnel transistors. *IEEE Electron Device Lett.* 27 (4): 297–300.
9 Khatami, Y. and Banerjee, K. (2009). Steep subthreshold slope n- and p-type tunnel-FET devices for low-power and energy-efficient digital circuits. *IEEE Trans. Electron Devices* 56 (11): 2752–2761.
10 Ionescu, A.M. and Riel, H. (2011). Tunnel field-effect transistors as energy-efficient electronic switches. *Nature* 479 (7373): 329–337.
11 Datta, S., Liu, H., and Narayanan, V. (2014). Tunnel FET technology: a reliability perspective. *Microelectron. Reliab.* 54 (5): 861–874.
12 Cao, W., Jiang, J., Kang, J. et al. (2015). Designing band-to-band tunneling field-effect transistors with 2D semiconductors for next-generation low-power VLSI. *IEEE International Electron Devices Meeting*, pp. 305–308.
13 Cao, W., Sarkar, D., Khatami, Y. et al. (2014). Subthreshold-swing physics of tunnel field-effect transistors. *AIP Adv.* 4 (6): 067141-1–067141-9.
14 Sarkar, D., Xie, X., Liu, W. et al. (2015). A subthermionic tunnel field-effect transistor with an atomically thin channel. *Nature* 526 (7571): 91–95.
15 Sarkar, D., Krall, M., and Banerjee, K. (2010). Electron-hole duality during band-to-band tunneling process in graphene-nanoribbon tunnel-field-effect-transistors. *Appl. Phys. Lett.* 97 (26): 263109-1–263109-3.
16 Solomon, P.M. (2010). Inability of single carrier tunneling barriers to give subthermal subthreshold swings in MOSFETs. *IEEE Electron Device Lett.* 31 (6): 618–620.
17 Sarkar, D. and Banerjee, K. (2012). Proposal for tunnel-field-effect-transistor as ultra-sensitive and label-free biosensors. *Appl. Phys. Lett.* 100 (14): 143108-1–143108-4.

18 Yan, R., Ourmazd, A., and Lee, K.F. (1992). Scaling the Si MOSFET : from bulk to SO1 to bulk. *IEEE Trans. Electron Devices* 39 (7): 704–710.

19 Moll, J.L. (1964). *Physics of Semiconductors*, 249. New York: McGraw-Hill.

20 Cao, W., Yao, C.J., Jiao, G.F. et al. (2011). Improvement in reliability of tunneling field- effect transistor with p-n-i-n structure improvement in reliability of tunneling field-effect transistor with p-n-i-n structure. *IEEE Trans. Electron Devices* 58 (7): 2122–2126.

21 Tomioka, K., Yoshimura, M., and Fukui, T. (2012). Steep-slope tunnel field-effect transistors using III-V nanowire / Si heterojunction. *Symposium on VLSI Technology* (2010): 47–48.

22 Gandhi, R., Chen, Z., Singh, N. et al. (2011). Vertical Si-nanowire n -type tunneling FETs with low subthreshold swing (≤ 50 mV/decade) at room temperature. *IEEE Electron Device Lett.* 32 (4): 437–439.

23 Gandhi, R., Chen, Z., Singh, N. et al. (2011). CMOS-compatible vertical-Silicon-nanowire gate-all-around p-type tunneling FETs with ≤ 50 -mV / decade subthreshold swing. *IEEE Electron Device Lett.* 32 (11): 1504–1506.

24 Jeon, K., Loh, W., Patel, P. et al. (2010). Si tunnel transistors with a novel silicided source and 46mV / dec swing. *Symposium on VLSI Technology* 1: 2009, 2009–2010.

25 Kim, S.H., Kam, H., Hu, C., and Liu, T.K. (2009). Germanium-source tunnel field effect transistors with record high I ON/I OFF. *Symposium on VLSI Technology* 178–179.

26 Mayer, F., Le Royer, C., Damlencourt, J.-F. et al. (2008). Impact of SOI, $Si1_{-x}Ge_{x}OI$ and GeOI substrates on CMOS compatible tunnel FET performance. *2008 IEEE International Electron Devices Meeting* 4: 1–5.

27 Choi, W.Y., Park, B., Lee, J.D., and Liu, T.K. (2007). Tunneling field-effect transistors (TFETs) with subthreshold swing (SS) less than 60 mV/dec. *IEEE Electron Device Lett.* 28 (8): 743–745.

28 Appenzeller, J., Lin, Y.-M., Knoch, J., and Avouris, P. (2004). Band-to-band tunneling in carbon nanotube field-effect transistors. *Phys. Rev. Lett.* 93 (19): 196805.

29 Knoll, L., Zhao, Q., Nichau, A. et al. (2013). Inverters with strained Si nanowire complementary tunnel field-effect transistors. *IEEE Electron Device Lett.* 34 (6): 813–815.

30 Krishnamohan, T., Kim, D., Raghunathan, S., and Saraswat, K. (2008). Double-gate strained-Ge heterostructure Tunneling FET (TFET) with record high drive currents and <60mV/dec subthreshold slope. *IEEE International Electron Devices Meeting* 67 (2006): 7–9.

31 Huang, Q., Huang, R., Wu, C. et al. (2014). Comprehensive performance reassessment of TFETs with a novel design by gate and source engineering from device / circuit perspective. *IEEE International Electron Devices Meeting* 335–338.

32 Villalon, A., Royer, C.Le., Cassé, M. et al. (2012). Strained tunnel FETs with record I ON : first demonstration of ETSOI TFETs with SiGe channel and RSD. *Symposium on VLSI Technology* 44510 (2011): 2011–2012.

33 Ganjipour, B., Wallentin, J., Borgstro, M.T. et al. (2012). Tunnel field-effect transistors based on InP-GaAs heterostructure nanowires. *ACS Nano* 6 (4): 3109–3113.

34 Kim, M., Wakabayashi, Y., Nakane, R. et al. (2014). High I on/I off Ge-source ultrathin body strained-SOI unnel FETs - impact of channel strain, MOS interfaces and back gate on the electrical properties. *IEEE International Electron Devices Meeting* 331–334.

35 Dewey, G., Chu-Kung, B., Boardman, J. et al. (2011). Fabrication, characterization, and physics of III–V heterojunction tunneling field effect transistors (H-TFET) for steep sub-threshold swing. *Int. Electron Devices Meet.* 3: 33.6.1–33.6.4.

36 Mak, K.F., Lee, C., Hone, J. et al. (2010). Atomically thin MoS_2: a new direct-gap semiconductor. *Phys. Rev. Lett.* 105 (13): 136805.

37 Liu, W., Kang, J., Cao, W. et al. (2013). High-performance few-layer-MoS_2 field-effect-transistor with record low contact-resistance. *IEEE International Electron Devices Meeting* 499–502.

38 Sarkar, D., Gossner, H., Hansch, W., and Banerjee, K. (2013). Tunnel-field-effect-transistor based gas-sensor: introducing gas detection with a quantum-mechanical transducer. *Appl. Phys. Lett.* 102 (2).

39 Srinivas, P.R., Kramer, B.S., and Srivastava, S. (2001). Trends in biomarker research for cancer detection. *Lancet Oncol.* 2: 696–704.

40 Galasko, D. (2005). Biomarkers for Alzheimer's disease – clinical needs and application. *J. Alzheimer's Dis.* 8: 339–346.

41 de Jong, D., Kremer, B.P.H., Olde Rikkert, M.G.M., and Verbeek, M.M. (2007). Current state and future directions of neurochemical biomarkers for Alzheimer's disease. *Clin. Chem. Lab. Med.* 45 (11): 1421–1434.

42 Barletta, J.M., Edelman, D.C., and Constantine, N.T. (2004). Lowering the detection limits of HIV-1 viral load using real-time immuno-PCR for HIV-1 p24 antigen. *Am. J. Clin. Pathol.* 122 (1): 20–27.

43 Bergveld, P. (1970). Development of an ion-sensitive solid-state device for neurophysiological measurements. *IEEE Trans. Biomed. Eng.* 17: 70–71.

44 Caras, S. and Janata, J. (1980). Field effect transistor sensitive to penicillin. *Anal. Chem.* 52: 1935–1937.

45 Souteyrand, E., Cloarec, J.P., Martin, J.R. et al. (1997). Direct detection of the hybridization of synthetic homo-oligomer DNA sequences by field effect. *J. Phys. Chem. B* 101: 2980–2985.

46 Cui, Y., Wei, Q., Park, H., and Lieber, C.M. (2001). Nanowire nanosensors for highly sensitive and selective detection of biological and chemical species. *Science* 293: 1289–1292.

47 Sarkar, D., Liu, W., Xie, X. et al. (2014). MoS_2 field-effect transistor for next-generation label- free biosensors. *ACS Nano* 8 (4): 3992–4003.

48 Sarkar, D. and Banerjee, K. (2012). Fundamental limitations of conventional-FET biosensors: quantum-mechanical-tunneling to the rescue. Device Research Conference - Conference Digest, DRC, 2012, pp. 83–84.

49 Nair, P.R. and a Alam, M. (2008). Screening-limited response of nanobiosensors. *Nano Lett.* 8 (5): 1281–1285.

50 McQuarrie, D. (1976). *Statistical Mechanics*. New York: Harper & Row.

51 a Gao, X.P., Zheng, G., and Lieber, C.M. (2010). Subthreshold regime has the optimal sensitivity for nanowire FET biosensors. *Nano Lett.* 10 (2): 547–552.

52 Nair, P.R. and Alam, M.a. (2006). Performance limits of nanobiosensors. *Appl. Phys. Lett.* 88 (23): 233120.

7

Energy-Efficient Computing with Negative Capacitance

Asif I. Khan

Georgia Institute of Technology, School of Electrical and Computer Engineering, Atlanta, GA 30309, USA

7.1 Introduction

Our appetite for generating and consuming digital data is going to only increase with time. "How much information is there stored by the mankind and what is the digital computation capability of our civilization?" – the answer to such questions yielded staggering estimations such as 2.9×10^{20} optimally compressed bits and 6.4×10^{18} instructions per second, respectively, for the year 2007 [1]. Close to 2.5 billion people–a number that has grown 566% since the year 2000–are online around the globe and nearly 70% of them use the Internet every day [2]. Every 60 seconds, 204 million emails are exchanged, 5 million searches are made on Google, 1.8 million "Like"s are generated on Facebook, 350,000 tweets are made on Twitter, $272 000 of merchandise is sold on Amazon, and 15,000 tracks are downloaded in iTunes [2]. With the "Internet of Things," (IoT) where intelligence in the form of tiny stand-alone devices and sensors numbering in trillions will be seamlessly integrated into our environment. Within the upcoming paradigm of the "Internet of Things," the number of data generating devices connected to the Internet is expected to grow to 20 billion by 2020 [3].

In other words, the explosive growth of the information paradigm over the past several decades has followed the path of accelerating returns–thanks to the innovation in all levels of the computing hierarchy starting from the materials and devices all the way to circuits, systems, algorithms, and softwares. What underlies the information ecosystem is the massive data centers and cloud computing infrastructures in tandem with personal computing and IoT devices.

All told, we are now faced with a serious challenge in dealing with energy in computing. Power dissipation and management issues in the information infrastructure threatens the cadence of the revolutions in the coming years. On a macro-scale, the 12 million servers in 3 million data centers that drive all the online activities in the United States consume 76 TW-hr of energy per year [2]. The total electric energy consumption of the United States in 2011 was 4113 TW-hr [4]; data centers constitute ~2% of the U.S. electricity consumption. If the current exponential growth pattern of information paradigms continues in the future, the power consumption will reach an unmanageable level in the near

Advanced Nanoelectronics: Post-Silicon Materials and Devices,
First Edition. Edited by Muhammad Mustafa Hussain.

future [5]. Along with the data center, the overall energy usage of the information infrastructure could reach one-third of the total U.S. energy consumption by the year 2025 [6, 7]. The carbon footprint of all the data centers worldwide is already equivalent to that of a mid-sized country such as Malaysia or the Netherlands [5].

So why cannot we make the computing hardware more efficient? The answer lies in how complementary metal-oxide-semiconductor (CMOS) technology and its downscaling over the years have progressed. The downscaling of CMOS transistors over the past four decades has been the core driver for the information revolution. The transistor density has approximately doubled every 18 months in accordance with the Moore's law [8]. However, as we are scaling down our transistors, we could not make them as energy efficient as we wanted them to be–the power supply voltage has been stuck at around 1 V for almost 15 years. The increase of microprocessor operating frequency resulted in the power density reaching 100 W cm^{-2} level, which is an order of magnitude higher than a typical hot plate. Increase of the power density beyond that level is unmanageable; as a result, the clock frequency stopped increasing beyond 3 GHz after 2005 [9]. In fact, it has been predicted that at sub-10-nm nodes, more than 50% of a fixed-size microprocessor chip must be powered off at a given time due to power management challenges–the age of the so-called Dark Silicon has already begun [10].

The reason why the power supply voltage in microprocessors has not scaled at par with the transistor dimensions originates from the fundamental physics of transistor operation. In a transistor, the carriers in the semiconductor channel follow the Boltzmann distribution. The Boltzmann distribution dictates that, to increase the drain current by an order of magnitude at room temperature, the gate voltage needs to be increased by at least $k_B T \log 10 = 60$ mV, k_B and T being the Boltzmann constant and the temperature, respectively. In other words, how sharply a transistor can be turned on has a lower limit of 60 mV/decade. To maintain a good on–off ratio of the current, ~1 V needs to be applied at the gate. This limitation is a fundamental physical one. However much engineering is put into a transistor design, the subthreshold slope cannot be lowered below 60 mV/decade. It is now generally agreed that, without introducing new physics into the transistor operation, this limitation cannot be overcome. As a result, there have been calls for fundamental innovations in transistors such that they operate on the basis of new principles [11–14].

Negative capacitance field-effect transistors (NCFETs)–first proposed by Salahuddin and Datta [15]–belong to a new class of devices that can overcome the Boltzmann barrier in the CMOS technology. It can do so by introducing a novel physical mechanism called the negative capacitance effect into the transistor operation. The device structure of an NCFET is very similar to that of a CMOS transistor (be it a planar bulk or silicon-on-insulator (SOI) FET or a Fin field-effect transistor (FinFET) or a nanowire FET, for example)–the only difference is that the gate oxide stack in an NCFET also includes an insulator that exhibits a differential negative capacitance. Over the past 10 years, activities in NCFETs have gained deep roots in materials science and physics of negative capacitance, device engineering, modeling, and simulation and large-scale circuit and system performance analysis. In this chapter, we discuss a selection of these activities. In explaining the operation of an NCFET, the two most important questions that need to be addressed first are (i) how a negative capacitance gate oxide reduces the subthreshold swing and (ii) how mechanisms in a certain class of materials called ferroelectrics can result in an effective differential negative capacitance. We start by elaborating these two points in the next two sections.

7.2 How a Negative Capacitance Gate Oxide Leads to Sub-60- Millivolt/Decade Switching

The key to the sub-60-mV/decade switching characteristics in an NCFET is the passive amplification of the gate voltage at the oxide–semiconductor interface through the negative capacitance effects of the gate oxide stack. To illustrate the operation of an NCFET, we resort to a simplistic description of MOSFET physics. A MOSFET (Figure 7.1a), can be represented by an equivalent capacitor network as shown in Figure 7.1b, where C_{ox}, C_S, C_D, and C_{Si} are the gate oxide, source-to-channel, drain-to-channel, and the semiconductor channel capacitance, respectively. The gate voltage V_G and the electrostatic potential at the oxide–semiconductor interface (i.e. surface potential) ψ_S are related by the following equation.

$$\frac{\partial \psi_s}{\partial V_G} = \frac{C_{ox}}{C_{ox} + C_{eq}} \tag{7.1}$$

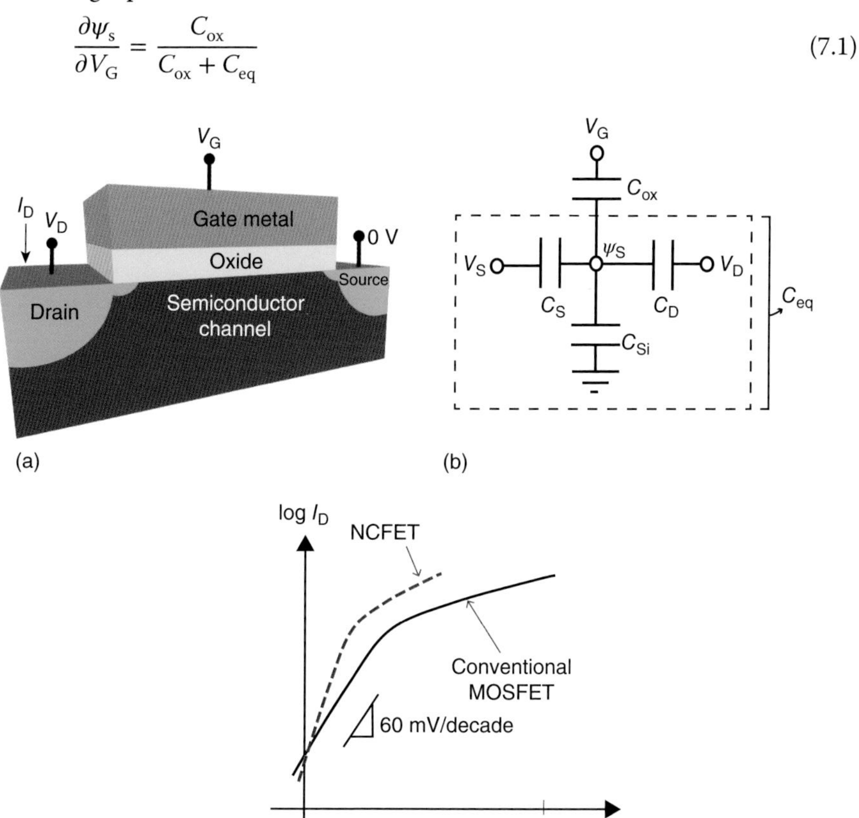

Figure 7.1 Schematic diagram of a (a) metal-oxide-semiconductor field-effect-transistor (MOSFET). (b) The equivalent circuit diagram of a MOSFET. (c) The output characteristics of a MOSFET.

Here, $C_{eq} = C_S + C_D + C_{Si}$. In a conventional MOSFET, all the capacitances are positive for which $\partial\psi_s/\partial V_G < 1$. At the ultimate limit where the oxide thickness t_{ox} is zero (i.e. $C_{ox} \to \infty$), $\partial\psi_s/\partial V_G = 1$. Now let us consider the case when $C_{ox} < 0$ and $|C_{ox}| > C_{eq}$. In this case, $\partial\psi_s/\partial V_G > 1$, which means that a smaller change in V_G can lead to a larger change in ψ_s. In other words, the gate voltage is differentially amplified at the oxide–semiconductor interface.

The carrier distribution in a semiconductor follows the Boltzmann statistics. This leads to the carrier density in the channel being an exponential function of the surface potential–i.e. the carrier density in the semiconductor $Q \propto \exp(q\psi_S/k_B T)$. Hence, to affect an order of magnitude change in the carrier density, the surface potential ψ_s needs to change by $k_B T \log 10 \approx 60$ mV. The drain current is directly proportional to the channel charge, and hence has the same dependence on ψ_s. Since the change in ψ_s is always smaller than that in V_G in a conventional MOSFET, larger than 60-mV change in V_G is required to the increase the drain current by 10 times in such a case. The advantage of differential amplification in an NCFET is evident right away. Less than 60-mV change in the gate voltage can lead to a 60-mV change in the surface potential, and one order of magnitude in drain current. In other words, the same amount of change in the gate voltage can create a larger change in the drain current in an NCFET than fundamentally possible in a conventional MOSFET.

Mathematically, subthreshold swing S is defined as follows.

$$S = \frac{\partial V_G}{\partial \log_{10} I_D} = \left(\frac{\partial V_G}{\partial \psi_s}\right)\left(\frac{\partial \psi_s}{\partial \log_{10} I_D}\right) \tag{7.2}$$

The second term in Eq. (7.2) in an NCFET is equal to 60 mV/decade as it is in a conventional MOSFET. The negative capacitance effect alters the first term in Eq. (7.2), which is also called the body factor m. The electrostatic amplification in an NCFET results in a body factor smaller than one leading to $S < 60$ mV/decade, whereas, in a conventional MOSFET, $m > 1$ resulting in $S > 60$ mV/decade.

It is important to note that, in an NCFET, the mode of carrier transport between the source and drain through the channel is not altered–rather, the electrostatics is altered keeping the transport mechanism the same as in a conventional MOSFET. This is in contrast with other proposals for sub-60-mV/decade transistors such as TFET [16, 17], impact ionization metal-oxide-semiconductor field-effect-transistors (IMOSFETs) [18], and nanoelectromechanical field-effect transistors (NEMFETs) [19, 20], where the mode of transport is changed.

7.3 How a Ferroelectric Material Acts as a Negative Capacitor

A capacitor is a device that stores charge. Capacitance of a device C is defined as the rate of increase of the charge Q with the voltage V ($C = \mathrm{d}Q/\mathrm{d}V$). Hence, by definition, for a negative capacitor, Q decreases as V is increased (see Figure 7.2). Alternatively, capacitance can also be defined in terms of the free energy U. For a

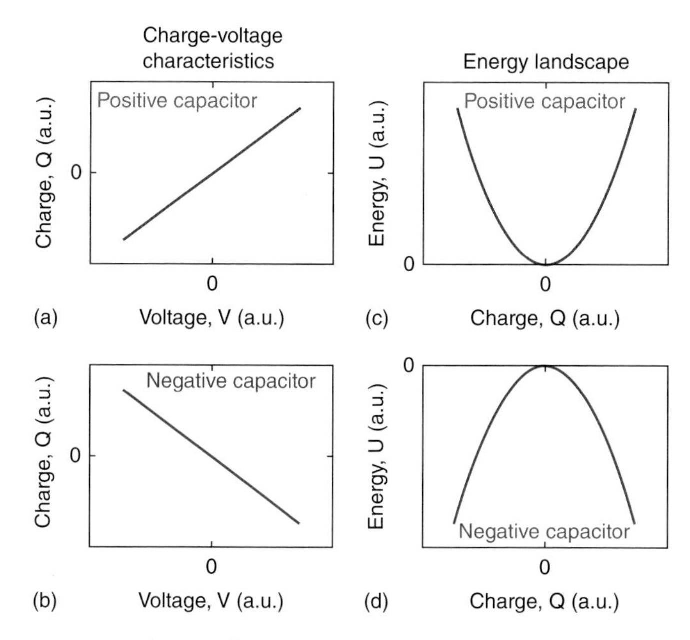

Figure 7.2 Charge-voltage characteristics and energy landscapes of (a,c) a positive capacitance and (b,d) a negative capacitor.

negative capacitor, the energy landscape is an inverted parabola (see Figure 7.2). For a linear capacitor, $U = Q^2/2C$. In terms of free energy, the capacitance can be defined as follows:

$$C = \left[\frac{d^2U}{dQ^2}\right]^{-1} \tag{7.3}$$

The same relation holds also for a nonlinear capacitor. In other words, the negative curvature region in the energy landscape of an insulating material corresponds to a negative capacitance.

To understand how a ferroelectric material acts a negative capacitor, we plot its generic energy landscape in Figure 7.3a, which has a double-well shape. A ferroelectric is an insulating material with two or more discrete stable or metastable states of different nonzero electric polarization in zero applied electric field, referred to as "spontaneous" polarization. In this chapter, we use the terms polarization and charge interchangeably for the sake of clarity. The stable polarization states in the absence of an applied electric field are noted by the points A and B in Figure 7.3a. For a material to be considered ferroelectric, it must be possible to switch between these states with an applied electric larger than the coercive field, which changes the relative energy of the states through the coupling between the electric field and the polarization. As such, there are hysteretic jumps in the charge-voltage characteristic of a ferroelectric material, as shown in Figure 7.3b. If we compare the characteristic ferroelectric energy landscape (Figure 7.3a) with that of an ordinary capacitor shown in Figure 7.2a, we see that, at the stable polarization states A and B, the ferroelectric energy

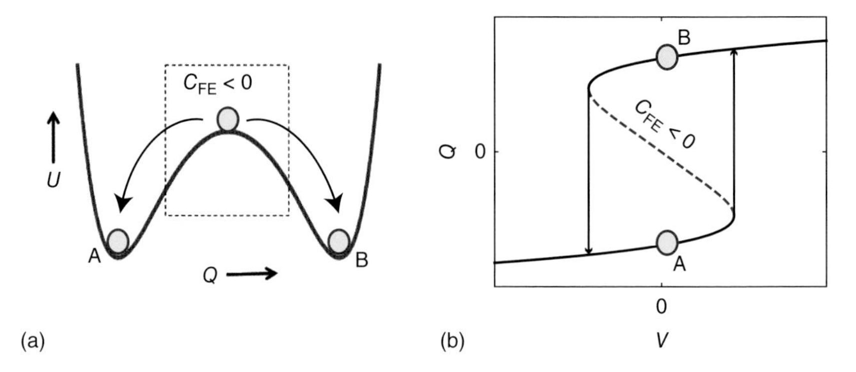

An LCR meter cannot directly measure a ferroelectric negative capacitance

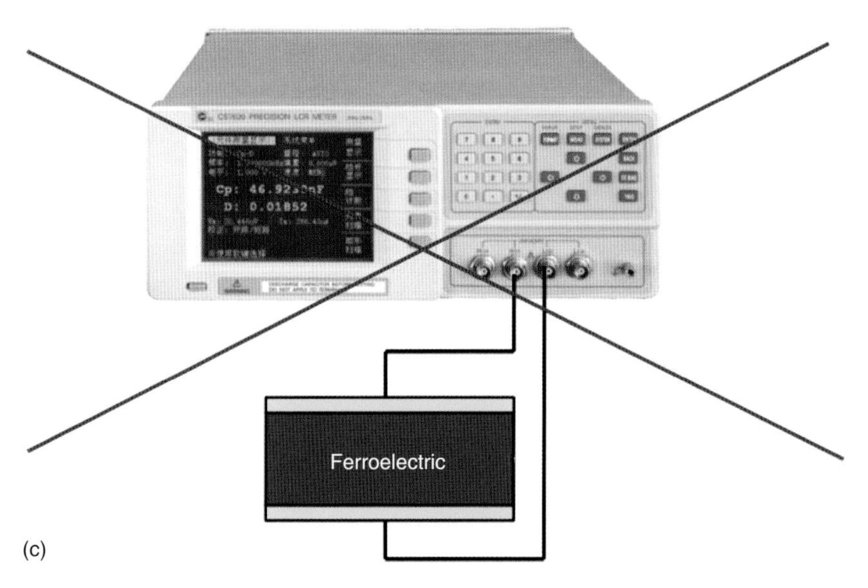

Figure 7.3 (a) Energy landscape of a ferroelectric material. The region under the dashed box corresponds to the negative capacitance state. Negative capacitance state is unstable and the polarization spontaneously rolls downhill from a negative capacitance state to one of the minima making a direct measurement of the phenomenon experimentally difficult. (b) Charge (or polarization)-voltage characteristics of a ferroelectric material, according to the Landau theory. The capacitance is negative in a certain region of charge and voltage, which is indicated by the dotted line. (c) An LCR meter cannot directly measure a ferroelectric negative capacitance.

landscape has a positive curvature, pointing to the fact that the capacitance of the ferroelectric is positive in these states. On the other hand, the curvature around $Q = 0$ of a ferroelectric is just the opposite of that of an ordinary capacitor. Remembering that the energy of an ordinary capacitor is given by $(Q^2/2C)$, this opposite curvature hints at a negative capacitance for the ferroelectric material around $Q = 0$. Therefore, around this point, a ferroelectric material could provide a negative capacitance.

The formal definition of negative capacitance in ferroelectric materials originates from the Landau–Devonshire theory. According to this theory, the free energy U of a ferroelectric material can be represented as an even-order polynomial of the polarization Q, which is as follows.

$$U = \alpha Q^2 + \beta Q^4 + \gamma Q^6 - EQ \tag{7.4}$$

Here, $E = V/d$ is the applied electric field; V and d are the voltage applied across the ferroelectric and the ferroelectric thickness, respectively. α, β, and γ are anisotropy constants. β and γ are temperature independent. γ is a positive quantity; β is positive and negative, respectively, for second-order and first-order phase transition. $\alpha = a_\circ(T - T_C)$, where $a_\circ$ is a temperature-independent positive quantity and T and T_C are the temperature and the Curie temperature, respectively. As a result, $\alpha < 0$ below the Curie temperature which results in the negative curvature of the energy landscape of a ferroelectric around $Q = 0$ and hence the double-well energy landscape.

Combining Eqs. (7.3) and (7.4), the following equation for capacitance around $Q = \sim 0$ and at $T < T_C$ is obtained.

$$C = \frac{1}{2\alpha} = \frac{1}{2a_\circ(T - T_C)} < 0 \tag{7.5}$$

Furthermore, at equilibrium, $\mathrm{d}U/\mathrm{d}Q = 0$, which, combined with Eq. (7.4), results in the following relation.

$$E = 2\alpha Q + 4\beta Q^3 + 6\gamma Q^5 \tag{7.6}$$

Figure 7.3b shows the polarization-voltage characteristics of a ferroelectric capacitor obtained using Eq. (7.6). We note in Figure 7.3b that, in accordance with the Landau theory of ferroelectrics, a ferroelectric capacitor has a nonlinear charge-voltage characteristic in which a negative capacitance can be obtained in a certain range of charge and voltage indicated by the red dashed curve.

Ferroelectricity is an established discipline in physics and materials science, with its origin back in the 1920s. The Landau theory of ferroelectricity has also been researched actively since the 1930s. And, ferroelectricity is a very active field of research with a publication rate of the order of 10,000 per year. Hence, it begs the question "why has the negative capacitance phenomenon never been explicitly observed in ferroelectric materials until now?" One of the reasons is the unstable nature of the negative capacitance in a ferroelectric capacitor. If the polarization is placed in the unstable region of the energy landscape, as shown in Figure 7.3a, the ferroelectric capacitor spontaneously self-charges and the polarization rolls downhill to one of the minima. This is also why, in the conventional experimental measurement of polarization-voltage characteristics where voltage is the control variable (see Figure 7.3b), the negative capacitance region is masked by a hysteresis region and sharp transitions between the two polarization states occur. As a result, a negative capacitance cannot be directly measured by connecting a ferroelectric capacitor to an LCR meter as shown in Figure 7.3c. It requires specialized experimental setup to directly measure the negative capacitance in a ferroelectric capacitor, which is the topic of the next section.

7.4 Direct Measurement of Negative Capacitance in Ferroelectric

To understand how the unstable negative capacitance states in a ferroelectric capacitor can be accessed, we resort to the energy landscape picture of its switching dynamics [21]. Figure 7.4 shows the evolution of the energy landscape with different applied voltages. Starting from an initial state P, as a voltage is applied across the ferroelectric capacitor, the energy landscape is tilted and the polarization will move to the nearest local minimum. Figure 7.4b shows this transition for a voltage which is smaller than the coercive voltage V_c. If the voltage is larger than V_c, one of the minima disappears and Q_F moves to the remaining minimum of the energy landscape (Figure 7.4c). Notably as the polarization rolls downhill in Figure 3.1c, it passes through the region where $C = [d^2U/dQ_F^2]^{-1} < 0$. Therefore, while switching from one stable polarization to the other, a ferroelectric material crosses through a region where the differential capacitance is negative.

A negative C_{FE} could be directly measured during the transit of Q_F through the barrier of the energy landscape by constructing an R – FE circuit as shown in Figure 7.5a,b. An epitaxially grown 60-nm $Pb(Zr_{0.2}Ti_{0.8})O_3$ (PZT) with metallic $SrRuO_3$ bottom and circular Au top electrode (area A=π(15 μm)^2) is used as the ferroelectric in the experiments. We note that if a step voltage pulse V_S is applied in a classical analog to the circuit shown in Figure 7.5b–a

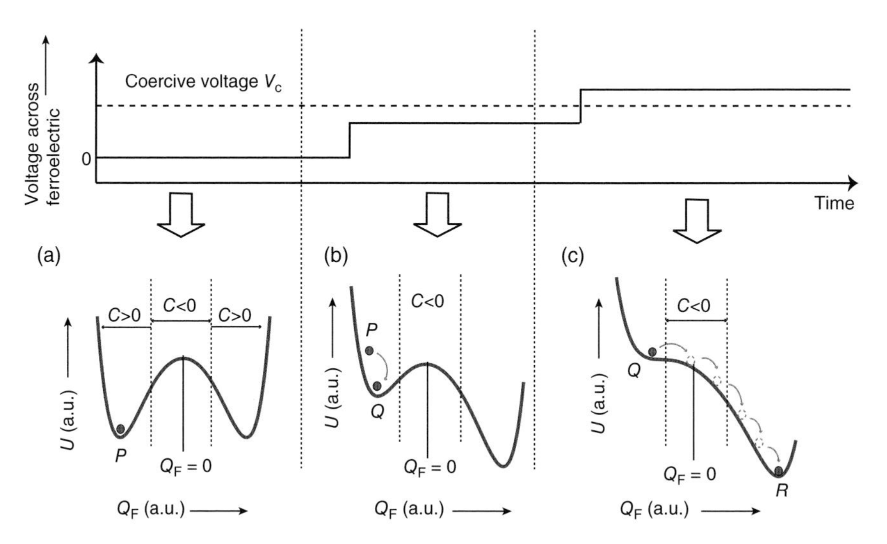

Figure 7.4 Energy landscape description of ferroelectric negative capacitance. (a) The energy landscape U of a ferroelectric capacitor in the absence of an applied voltage. The capacitance C is negative only in the barrier region around charge $Q_F = 0$. (b,c) The evolution of the energy landscape upon the application of a voltage across the ferroelectric capacitor that is smaller (b) and larger than the coercive voltage V_c (c). If the voltage is larger than the coercive voltage, the ferroelectric polarization rolls downhill through the negative capacitance states. P, Q, and R represent different polarization states in the energy landscape.

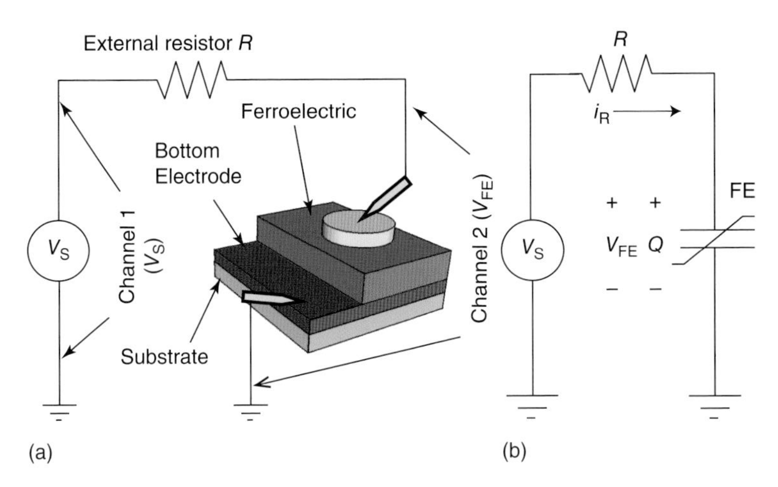

Figure 7.5 (a) Schematic diagram of the experimental setup. A 60-nm-thick ferroelectric $Pb(Zr_{0.2}Ti_{0.8})O_3$ with an effective top electrode diameter of 30 μm is connected to a voltage source V_s through an external resistor R. The oscilloscope connections are also shown. (b) Equivalent circuit diagram of the R – FE circuit.

regular $R - C$ circuit–the capacitor voltage V_c and the charge Q would increase with a time constant RC given by the relations: $V_c(t) = V_s(1 - e^{-t/RC})$ and $Q(t) = CV_s(1 - e^{-t/RC})$. As a result, at any time t during the transients, $dQ(t)/dV_c(t)$ would turn out to be a positive number equal to the capacitance of the capacitor. Figure 7.6a shows the transients corresponding to the ferroelectric voltage V_{FE}, the current i_R, and the charge $Q = \int i_R(t)dt$ in the R – FE circuit ($R = 50\ k\Omega$) upon the application of a step voltage pulse V_S: $-4.5\ V \rightarrow +4.5\ V \rightarrow -4.5\ V$. In Figure 7.6a, we observe that during the initial period OA, both V_{FE} and Q increase; however, during the subsequent period AB, V_{FE} decreases while Q increases. As a result, during AB, $C_{FE} = dQ(t)/V_{FE}(t) < 0$. The behavior of the ferroelectric during AB in Figure 7.6a is nontrivial; this is exactly opposite to what happens in a regular $R - C$ circuit, where both Q and V_c monotonically increase or decrease in response to a step V_s pulse. This points to the fact that, starting from point Q in Figure 7.4c, the ferroelectric charge enters into the negative capacitance state of the energy landscape at time A. Afterwards during BC in Figure 7.6a, V_{FE} and Q again change in the same direction, which indicates that the ferroelectric has returned to a positive capacitance state ($C_{FE} > 0$) near the point R in Figure 7.4c at time B. Similarly, during the time segment CD in the negative voltage pulse cycle in Figure 7.6a, a negative $dQ(t)/V_{FE}(t)$ is observed. During CD, the ferroelectric polarization transits from the state R back to the state P. In Figure 7.6b, we plot the ferroelectric polarization $P(t) = Q(t)/A$ as a function of $V_{FE}(t)$. The $P(t) - V_{FE}(t)$ curve clearly shows that, in regions corresponding to AB and CD periods, the slope is negative.

The time-dependent technique has recently been utilized to understand the negative capacitance phenomena in HfO_2-based ferroelectrics [22, 23] and organic ferroelectrics [24].

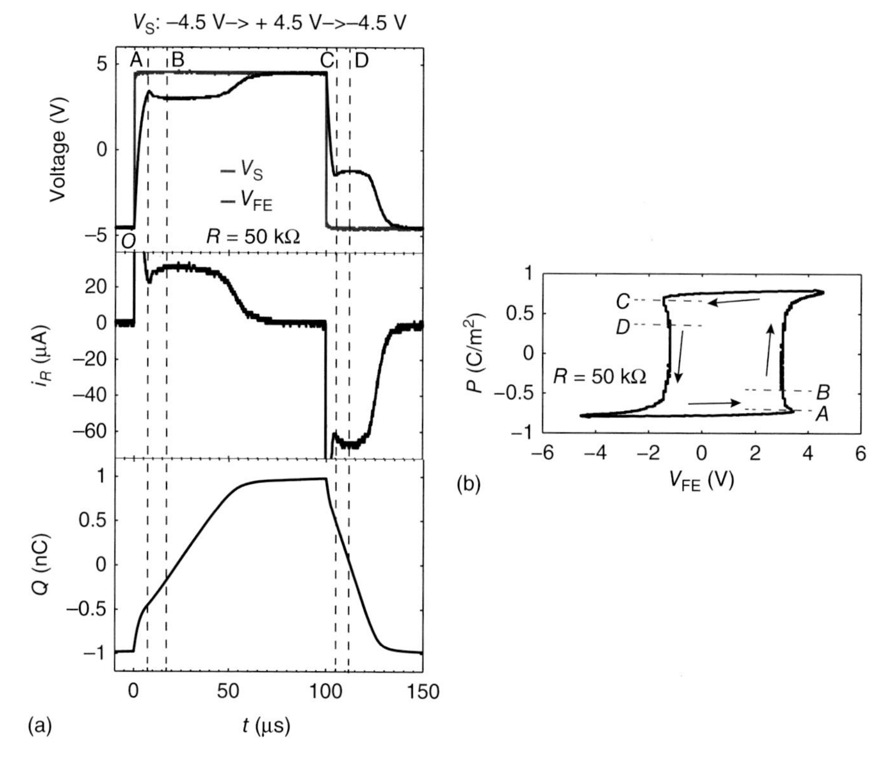

Figure 7.6 (a) Transients corresponding to the source voltage V_s, the ferroelectric voltage V_{FE}, the loop current i_R, and the ferroelectric charge Q in response to a step voltage pulse V_s:−4.5 V→+4.5 V → −4.5 V. R=50 kΩ. (b) Ferroelectric polarization $P(t)$ as a function of $V_{FE}(t)$ plotted using the results shown in Panel (a). In sections AB and CD, the slope of the $P(t) - V_{FE}(t)$ curve is negative, indicating a negative capacitance in these regions.

7.5 Properties of Negative Capacitance FETs: Modeling and Simulation

A significant amount of work has been reported on the modeling and simulation of NCFETs [25–41]. The simulation framework of NCFETs, as proposed first in [25], is based on a simple capacitance-based equivalent device model as shown in Figure 7.7a. Here, C_{FE} represents the ferroelectric capacitance as a nonlinear function of charge and voltage and is negative for a certain range of charge and voltage; and C_{MOS} is the equivalent capacitance of all the positive capacitances in the baseline device. When designed properly, a negative C_{FE} results in the voltage amplification at the oxide–semiconductor interface, leading to sub-60-mV/decade switching characteristics. The condition $|C_{FE}| \approx C_{MOS}$ and $|C_{FE}| > C_{MOS}$ results in the steepest turn-on characteristics without a hysteresis. On the other hand, $|C_{FE}| < C_{MOS}$ results in a hysteresis in the $I_D - V_G$ characteristics of NCFET. To the first order, C_{FE} is related to the ferroelectric parameters: remnant polarization P_o, coercive field E_C, and the thickness t_{FE} as follows [33].

$$C_{FE} \approx -\frac{2}{3\sqrt{3}}\frac{P_o}{E_c t_{FE}} \tag{7.7}$$

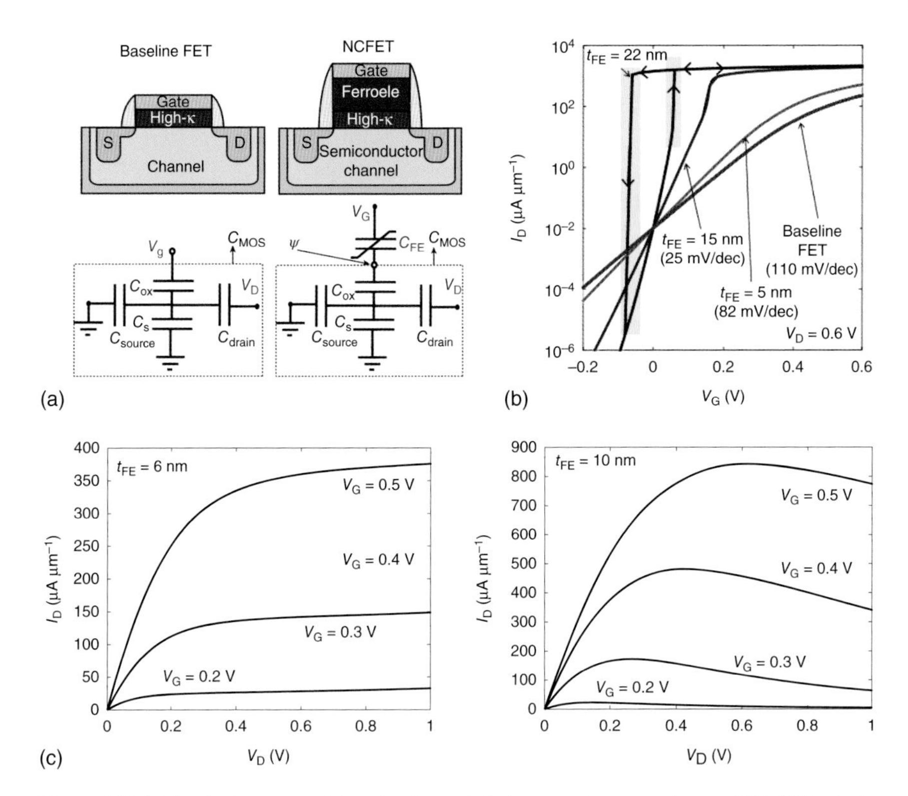

Figure 7.7 (a) Equivalent capacitive circuit model of a negative capacitance FET. (b) Evolution of $I_D - V_G$ characteristics of an NCFET with respect to the ferroelectric thickness t_{FE}. (c) Evolution of $I_D - V_D$ characteristics of an NCFET with respect to the ferroelectric thickness t_{FE}.

Figure 7.7 shows how the NCFET characteristics evolve as the thickness of the ferroelectric changed. The baseline FET is an Intel 45-nm n-type bulk FET, which is modeled using the MIT Virtual Source (MVS) compact device model [42]. P_o and E_c were assumed to be 800 kV cm^{-1} and 5 μC cm^{-2}. The baseline FET has a subthreshold slope of ~110 mV/decade. As the ferroelectric thickness is changed to 5 and 15 nm, the subthreshold slope improves to 82 mV/decade and ~25 mV/decade, respectively. Increasing t_{FE} decreases the magnitude of C_{FE} bringing it closer to C_{MOS}; $|C_{FE}|$ is still larger than C_{MOS} in this t_{FE} range. However, as t_{FE} is increased to 22 nm, the magnitude of C_{FE} actually becomes smaller than C_{MOS}, which results in a hysteresis in the corresponding $I_D - V_G$ curve.

Another interesting signature of the negative capacitance effects is the emergence of negative output differential resistance. For example, Figure 7.7c shows the $I_D - V_D$ characteristics of NCFETs simulated on the basis of the same framework (Intel 45 nm bulk FET and MIT MVS model) for two different ferroelectric thicknesses. As t_{FE} is increased from 6 to 10 nm, the output characteristics start exhibiting negative output differential resistance. Such a behavior can be understood by considering the circuit model shown in Figure 7.7c. Because of the drain-to-channel coupling capacitance C_{drain}, the surface potential ψ can change with the change of V_D. Due to C_{FE} being negative, it can be shown that under certain conditions $d\psi/dV_D$ can actually be negative. The negative output differential resistance in NCFETs was discussed in detail in Ref. [43]. This phenomenon is beneficial for short channel devices; this is because of the fact that negative capacitance behavior can be tuned through device design such that, in ultrashort channel lengths, the $I_D - V_D$ characteristics in the saturation regime can be made completely flat, leading to an infinite output resistance.

Often, an intermediate metallic layer is included between the ferroelectric layer and the interfacial oxide [25]. The purpose of this layer is to average out the nonuniform potential profile along the source-drain direction as well as any charge nonuniformity due to domain formation in the ferroelectric layer. Thus, as far as the MOSFET is concerned, the FE looks like a monodomain dipole. From an experimental point of view, this metallic layer provides interesting opportunities for understanding the physics of operation of NCFETs due to the access to the internal node voltage [44–51]. However, recent theoretical analysis shows that such an intermediate metallic layer can change the dynamics of the system such that the negative capacitance state in the ferroelectric cannot be stabilized [46, 52]. This imposes a timescale limitation on the NCFET operation–i.e. NCFETs with internal metallic layers need to operate faster than a characteristics timescale such that the voltage amplification and the associated steep switching and on-current boost can be obtained.

7.6 Experimental Demonstration of Negative Capacitance FETs

Over the past couple of years, there have been a significant number of experimental demonstrations on sub-60-mV/decade characteristics in NCFETs. The NCFETs varied in terms of choice of ferroelectric material (perovskite, organic polymer, or HfO_2-based ones), structures (bulk, SOI, FinFET), channel material (Si, Ge, 2-D), channel length (short channel to long channel). However, those experimental structures can be categorized in two broad types: ones with an intermediate metallic layer between the ferroelectric layer and the interfacial oxide (Figure 7.8b,c) and the ones without (Figure 7.8a). For NCFETs with an

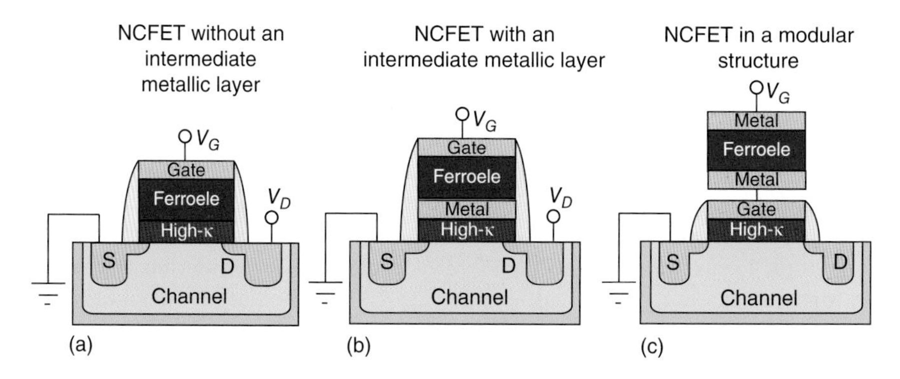

Figure 7.8 Different structures of negative capacitance FETs: (a,b) with and without an intermediate metallic layer and (c) the modular structure.

intermediate metallic layer, the ferroelectric layer can be fully integrated into the gate stack (Figure 7.8b) or it can be externally connected (Figure 7.8c). Note that the modular structure provides interesting opportunities for understanding the physics of operation of NCFETs, while the integrated structures are essential for wafer-scale integration of NCFETs.

The earliest reports of sub-60-mV/decade switching characteristics in a ferroelectric gated FET are due to Adrian Ionescu's group at EPFL [44, 45]. In 2008 at IEDM, they reported measurements of less than 60 mV/decade subthreshold swing in a long channel (L_G = 50 μm) bulk Si FET with 40-nm-thick organic polymer ferroelectric: poly[(vinylidenefluoride-co-trifluoroethylene] (P(VDF-TrFE))/10 nm SiO_2 used as the gate stack [44]. However, the reduction of subthreshold swing was observed mostly at sub-pA current levels where noise and leakage effects are dominant. Building on this work, the same group reported bulk Si FET with different gate lengths at IEDM in 2010 [45], where particular attention was paid to isolation techniques (such as optimized well and shallow trench isolation) to curb leakage from different sources. In one of their best devices, they reported a minimum subthreshold swing of 50 mV/decade with an average swing of 57 mV/decade over ~3 orders of magnitude of current (1 pA to 1 nA). However, a sub-60 mV/decade characteristic was observed in only one branch of the hysteresis. Nonetheless, their study could correlate the steep turn-on behavior with voltage amplification due to negative capacitance–thanks to their device structure which had a 50-nm metallic AlSi layer introduced between the 40-nm ferroelectric P($VDF_{0.7}$-$TrFE_{0.3}$) and 10-nm SiO_2, allowing them to directly access the internal node voltage. They were able to show a voltage amplification>1 in the region where a sub-60-mV/decade characteristic was observed, and, in addition, they reported a negative slope of the extracted charge-voltage characteristics of the ferroelectric layer.

Modular NCFET where a ferroelectric capacitor is connected externally to the gate terminal of a MOSFET was reported by multiple groups. Khan et al. [46] used 100-nm gate length P- and N-type FinFETs as baseline MOSFETs and an epitaxial 300-nm-thick $BiFeO_3$ capacitor as the ferroelectric capacitor, and showed in the modular structure, as low as 8.5 mV/decade of swing over as high as 8 orders of magnitude of drain current can be obtained while the baseline MOSFET has a minimum subthreshold swing of 65 mV/decade. In this report, the I_D-V_G characteristics were hysteretic with 4–6 V memory windows, and sub-60-mV/decade switching was observed in both the branches of the hysteresis loop. The key advantage of the modular structure is that it allows for independent investigation of the characteristics of the baseline MOSFET and the NCFET. Hence, how the addition of the ferroelectric changes the device characteristics can be identified and understood in a clean manner. They were able to backtrack the charge-voltage characteristics of the ferroelectric capacitor, which clearly showed a negative slope and confirmed that the sub-60-mV/decade characteristic was due to the negative capacitance effect. Changhwan Shin's group at University of Seoul, in one of their first studies in 2015 [47], reported that starting with a 1-μm channel length baseline N-MOSFET with a S_{min}=~92 mV/decade, adding a ~20 nm thick ferroelectric P($VDF_{0.75}$-$TrFE_{0.25}$) capacitor to the gate terminal reduces the subthreshold swing to ~20 mV/decade. In 2016, using a 180-nm planar bulk baseline P-MOSFET and a similar P(VDF$_{0.75}$-TrFE$_{0.25}$) capacitor, the same group showed ~50 mV/decade switching with a very small hysteresis (less than 100 mV) at $V_{DS} \leq 2$ V. An interesting observation they made in Ref. [48] is that a hysteresis opens up for $|V_{DS}| \geq 3$ V with sub-60-mV/decade swing preserved in both the branches of the hysteresis. The fact that in a modular structure any of the components: baseline MOSFET and the ferroelectric capacitor can be changed independently without having to fabricate completely new devices enables the quick study of the evolution of the device characteristics with respect to different device parameters. Shin's group later in 2017 constructed modular NCFETs using short-channel P-FinFETs with different source-drain extension lengths as their baseline MOSFETs and an epitaxial 60-nm $Pb(Zr_{0.2}Ti_{0.8})O_3$ (PZT) thin-film capacitor as the ferroelectric [49]. In this report, they showed that as the extension length is increased, the hysteresis window narrows; in all cases, the average S during the forward sweep direction remains below 20 mV/decade, while during the reverse sweep they average S increases with an increased extension length. It must be noted that, in all these studies, the area of the ferroelectric capacitor was much larger than the effective area of the baseline MOSFETs leading. As a result, due to charge balance conditions, the charge in the ferroelectric is controlled by the baseline MOSFET and not vice versa. For example, in Ref. [46], it was reported that less than 0.25% of the ferroelectric charge switched during the switching.

Li et al. in IEDM 2015 [50] reported the 30-nm gate length integrated NCFET with a 5-nm ferroelectric $HfZrO_2$ gate oxide. By comparing the characteristics of the NCFET with that of the control FET, they were able to show for the first time that the negative capacitance effect can lead to an increase of the on-current for the same off-current for a given gate voltage swing without any hysteresis. What allowed them to perform this control study is that their structure, which was a bulk FinFET with 30-nm fin width, included an intermediate gate metal between the ferroelectric layer and the high-κ dielectric such that the device characteristics could be measured from both the internal and the external gate terminal. For both N- and P-type NC-FinFET, they showed that the addition of the ferroelectric layer in the measurement decreases the minimum subthreshold swing from ~88 mV/decade to below 60 mV/decade. By measuring the internal node voltage through the intermediate metal layer, they also reported a measured voltage amplification of ~1.5. In the same year in IEDM, Lee et al. [53] reported long channel bulk Si NCFET (30 μm), which showed minimum subthreshold swings of ~40 and ~30 mV/decade in the forward and reverse sweeps with a hysteresis window of ~200 mV. High-resolution transmission electron microscopy (HR-TEM) analysis of their gate oxide stack showed that there was a 0.3-nm intermediate layer of SiO_2 between a 5-nm annealed $HfZrO_2$ layer and the Si channel. In IEDM 2016, Lee et al. [54] reported ~52 mV/decade of minimum subthreshold swing without hysteresis in long channel Si NCFETs with 1.5 nm $HfZrO_2$ gate oxide. In IEDM 2016, Chang et al. [55] reported a low thermal budget process flow that makes La_2O_3 into a ferroelectric, utilizing which they demonstrated negative capacitance effect in InGaAs MOSFETs with 15-nm La_2O_3 used as the gate oxide in the temperature range: 150–250 K. They correlated the sub-60-mV/decade subthreshold swing at low temperature with their experimental observation that at low temperatures, ferroelectricity in their La_2O_3 films becomes stronger. At 150 K, they reported subthreshold swings below the theoretical limit for that temperature (~30 mV/decade) over ~2.5 orders of magnitude of drain current. Dasgupta et al. [56] in a 10-μm gate length bulk Si MOSFET used 100-nm sputter-deposited ferroelectric $PbZr_{0.52}Ti_{0.48}O_3$ layer with a 10-nm non-ferroelectric ALD HfO_2 buffer layer. The as-deposited $PbZr_{0.52}Ti_{0.48}O_3$ layer was amorphous, which was crystallized through a 90-s rapid thermal annealing in air at 620 °C. Their device exhibited minimum swings of 13 and 32 mV/decade in the forward and reverse sweep directions, respectively, with a hysteresis window of ~11 V. Interestingly, the sub-60-mV/decade characteristic was observed in strong inversion (~ 100 μm/μm current range) rather than in the subthreshold regime.

There have been multiple reports of NCFET that use alternative channel materials. In IEDM 2016, Zhou et al. demonstrated P-type Ge NCFETs using sub-10-nm thick ferroelectric $Hf_{0.5}Zr_{0.5}O_2$ layer with an intermediate TaN metallic layer. This 5-μm gate length P-type Ge NCFET showed average forward and reverse subthreshold swings of 47 and 43 mV/decade, respectively, with a hysteresis window of ~ 2 V. They performed detailed characterization of the ferroelectric layer, which showed clean hysteresis loops with remnant polarization of ~20 μC cm^{-3} and coercive field of ~1 MV cm^{-1}. One of the interesting features of their device was that in their measured $I_D - V_{DS}$ characteristics, they observed a negative output resistance in the reverse sweep of V_{DS} 5-nm $HfZrO_2$–a feature that was predicted on the basis of simulations. In IEDM 2017, Chung et al. from Peide Ye's group at Purdue University demonstrated Ge-based NC-FinFETs [57]. In their devices, ferroelectric $Hf_{0.5}Zr_{0.5}O_2$ film was integrated with Al_2O_3/GeO_x as the gate stack, and recessed source/drain and fin-channel process scheme was implemented. When they used 10-nm ferroelectric films, ~7 and ~17 mV/decade subthreshold slope was observed in the reverse and forward sweep direction, respectively, with a large voltage hysteresis window of −4.3 V. With ferroelectric thickness scaling down to 2 nm, Ge p-NC-FinFETs and n-NC-FinFETs show subthreshold swings of (forward/reverse) of 56/41 and 43/49 mV/dec, respectively, with negligible hysteresis windows (~4 mV for p-NC-FinFETs and ~17 mV for n-NC-FinFETs). Si et al. from the same group reported NCFET consisting of a 86-nm-thick β-Ga_2O_3 nanomembrane as the channel, a 20-nm ferroelectric $Hf_{0.5}Zr_{0.5}O_2$/3-nm dielectric Al_2O_3 bilayer as the gate dielectric stack, heavily n-doped (n++) silicon substrate as the gate electrode, and a Ti/Au source/drain as the metal contact [58]. They reported a minimum subthreshold slope of ~30 mV/dec at the reverse gate-voltage sweep and ~50 mV/dec at the forward gate-voltage sweep at V_{DS} = 0.5 V with a hysteresis window less than 0.1 V.

Noorbaksh et al. [59] from Palacios group at the Massachusetts Institute of Technology reported a MoS_2 FET with a gate oxide stack of a 10-nm ferroelectric HfO_2 layer doped with Al at 7.3% and a 10-nm undoped non-ferroelectric HfO_2 layer with an intermediate Ni layer. Their gate length was 3 μm. The minimum subthreshold swings of their NC and control MoS_2 FET were 57 and 67 mV/decade, respectively; both of these devices had a hysteresis window of less than 20 mV. The Al:HfO_2 layer showed a remnant polarization of 9.5 μC cm^{-3}

and a coercive field of 1.55 MV cm^{-1}. They also reported independent characterization of the negative capacitance behavior of the ferroelectric layer in a time-dependent measurement setup which clearly showed the characteristic negative capacitance transients. By accessing the internal node voltage through the intermediate Ni layer, they were able to deduce a voltage amplification of ~1.25. Franklin's group at the Duke University reported in 2016 sub-60-mV/decade operation in MoS_2 FETs with organic ferroelectric polymer: poly(vinylidene difluoride-trifluoroethylene) ($P(VDF_{0.7}\text{-}TrFE_{0.3})$) used as the negative capacitance material [60]. The reported two device structures: one with an intermediate metallic layer and the other without. A 20-nm Al_2O_3 layer was used as the top gate dielectric. In the ferroelectric devices without the intermediate metal layer, they reported a ~50% improvement of subthreshold swing; the control device had a minimum subthreshold slope of 252 mV/decade which improved to ~124 mV/decade after the addition of the ferroelectric layer. Despite the reduction in SS, the ferroelectric device was still well above the thermal limit of 60 mV/decade, which they attributed to the interfacial effects between the polymer and the dielectric. On the other hand, in the devices with an intermediate metal layer, they reported minimum subthreshold swings of 11.7 mV/dec over 2 orders and 14.4 mV/decade over ~4 orders for 1-μm and 500-nm channel lengths, respectively. The same group in 2017 reported similar NC MoS_2 FETs where they used a 12-nm-thick $Hf_{0.5}Zr_{0.5}O_2$ as the ferroelectric layer [51]. This time, their device structure included a 22.8-nm undoped HfO_2 as the dielectric layer and TiN as the intermediate metallic layer. The interesting feature they reported was the drain voltage dependence of the hysteresis loop in the $I_D - V_{GS}$ characteristics. They showed that the hysteresis window shifts to the right and becomes narrower as V_{DS} is increased, which was also predicted in earlier simulations. They observed a minimum subthreshold swing of 6.07 mV/decade with average of 8.03 mV/decade over 4 orders of magnitude in drain current in the reverse sweeps, although in the forward sweeps, the subthreshold swing was always larger than the thermal limit. Using the intermediate metal layer, they also measured a maximum voltage amplification of 28. Peide Ye's group at Purdue University demonstrated MoS_2-based steep slope NCFETs [61]. Their devices consisted of a monolayer up to a dozen layers of MoS_2 as the channel, 2-nm amorphous aluminum oxide (Al_2O_3) layer and 20-nm ferroelectric $Hf_{0.5}Zr_{0.5}O_2$ as the top gate dielectric stack, heavily doped silicon substrate as the gate electrode and nickel source/drain contacts. Apart from an improvement of the subthreshold swing in the NCFETs, they observed a significant on-current boost (five times larger than that in the control FET). Furthermore, they observed a negative output differential resistance in the NCFETs. They also performed detailed measurement and analysis to decouple the negative-capacitance-induced reverse-drain-induced-barrier-lowering effects from the self-heating effects.

Writing in IEDM 2017, Krivokapic et al. from GlobalFoundries demonstrated for the first time wafer-scale integration of NC-FinFETs at the state-of-the-art 14-nm node [62]. They not only showed a reduced subthreshold slope and an increased on-current of NC-FinFETs compared to those of the corresponding control devices at the same voltage swing but also reported 101-stage, fan-out three-ring oscillators based on NC-FinFETs operating at high frequencies. They also showed that NC-FinFET ring oscillators consistently dissipate ~40% less active power compared to that in the same based on control FinFETs while operating at the same frequency.

7.7 Speed of Negative Capacitance Transistors

There has been a significant discussion as to what the maximum operating frequency of an NCFET is. The discussion originates from the question: how does the switching speed limitations of ferroelectric oxides affect the operation of an NCFET? For example, the fastest switching time for a ferroelectric oxide in a capacitor structure was reported to be ~220 ps for micron-sized capacitors of ferroelectric Nb-doped $Pb(ZrTi)O_3$ [63]. On the other hand, for high-performance logic applications, the switching speed of transistors needs to be below 10 ps or so. However, it needs to be noted that, in such large dimensions (in microns), domain growth mechanisms dominate. On the other hand, in the relevant dimensions for scaled transistors (in single- or double-digit nanometers), domain nucleation mechanisms will dominate, for which the timescale of ferroelectric polarization switching would be much smaller than that for the case dominated by domain growth mechanisms. In fact, Chaterjee et al. has recently shown by extracting damping and mass terms from spectroscopic measurements for HfO_2 that the characteristic timescale for ferroelectric switching in the domain nucleation limit can be much smaller than 1 ps. Experimentally, the demonstration of high-frequency operation of NC-FinFET-based ring oscillators by Krivokapic et al. [62] also points to the fact that there is no roadblock for operating NCFETs at frequencies relevant for high-performance logic applications.

7.8 Conclusions

In summary, this chapter provides an overview of the developments in the field of NCFETs that happened over the past decade–starting with the initial proposal in 2008 to the recent experimental demonstrations of negative capacitance phenomena in ferroelectric capacitors and steep switching in advanced node transistors and power reduction in corresponding ring oscillators. It remains to be seen as to what the maximum amount of power saving is that can be achieved using this technology and what the associated device design guidelines are. Initial performance projections are also encouraging [41, 64, 65]. On the basis of full chip-level simulations, it has recently been projected that, using negative capacitance transistors, as high as 4× reduction in power dissipation is possible with respect to baseline transistor technology at iso-performance [64]. Going forward, the integration and reliability aspects of NCFETs need to be addressed.

References

1 Hilbert, M. and López, P. (2011). The world's technological capacity to store, communicate, and compute information. *Science* 332 (6025): 60–65.
2 Whitney, J. and Delforge, P. (2014). Data center efficiency assessment. National Resource Defense Council. http://www.nrdc.org/energy/files/data-center-efficiency-assessment-IP.pdf (Retrieved 5 May 2015).
3 Evans, D. (2011). The internet of things how the next evolution of the internet is changing everything. Internet Business Solutions Group. http://www.cisco.com/web/about/ac79/docs/innov/IoT_IBSG_0411 FINAL.pdf (Retrieved 12 May 2015).
4 Energy in the United States. http://en.wikipedia.org/wiki/Energy_in_the_United_States (retrieved 5 May 2015).
5 Kaplan, J.M., Forrest, W., and Kindler, N. (2008). Revolutionizing Data Center Energy Efficiency. Technical Report. McKinsey & Company.
6 Brown, R., Masanet, E., Nordman, B. et al. (2008). Report to Congress on Server and Data Center Energy Efficiency: Public law 109-431. Lawrence Berkeley National Laboratory.
7 Pop, E. (2010). Energy dissipation and transport in nanoscale devices. *Nano Res.* 3 (3): 147–169.
8 50 Years of Moore's Law. http://www.intel.com/content/www/us/en/silicon-innovations/moores-law-technology.html (retrieved 5 May 2015).
9 Danowitz, A., Kelley, K., Mao, J. et al. (2012). CPU DB: recording microprocessor history. *Commun. ACM* 55 (4): 55–63.
10 Esmaeilzadeh, H., Blem, E., St Amant, R. et al. (2011). Dark silicon and the end of multicore scaling. *ACM SIGARCH Comput. Archit. News* 39 (3): 365–376.
11 Theis, T.N. and Solomon, P.M. (2010). It's time to reinvent the transistor! *Science* 327 (5973): 1600–1601.
12 Theis, T.N. and Solomon, P.M. (2010). In quest of the "next switch": prospects for greatly reduced power dissipation in a successor to the silicon field-effect transistor. *Proc. IEEE* 98 (12): 2005–2014.
13 Cavin, R.K., Zhirnov, V.V., Hutchby, J.A., and Bourianoff, G.I. (2005). Energy barriers, demons, and minimum energy operation of electronic devices. *Fluctuation Noise Lett.* 5 (04): C29–C38.
14 Zhirnov, V.V., Cavin, R.K., Hutchby, J.A., and Bourianoff, G.I. (2003). Limits to binary logic switch scaling-a gedanken model. *Proc. IEEE* 91 (11): 1934–1939.
15 Salahuddin, S. and Datta, S. (2008). Use of negative capacitance to provide voltage amplification for low power nanoscale devices. *Nano Lett.* 8 (2): 405–410. pMID: 18052402. http://pubs.acs.org/doi/abs/10.1021/nl071804g.
16 Banerjee, S., Richardson, W., Coleman, J., and Chatterjee, A. (1987). A new three-terminal tunnel device. *IEEE Electron Device Lett.* 8 (8): 347–349.
17 Hu, C., Chou, D., Patel, P., and Bowonder, A. (2008). Green transistor - a VDD scaling path for future low power ICs. *International Symposium on VLSI Technology, Systems and Applications (VLSI-TSA)*, pp. 14–15.

18 Gopalakrishnan, K., Griffin, P.B., and Plummer, J.D. (2005). Impact ionization MOS (I-MOS)–Part I: device and circuit simulations. *IEEE Trans. Electron Devices* 52 (1): 69–76.

19 Kam, H. and Liu, T.-J.K. (2009). Pull-in and release voltage design for nano-electromechanical field-effect transistors. *IEEE Trans. Electron Devices* 56 (12): 3072–3082.

20 Enachescu, M., Lefter, M., Bazigos, A. et al. (2013). Ultra low power NEM-FET based logic. *2013 IEEE International Symposium on Circuits and Systems (ISCAS)*, pp. 566–569.

21 Khan, A.I., Chatterjee, K., Wang, B. et al. (2015). Negative capacitance in a ferroelectric capacitor. *Nat. Mater.* 14 (2): 182–186.

22 Hoffmann, M., Pešić, M., Chatterjee, K. et al. (2016). Direct observation of negative capacitance in polycrystalline ferroelectric Hfo_2. *Adv. Funct. Mater.* 26 (47): 8643–8649.

23 Kobayashi, M., Ueyama, N., Jang, K., and Hiramoto, T. (2016). Experimental study on polarization-limited operation speed of negative capacitance FET with ferroelectric Hfo_2. *2016 IEEE International on Electron Devices Meeting (IEDM).* IEEE, pp. 12.3.1–12.3.4.

24 Ku, H. and Shin, C. (2017). Transient response of negative capacitance in P(VDF 0.75-TrFE 0.25) organic ferroelectric capacitor. *IEEE J. Electron Devices Soc.* 5 (3): 232–236.

25 Khan, A.I., Yeung, C.W., Hu, C., and Salahuddin, S. (2011). Ferroelectric negative capacitance MOSFET: capacitance tuning & antiferroelectric operation. *Electron Devices Meeting (IEDM), 2011 IEEE International.* IEEE, pp. 11–3.

26 Yeung, C.W., Khan, A.I., Cheng, J.-Y. et al. (2012). Non-hysteretic negative capacitance FET with sub-30mV/dec swing over 106X current range and ion of 0.3 mA/μm without strain enhancement at 0.3 V VDD. *2012 International Conference on Simulation of Semiconductor Processes and Devices (SISPAD)*, pp. 257–259.

27 Yeung, C.W., Khan, A.I., Sarker, A. et al. (2013). Low power negative capacitance FETs for future quantum-well body technology. *2013 International Symposium on VLSI Technology, Systems, and Applications (VLSI-TSA).* IEEE, pp. 1–2.

28 Hu, C., Salahuddin, S., Lin, C.-I., and Khan, A.I. (2015). 0.2 V adiabatic NC-FinFET with 0.6 mA/μm I_{ON} and 0.1 nA/μm I_{OFF}. *2015 73rd Annual Device Research Conference (DRC)*, pp. 39–40.

29 Cano, A. and Jiménez, D. (2010). Multidomain ferroelectricity as a limiting factor for voltage amplification in ferroelectric field-effect transistors. *Appl. Phys. Lett.* 97 (13): 133509.

30 Frank, D.J., Solomon, P.M., Dubourdieu, C. et al. (2014). The quantum metal ferroelectric field-effect transistor. *IEEE Trans. Electron Devices* 61 (6): 2145–2153.

31 Rusu, A. and Ionescu, A.M. (2012). Analytical model for predicting sub-threshold slope improvement versus negative swing of S-shape polarization in a ferroelectric FET. *2012 Proceedings of the 19th International Conference on Mixed Design of Integrated Circuits and Systems (MIXDES).* IEEE, pp. 55–59.

32 Jimenez, D., Miranda, E., and Godoy, A. (2010). Analytic model for the surface potential and drain current in negative capacitance field-effect transistors. *IEEE Trans. Electron Devices* 57 (10): 2405–2409.

33 Lin, C.-I., Khan, A.I., Salahuddin, S., and Hu, C. (2016). Effects of the variation of ferroelectric properties on negative capacitance FET characteristics. *IEEE Trans. Electron Devices* 63 (5): 2197–2199.

34 Jain, A. and Alam, M.A. (2014). Stability constraints define the minimum subthreshold swing of a negative capacitance field-effect transistor. *IEEE Trans. Electron Devices* 61 (7): 2235–2242.

35 Karda, K., Jain, A., Mouli, C., and Alam, M.A. (2015). An anti-ferroelectric gated landau transistor to achieve sub-60 mV/dec switching at low voltage and high speed. *Appl. Phys. Lett.* 106 (16): 163501.

36 Karda, K., Mouli, C., and Alam, M. (2016). Switching dynamics and hot atom damage in landau switches. *IEEE Electron Device Lett.* 37 (6): 801–804.

37 Pahwa, G., Dutta, T., Agarwal, A., and Chauhan, Y.S. (2017). Compact model for ferroelectric negative capacitance transistor with MFIS structure. *IEEE Trans. Electron Devices* 64 (3): 1366–1374.

38 Kobayashi, M. and Hiramoto, T. (2016). On device design for steep-slope negative-capacitance field-effect-transistor operating at sub-0.2V supply voltage with ferroelectric Hfo_2 thin film. *AIP Adv.* 6 (2): 025113.

39 Qi, Y. and Rappe, A.M. (2015). Designing ferroelectric field-effect transistors based on the polarization-rotation effect for low operating voltage and fast switching. *Phys. Rev. Appl.* 4 (4). doi: 10.1103/physrevapplied.4.044014.

40 Aziz, A., Ghosh, S., Datta, S., and Gupta, S.K. (2016). Physics-based circuit-compatible spice model for ferroelectric transistors. *IEEE Electron Device Lett.* 37 (6): 805–808.

41 Khandelwal, S., Khan, A.I., Duarte, J.P. et al. (2016). Circuit performance analysis of negative capacitance FinFETs. *2016 Symposium on VLSI Technology and Circuits*, pp. 1–2.

42 Wei, L., Mysore, O., and Antoniadis, D. (2012). Virtual-source-based self-consistent current and charge FET models: from ballistic to drift-diffusion velocity-saturation operation. *IEEE Trans. Electron Devices* 59 (5): 1263–1271.

43 Dong, Z. and Guo, J. (2017). A simple model of negative capacitance FET with electrostatic short channel effects. *IEEE Trans. Electron Devices* 64 (7): 2927–2934.

44 Salvatore, G.A., Bouvet, D., and Ionescu, A.M. (2008). Demonstration of subthrehold swing smaller than 60mV/decade in Fe-FET with P(VDF-TrFE)/sio_2 gate stack. *2008 IEEE International on Electron Devices Meeting. IEDM 2008.* IEEE, pp. 1–4.

45 Rusu, A., Salvatore, G.A., Jimenez, D., and Ionescu, A.M. (2010). Metal-ferroelectric-meta-oxide-semiconductor field effect transistor with sub-60mV/decade subthreshold swing and internal voltage amplification. *2010 IEEE International on Electron Devices Meeting (IEDM)*, pp. 16.3.1–16.3.4.

46 Khan, A., Chatterjee, K., Duarte, J.P. et al. (2016). Negative capacitance in short channel FinFETs externally connected to an epitaxial ferroelectric capacitor. *IEEE Electron Device Lett.* 37 (1): 111–114.

47 Jo, J., Choi, W.Y., Park, J.-D. et al. (2015). Negative capacitance in organic/ferroelectric capacitor to implement steep switching MOS devices. *Nano Lett.* 15 (7): 4553–4556.
48 Jo, J. and Shin, C. (2016). Negative capacitance field effect transistor with hysteresis-free sub-60-mV/decade switching. *IEEE Electron Device Lett.* 37 (3): 245–248.
49 Ko, E., Lee, J.W., and Shin, C. (2017). Negative capacitance FinFET with sub-20-mV/decade subthreshold slope and minimal hysteresis of 0.48 V. *IEEE Electron Device Lett.* 38 (4): 418–421.
50 Li, K.-S., Chen, P.-G., Lai, T.-Y. et al. (2015). Sub-60mV-swing negative-capacitance FinFET without hysteresis. *2015 IEEE International on Electron Devices Meeting (IEDM).* IEEE, pp. 22–6.
51 McGuire, F.A., Lin, Y.-C., Price, K. et al. (2017). Sustained sub-60 mV/decade switching via the negative capacitance effect in MoS_2 transistors. *Nano Lett.* 17 (8): 4801–4806.
52 Khan, A.I., Radhakrishna, U., Salahuddin, S., and Antoniadis, D. (2017). Work function engineering for performance improvement in leaky negative capacitance FETs. *IEEE Electron Device Lett.* 38 (9): 1335–1338.
53 Lee, M., Chen, P.-G., Liu, C. et al. (2015). Prospects for ferroelectric HfZrOx FETs with experimentally CET=0.98nm, SSfor=42mV/dec, SSrev=28mV/dec, switch-off <0.2 V, and hysteresis-free strategies. *2015 IEEE International on Electron Devices Meeting (IEDM).* IEEE, pp. 22.5.1–22.5.4.
54 Lee, M., Fan, S.-T., Tang, C.-H. et al. (2016). Physical thickness 1.x nm ferroelectric HfZrOx negative capacitance FETs. *2016 IEEE International on Electron Devices Meeting (IEDM).* IEEE, pp. 12.1.1–12.1.4.
55 Chang, C.-Y., Endo, K., Kato, K. et al. (2016). Impact of La_2O_3/ingaas MOS interface on InGaAs MOSFET performance and its application to InGaAs negative capacitance FET. *2016 IEEE International on Electron Devices Meeting (IEDM).* IEEE, pp. 12.5.1–12.5.4.
56 Dasgupta, S., Rajashekhar, A., Majumdar, K. et al. (2015). Sub-kT/q switching in strong inversion in $PbZr_{0.52}Ti_{0.48}O_3$ Gated negative capacitance FETs. *IEEE J. Explor. Solid-State Comput. Devices Circuits* 1, 43–48.
57 Chung, W., Si, M., and Ye, P.D. (2017). Hysteresis-free negative capacitance germanium CMOS FinFETs with Bi-directional Sub-60 mV/dec. *2017 IEEE International on Electron Devices Meeting (IEDM).* IEEE.
58 Si, M., Yang, L., Zhou, H., and Ye, P.D. (2017). β-Ga_2O_3 nanomembrane negative capacitance field-effect transistors with steep subthreshold slope for wide band gap logic applications. *ACS Omega* 2 (10): 7136–7140.
59 Nourbakhsh, A., Zubair, A., Joglekar, S. et al. (2017). Subthreshold swing improvement in MoS_2 transistors by the negative-capacitance effect in a ferroelectric al-doped-HFO_2/HFO_2 gate dielectric stack. *Nanoscale* 9 (18): 6122–6127.
60 McGuire, F.A., Cheng, Z., Price, K., and Franklin, A.D. (2016). Sub-60 mV/decade switching in 2D negative capacitance field-effect transistors with integrated ferroelectric polymer. *Appl. Phys. Lett.* 109 (9): 093101.
61 Si, M., Su, C.J., Jiang, C. et al. (2018). Steep-slope hysteresis-free negative capacitance MoS_2 transistors. *Nat. Nanotechnol.* 13 (1): 24

62 Krivokapic, Z., Rana, U., Galatage, R. et al. (2017). 14nm ferroelectric FinFET technology with steep subthreshold slope for ultra low power applications. *2017 IEEE International on Electron Devices Meeting (IEDM)*. IEEE.

63 Li, J., Nagaraj, B., Liang, H. et al. (2004). Ultrafast polarization switching in thin-film ferroelectrics. *Appl. Phys. Lett.* 84 (7): 1174–1176.

64 Samal, S.K., Khandelwal, S., Khan, A.I. et al. (2017). Full chip power benefits with negative capacitance FETs. *2017 IEEE/ACM International Symposium on Low Power Electronics and Design (ISLPED)*. IEEE, pp. 1–6.

65 Dutta, T., Pahwa, G., Trivedi, A.R. et al. (2017). Performance evaluation of 7-nm node negative capacitance FinFET-based SRAM. *IEEE Electron Device Lett.* 38 (8): 1161–1164.

8

Spin-Based Devices for Logic, Memory, and Non-Boolean Architectures

Supriyo Bandyopadhyay

Virginia Commonwealth University, Department of Electrical and Computer Engineering, 601 W. Main Street, Richmond, VA 23284, USA

8.1 Introduction

At the heart of nearly all electronic computing and signal processing circuits today is a device known as the "field-effect-transistor." It is a three-terminal device where the current flowing between two of the terminals ("source" and "drain") is modulated with a current or potential applied to the third terminal ("gate"). It has all the desirable characteristics of a digital "switch": the potential at the third terminal can change the current between the other two by several orders of magnitude to yield two well-separated conductance states that can encode the binary bits 0 and 1, switching between the two conductance states can be accomplished rapidly (in a few tens of picoseconds), the energy dissipated to switch can be small ($\sim$100 pJ at room temperature), and there is isolation between the third terminal and the other two to allow unidirectional steering of information from an input stage to an output stage. These wonderful properties have ensured that the transistor rules the roost when it comes to digital computing or signal processing. Yet, the electronic device community is concerned that the heyday of the transistor might be ending. This fear stems from the realization that when the transistor size scales down, the energy dissipated to switch it between its two conductance states may not scale down proportionately, which means that the heat generated per unit area will increase. As a result, thermal management on a chip can be threatened to the point where chip meltdown becomes a real possibility.

Let us examine this scenario more closely. Consider, for example, the Intel® Core™ i7-6700K processor built with 14-nm FinFET technology and introduced in 2015. It dissipates 91 W of power while operating at a clock frequency of 4 GHz, with a power supply voltage of 1.2 V, and contains roughly 1.75 billion transistors (https://ark.intel.com/products/88195/Intel-Core-i7-6700K-Processor-8M-Cache-up-to-4_20-GHz). We can assume that approximately 20% of them switch at any given time. Therefore, the number of transistors N that are switching at any given time is $N = 1.75 \times 10^9 \times 0.2 = 3.5 \times 10^8$. The energy dissipated

Advanced Nanoelectronics: Post-Silicon Materials and Devices,
First Edition. Edited by Muhammad Mustafa Hussain.

per transistor is $E_d = P_d/(Nf)$, where P_d is the power dissipation (91 W) and f is the clock frequency (4 GHz). Putting these numbers together, we get

$$E_d = \frac{P_d}{Nf} = \frac{91}{3.5 \times 10^8 \times 4 \times 10^9} \approx 65 \text{ aJ} \tag{8.1}$$

which means that a state-of-the-art transistor today dissipates about 65 aJ to switch. This estimate is a little pessimistic since we have ignored power dissipation in the interconnection wires, and assumed that all the dissipation comes from the switching transistors. Despite these small uncertainties, we can assert with some confidence that the ballpark figure for the energy dissipation in a transistor of c. 2017 is ~50 aJ.

Enter Moore's law [1] which envisions that the transistor density on a chip will roughly double every two years. If that trend were to continue, then by the year 2025, the transistor count in a chip like the i7-6700K will increase by $2^5 = 32$ times. If, at the same time, the clock frequency were to increase to 10 GHz (2.5 times), then the power dissipation in the chip will have increased by $32 \times 2.5 = 80$ times (assuming that the energy dissipation per transistor remains the same) to become 7.28 kW! That kind of dissipation may overwhelm all existing heat sinking technologies and portend chip meltdown. The only way to avoid that will be to reduce the energy dissipation in a transistor when it switches (and the dissipation in the connecting wires, if possible).

We can carry out a projective estimate of how low the energy dissipation in a transistor can be. This is an order estimate, not necessarily very accurate, but it will provide a glimpse of how far we can go with the transistor paradigm in electronic circuits. We need to understand first where the energy dissipation during switching comes from. The transistor is a charge-based device where the two conductance states are demarcated by the amount of charge stored in the channel of the transistor. More charge in the device takes it to one state and less charge to the other state. Charge is a scalar quantity that has a magnitude and no other attribute (like direction in the case of a vector). Therefore, the two states *must* be encoded by two different amounts of charge Q_1 and Q_2. Every time the transistor switches, the amount of charge changes from Q_1 to Q_2, or vice versa [2, 3]. If this change takes place in a time Δt, then there is an accompanying current flow somewhere in the transistor circuitry which is given by

$$I = \frac{|Q_1 - Q_2|}{\Delta t} = \frac{\Delta Q}{\Delta t} \tag{8.2}$$

with an accompanying power dissipation I^2R and energy dissipation

$$E_d = I^2 R \Delta t = \Delta Q(IR) = \Delta Q \Delta V \tag{8.3}$$

where R is the resistance in the path of the current and $\Delta V = IR$. Essentially, ΔV is the potential needed to move the amount of charge ΔQ in the channel of the transistor.

Note that for a given ΔQ, the corresponding ΔV is *not* independent of the speed with which ΔQ amount of charge is moved. Since $\Delta V = IR = (\Delta Q/\Delta t)R$, it is clear that ΔV is inversely proportional to the switching duration Δt. Hence, faster switching will always cause larger energy dissipation.

We can estimate ΔQ for the 14-nm FinFET in the i7-6700K processor. The power supply voltage is $\Delta V = 1.2$ V. The energy dissipation E_d is 65 aJ. Therefore, $\Delta Q = \frac{E_d}{\Delta V} = 5.4 \times 10^{-17}$ C = 336 electronic charge. That means 300–400 electrons are moved in the channel of a modern transistor to make it switch.

We would not want the number of electrons moved to be too few lest we are overwhelmed by background charge fluctuations which have been the nemesis of single-electron transistors. Let us be safe and assume that the minimum number is 20 electrons, which will make $\Delta Q|_{min} = 3.22 \times 10^{-18}$ C. We will also assume for the sake of simplicity that the voltage needed to move charge into or out of the transistor channel scales linearly with the amount of charge moved. If 1.2 V moves 336 electrons, then the voltage required to move 20 electrons (everything else being the same) is 71 mV. In that case, the minimum energy dissipation that we can expect in the FinFET-type transistor will be

$$\Delta Q|_{min} \times \Delta V|_{min} = 3.22 \times 10^{-18} \times 0.071 = 0.23 \text{ aJ} \tag{8.4}$$

If we wish to limit our energy dissipation in the chip to 100 W and work with a clock frequency of 10 GHz, then this energy dissipation will limit our transistor count in the chip to

$$N = \frac{P_d}{E_d f} = \frac{100}{2.3 \times 10^{-19} \times 10^{10}} = 43.5 \text{ billion} \tag{8.5}$$

The fact that the maximum number of transistors on a chip is limited by the maximum tolerable power dissipation and the minimum tolerable clock speed is a matter of concern. It tells us that we cannot continue to increase the transistor count on a chip indefinitely and Moore's law must, ultimately, come to an end. The Semiconductor Industry Association and the Semiconductor Research Corporation (a consortium of chip manufacturers) has called this the "red brick wall" in the way of progress in the semiconductor industry. It is a real menace that has prompted the search for alternate state variables (something other than "charge") to encode bit information, with the hope that a different state variable may be able to drastically reduce E_d. While there are a number of likely state variables to choose from, the one that has attracted the most attention is the electron's quantum mechanical "spin," which can be easily adapted to host bit information.

8.2 Spin-Based Devices

The basic problem with charge is that it is a *scalar* quantity and therefore has a magnitude and no other attribute. If we wish to store the bits 0 and 1 in the form of charge, then we must store them in two different amounts (magnitudes) of charge. We have no other option because "charge" has only magnitude. More charge could represent bit 1 and less charge bit 0, or vice versa. Whenever we wish to switch, we must *change the magnitude of charge* ($\Delta Q \neq 0$), which will invariably cause the $\Delta Q \Delta V$ dissipation.

In contrast, we can (classically) think of spin as a vector that has both a magnitude and a direction. In fact, we can view it as a tiny magnetic moment attached to each electron. It has a constant magnitude of $\hbar/2$ and a variable direction.

We can encode information in the "direction" as opposed to the "magnitude," which would not have been possible with "charge." This could yield an energy advantage. Switching bits would entail simply flipping the spin (i.e. reversing the direction of the vector), without moving the electron in space and causing current flow. This would make $\Delta Q = 0$ and eliminate the $\Delta Q \Delta V$ dissipation altogether. Although this is an enticing possibility, it actually becomes beneficial only if the energy expended (and ultimately dissipated) to flip the spin is considerably less than the $\Delta Q \Delta V$ dissipation. That is the most important question which will ultimately decide if spin-based devices can be more energy efficient than charge-based devices. In this chapter, we explore this question.

There are two other (perhaps slightly less important) issues of switching speed and device reliability when it comes to digital switches. Faster devices are always preferred, but slower devices can be accommodated if massively parallel architectures can be built around them to overcome the speed limitation. A good example is the human brain. Neurons fire in about 1 ms (7–8 orders of magnitude slower than a transistor), but the massive interconnection between neurons and the way the human brain computes, make humans beat the best digital computers in many tasks (e.g. face recognition). At this time, spin-based switches are at least an order of magnitude slower than transistor switches, which tells us that they might need special (perhaps neuromorphic) architectures to be competitive. Spin-based switches are also less reliable (switching error probability at room temperature may be 7–10 *orders of magnitude* larger than in transistors), but that can be ameliorated by using them in architectures such as neuromorphic, stochastic, or Bayesian inference engines which are much more forgiving of individual device errors than Boolean logic. Spin-based devices can also be used in (non-Boolean) collective computational architectures where the collective activity of many devices working in unison elicit the computational activity and the failure of a large fraction of the devices does not impair the circuit function. An example of this is computing with coupled spin torque oscillators [4, 5].

The most important advantage of spin-based devices over charge-based devices is the *non-volatility*. Information written into a spin-based device (such as a nanomagnetic switch) can be retained for decades or centuries after the device has been powered off, while charge-based devices are *volatile* and the information stored in them is lost almost immediately after the device has been powered off. While it is obvious that non-volatility would be an advantage for memory applications, it is not so obvious that it can be a serious advantage for logic operations as well. *Non-volatile logic* is an emerging area where a logic unit is fashioned out of nanomagnets which enable retaining the output of the logic operation in the gate itself after powering off. The gate therefore doubles as both a logic processor and memory. This saves the trouble of having to store the result in a remote memory and then fetching it from there when the next logic step is to be executed. Such a feature allows one to easily build non–von-Neumann Boolean architectures and a host of other intriguing circuit topologies that reduce device count and lower the energy-delay product associated with specific tasks. This is where spin-based devices may make an inroad. They may not be able to eclipse charge-based devices in traditional Boolean computing, but may do so in nontraditional and non-Boolean computing paradigms which

Table 8.1 Comparison between charge- and spin-based devices.

	Charge-based devices	Spin-based devices
Energy dissipation	?	?
Switching speed	Superior	Inferior
Reliability	Superior	Inferior
Retention	Volatile	Non-volatile

are becoming increasingly important in the context of deep and wide learning algorithms, and "Internet of Things."

The above table makes a rough comparison between charge- and spin-based devices which is valid at the time of writing this chapter. These comparisons change rapidly with time and may change in the near future; so, they should be imbibed with a certain degree of circumspection (Table 8.1).

8.2.1 Spin Field-Effect Transistor (SPINFET)

An intriguing spin-based device concept (that was not necessarily motivated by the desire to avoid excessive energy dissipation or any other concern) is the *Spin Field-Effect Transistor* (SPINFET) [6] where the transistor is switched on and off by *not* moving charge into or out of the channel of the transistor, but by changing their spin polarization instead. At first glance, it may appear that this will make $\Delta Q = 0$, but we show later that this is not true.

The basic structure of the SPINFET is identical to that of a traditional depletion-mode n-channel metal-oxide semiconductor field-effect transistor (n-MOSFET) and is shown in Figure 8.1. The only difference is that the source and drain contacts are ferromagnets that are permanently magnetized along the direction of current flow. For the sake of simplicity, we will assume that the ferromagnets are perfect spin polarizers and analyzers, i.e. the source contact injects only its majority spins into the channel at the complete exclusion of any minority spin, and the drain contact completely blocks those electrons (impinging on it from the channel) whose spins are antiparallel to its own majority spins. Although the transistor's channel can be either one-, two-, or three-dimensional, the operation of the transistor will be explained by assuming it to be one-dimensional (a quantum wire), where only the lowest transverse subband in the conduction band is occupied by carriers. Extension of the theory to polydimensional channels, or even the one-dimensional channel with multiple subbands occupied, is not trivial and is not addressed here, but the interested reader can find it in [7]. Bandyopadhyay and Cahay [7] also contains an analysis of SPINFET, whose source and drain contacts are not perfect spin polarizers and analyzers.

The two ferromagnetic source and drain contacts in the SPINFET are magnetized along the direction of current flow in the channel and their magnetizations are mutually parallel. The source injects electrons into the channel with their spins aligned exclusively along the direction of the source's magnetization, which, in this case, is the $+x$ direction. Every injected carrier has its spin aligned in the $+x$

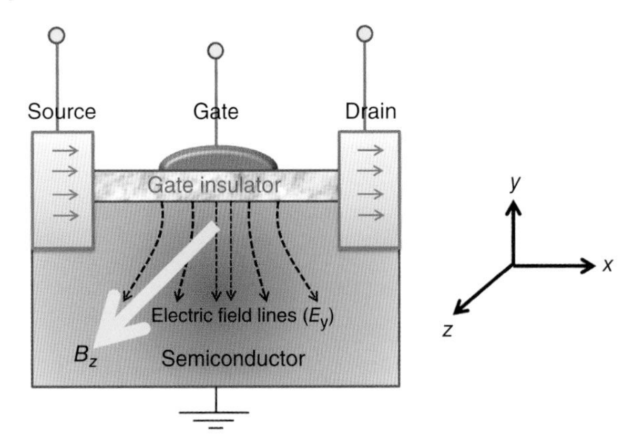

Figure 8.1 The basic structure of a SPINFET proposed in Ref. [6]. A gate voltage, applied between the gate terminal and ground, induces a *y*-directed transverse electric field in the channel region underneath the gate insulator where it induces a Rashba-type spin–orbit interaction. Because of this interaction, the electrons experience an effective magnetic field in the *z*-direction. The source and drain contacts are assumed to be perfect spin injectors and detectors, and they are magnetized in the $+x$-direction. Hence, electrons are injected into the channel by the source with their spins polarized in the $+x$-direction. When there is no gate voltage, the *z*-directed effective magnetic field is absent and the injected spins do not precess in the channel. If there is no spin relaxation in the channel, then they arrive as $+x$-polarized at the drain contact and are fully transmitted. When the gate voltage is nonzero, the *z*-directed effective magnetic field is present and makes the injected spin precess. When they arrive at the drain, they are filtered by their spin polarization. If the precession has been through an odd multiple of π radians, then the spins are $-x$-polarized and blocked by the drain, whereas if the precession has been through an even multiple of π, the spins are $+x$-polarized and transmitted. The precession angle depends on the strength of the *z*-directed effective magnetic field, which, in turn, depends on the *y*-directed electric field and hence the gate voltage. The source-to-drain current is proportional to the transmission through the drain contact. Thus, changing the gate voltage modulates the source-to-drain current and realizes transistor action.

direction. If there is no spin–orbit interaction in the channel and one can neglect the effect of any fringing magnetic field caused by the magnetized contacts, then the electron spins do not precess as the electrons travel from the source to the drain under a source-to-drain bias V_{DS}. Now, if there is no spin relaxation in the channel due to imperfections such as magnetic impurities, spin flip events, hyperfine interaction with nuclear spins, etc., then the injected carriers arrive at the "drain" contact with their spins still aligned in the original ($+x$) direction. Since all the arriving carriers have their spins aligned parallel to the drain's magnetization, the drain transmits all of them and the maximum possible current flows between the source and drain contacts. This is the ON state of the transistor.

When an electrostatic potential V_G is applied between the gate terminal and the grounded substrate, it generates an electric field transverse to the channel (in the *y*-direction). This electric field induces Rashba spin–orbit interaction [8] in the channel due to structural symmetry breaking and that produces an effective magnetic field oriented in a direction mutually perpendicular to the direction of current flow and the gate-induced electric field. Since the channel is strictly one-dimensional, current flows only in the *x*-direction. Therefore, the effective magnetic field of flux density B_{Rashba} is directed along the z-direction. Because of the one-dimensionality of the channel, the axis of this effective magnetic field is *fixed* and always points along the z-axis.

The strength of this magnetic field depends on the carrier's velocity and is given by [7]

$$B_{\text{Rashba}}(v) = \frac{2m^* a_{46}}{g\mu_B \hbar} E_y v \tag{8.6}$$

where v is the carrier velocity, E_y is the gate-induced electric field causing the Rashba interaction, m^* is the carrier effective mass, a_{46} is a material constant, g is the g-factor of the surrounding medium, μ_B is the Bohr magnetron, and $\hbar$ is the reduced Planck constant.

The spins of the injected carriers execute Larmor precession about the effective magnetic field $\vec{B}_{\text{Rashba}}(v)$ with a frequency Ω given by the Larmor relation:

$$\Omega(v) = \frac{g\mu_B B_{\text{Rashba}}(v)}{\hbar} = \frac{2m^*}{\hbar^2} a_{46} E_y v \tag{8.7}$$

This precession takes place on the x–y plane since the axis of the magnetic field is along the z-direction.

At this point, it is necessary to assume that there is no "damping" in the system, meaning that there is no energy relaxation. In other words, inelastic processes are absent. If energy relaxation is allowed, then the electron spins will ultimately tend to align along the magnetic field $\vec{B}_{\text{Rashba}}(v)$, since that is the lowest energy state. This will result in most spins pointing in the z-direction. Of course, not all spins will point along the z-direction because the lowest energy state will be occupied according to Fermi–Dirac probability, which is not unity at room temperature, but this is not important for this discussion. We need to prevent this eventuality from materializing, and for that reason all energy relaxation processes must be eliminated. That requires the channel length to be much shorter than the inelastic mean free path of electrons, which may be a few tens of nanometers at best if the device is operated at room temperature. At low temperatures (~4.2 K), the inelastic mean free path in a pristine one-dimensional semiconductor channel can be a few tens of micrometers.

If no damping is present (no energy relaxation), then the electron spins will precess continuously about $\vec{B}_{\text{Rashba}}(v)$ since that field is in the z-direction while the spins are polarized in the x-direction. Therefore, the *spatial* rate at which spin rotates when an electron travels through the SPINFET's channel can be obtained from Eq. (8.7) as

$$\begin{aligned} \Omega(v) &= \frac{\mathrm{d}\phi}{\mathrm{d}t} = \frac{\mathrm{d}\phi}{\mathrm{d}x}\frac{\mathrm{d}x}{\mathrm{d}t} = \frac{\mathrm{d}\phi}{\mathrm{d}x} v = \frac{2m^*}{\hbar^2} a_{46} E_y v \\ \Rightarrow \underbrace{\frac{\mathrm{d}\phi}{\mathrm{d}x}}_{\text{spatial rate}} &= \frac{2m^*}{\hbar^2} a_{46} E_y \end{aligned} \tag{8.8}$$

where ϕ is the angle through which spin rotates.

Note that the spatial rate $\mathrm{d}\phi/\mathrm{d}x$ is *independent* of the carrier velocity. Therefore, every electron, regardless of its injection velocity and regardless of any elastic momentum randomizing collision that it suffers in the channel, rotates by exactly the *same* angle as it traverses the distance between the source and drain. This angle is given by

$$\Phi_{\text{Rashba}} = \frac{2m^*}{\hbar^2} a_{46} E_y L \tag{8.9}$$

where L is the source-to-drain separation (or the channel length). Thus, if every electron was injected by the source with the same spin polarization (say, $+x$-polarized), then every electron that arrives at the drain will also have the same spin polarization, although it may be different from the initial one. In other words, the magnitude of the ensemble-averaged spin *does not change* and there is no decay of spin polarization. There is no randomization of spin polarization in the channel, *no matter how much momentum randomizing elastic scattering there is,* because the spin precession angle is a constant independent of carrier velocity and therefore is the same for every electron regardless of its injection velocity and scattering history! This is a remarkable result for the strictly one-dimensional SPINFET.

Now, if the electric field E_y is such that $\Phi_{\text{Rashba}} = (2m+1)\pi$, where m is an integer, then the spin polarizations of the carriers arriving at the drain would have rotated by an odd multiple of 180° in traversing the channel. In this case, their spins would be antiparallel to the drain's magnetization and hence they would be blocked by the drain from transmitting. On the other hand, if $\Phi_{\text{Rashba}} = 2m\pi$, then the arriving carriers will have their spins aligned parallel to the drain's polarization and will be fully transmitted. Thus, by changing E_y with a gate potential, one can change Φ_{Rashba} and modulate the source-to-drain current. *This realizes the field-effect transistor action.*

Note that this one-dimensional SPINFET can operate without any degradation at elevated temperatures as long as the elevated temperature does not introduce too many inelastic collisions and damping. Higher operating temperatures will cause a larger thermal spread in the electron velocity and perhaps also increase the rate of elastic collisions that change an electron's velocity randomly, but none of this matters. Since Φ_{Rashba} is independent of electron velocity, a larger spread in the velocity due to higher temperature and more frequent elastic collisions will have no effect on Φ_{Rashba}. Therefore, raising the temperature does not degrade the performance of the one-dimensional SPINFET as long as the temperature is not so high as to make the channel length exceed the inelastic mean free path.

The reader would have understood that the SPINFET is not turned on and off by changing the *number* of electrons in the channel. Instead, it is switched by changing the spin polarization of the electrons in the channel. At first glance, this would give the impression that $\Delta Q = 0$ and hence the energy dissipation will vanish. This is wrong. Even though no charge is moved in the channel, some charge has to be moved in some external circuit to charge the gate and change the gate voltage. We can get an estimate of this ΔQ as follows.

From Eq. (8.9), we see that the transverse electric field needed to change Φ_{Rashba} by $\pi/2$ and turn the SPINFET off (from the normally on state) is

$$E_y^{\text{off}} = \frac{\pi\hbar^2}{4m^* a_{46} L} \tag{8.10}$$

The gate voltage needed to produce this transverse electric field is obtained from standard MOSFET theory as [9]

$$V_g = \Delta V = \frac{\varepsilon_s}{\varepsilon_i} \frac{\pi \hbar^2 d}{4 m^* a_{46} L} \tag{8.11}$$

where ε_s is the dielectric constant of the semiconductor channel and ε_i is the dielectric constant of the gate insulator.

The amount of charge moved in the gate circuit to change the gate voltage by V_g is $\Delta Q = C_g V_g$, where C_g is the gate capacitance. The associated energy dissipation will be $E_d = \Delta Q \Delta V = C_g V_g^2$.

We can estimate the gate capacitance of a 14-nm SPINFET from that of the 14-nm FinFET. They are not structurally identical; but for an order estimate, we can ignore this difference. For the FinFET, $C = \frac{\Delta Q}{\Delta V} = \frac{5.4\times 10^{-17}}{1.2} = 45$ aF. Assuming the same C for the SPINFET, we get

$$E_d^{\text{SPINFET}} = C\left(\frac{\varepsilon_s}{\varepsilon_i} \frac{\pi \hbar^2 d}{4 m^* a_{46} L} \right)^2 = 231 \text{ aJ} \tag{8.12}$$

where we assumed that the channel material is InAs (it has a large value of $a_{46} = 10^{-38}$ C-m^2 [10]), the gate insulator is InAlAs, $L = 14$ nm, $d = 1$ nm, and m^* is 0.03 times the free electron mass [11]. The ratio $\varepsilon_s/\varepsilon_i \approx 1$.

Clearly, the SPINFET is more dissipative than an equivalent MOSFET (or its various avatars like the FinFET) and does not produce an energy advantage. Even worse, it does not possess the most desirable property of spin-based devices, namely the non-volatility. If we turn the power off to the gate, the Rashba interaction will vanish and the SPINFET will revert to the "on-state" and all memory about its previous state will be lost. In addition, the conductance on/off ratio of a SPINFET cannot exceed $(1 + \eta_S\eta_D)/(1 - \eta_S\eta_D)$ [7], where η_S is the actual spin injection efficiency of the source contact and η_D is the actual spin detection efficiency of the drain contact. These efficiencies are not likely to far exceed 70% at room temperature [12] and hence the on/off ratio for a SPINFET is unlikely to exceed 3 : 1! In contrast, the same quantity for a MOSFET can be 10^6 : 1 at room temperature. The SPINFET's on/off ratio is actually extremely sensitive to the spin injection/detection efficiencies and drops off rapidly as the efficiencies decrease [13]. Thus, the SPINFET and its various clones [14–18] really offer no device advantage [13, 19], but the original proposal in Ref. [6] will remain an icon in the field of spintronics because it was among the first attempts to harness the spin degree of freedom of an electron to elicit device functionality.

8.2.2 Single Spin Logic Devices and Circuits

The reason that the SPINFET is not particularly energy efficient as a switch is because it did not attempt to exploit the advantage of spin over charge, namely, encoding information in the spin polarization (direction of the "vector"). Instead, it encodes information in the amount of charge transmitting through the drain (or, equivalently, the source-to-drain current). When more charge transmits through the drain, the transistor is on, and when less charge (or no charge) transmits through the drain, the transistor is off. Therefore, the transistor is switched by modulating the amount of charge exiting the drain, which means that $\Delta Q \neq 0$. This prompted us to investigate a paradigm that actually encodes bit information in spin polarization of a single electron – one polarization encoding the bit 0 and the opposite polarization the bit 1. We have to, of course, first make only these two spin polarizations stable (or metastable) so that we would be able to dispense with logic level restoration. In the following, we describe how this is achieved.

If we place a single spin in a magnetic field (assumed to be oriented in the z-direction) then the Hamiltonian describing the electron is

$$H = H_0 - (g/2)\mu_B \vec{B} \cdot \vec{\sigma} = H_0 - (g/2)\mu_B B_z \sigma_z$$
$$= \begin{bmatrix} H_0 - (g/2)\mu_B B_z & 0 \\ 0 & H_0 + (g/2)\mu_B B_z \end{bmatrix} \quad (8.13)$$

where B_z is the magnetic flux density and σ is the Pauli spin matrix. The eigenspinors for this Hamiltonian are clearly $\begin{bmatrix} 1 \\ 0 \end{bmatrix}$ and $\begin{bmatrix} 0 \\ 1 \end{bmatrix}$, which are spin polarizations directed parallel and antiparallel to the magnetic field. That means only spin polarizations parallel and antiparallel to the magnetic field are stable and metastable orientations. They can encode the binary bits 0 and 1. To switch between the bits, one needs to merely toggle the spin, without physically moving the electron in space and causing any current flow. In other words, $\Delta Q = 0$. Of course, some energy will have to be dissipated externally to flip the spin and some energy will be dissipated internally since the two spin states are not energetically degenerate (they are separated in energy by $g\mu_B B_z$), but this energy could conceivably be much lower than any $\Delta Q \Delta V$ energy.

This realization led to the proposal for *single spin logic* (SSL) [20–25], which encodes digital bit information in the direction of spin polarization. In SSL, single electrons are confined in semiconductor quantum dots that are placed close enough to each other on a substrate such that nearest neighbor spins interact via exchange while second nearest and more remote neighbors do not interact at all (or at least not sufficiently) since exchange interaction strength decays exponentially with distance. The wafer is placed in a dc magnetic field which defines the spin quantization axis and makes the spin polarization of every electron on the wafer either parallel or antiparallel to the global field. These two polarizations represent the logic bits 1 and 0, respectively. The global magnetic field has to be weak enough that the Zeeman splitting in every quantum dot is much weaker than the exchange splitting due to nearest neighbor exchange interactions. Exchange interaction favors the spins in two neighboring dots to be mutually antiparallel, while Zeeman interaction will tend to make them parallel and point along the magnetic field. If exchange interaction dominates, then the ordering will be antiparallel and this is what we want.

The array of interacting spins in a weak global magnetic field acts as a computing system. Input data are provided to the quantum dot array by aligning the spins in certain chosen quantum dots (designated as input ports) in desired directions (parallel or antiparallel to the global magnetic field) using external agents, such as local magnetic fields that could be generated with nanomachined, spin-polarized scanning tunneling microscope tips. This action "writes" the input bits into the array. The arrival of the input bit stream takes the interacting array into a many-body excited state. The system is then allowed to relax to the thermodynamic ground state by emitting magnons and phonons.

The system is engineered such that when the ground state is reached, the spin orientations in certain other chosen dots (designated as output ports) will represent the result of a specific computation in response to the input bits. The quantum dots are arranged on the wafer in such a way that the nature of the nearest neighbor interactions guarantees this occurrence. Just as a combinational or sequential Boolean circuit is implemented by connecting gates in a certain way with wires, computing units in SSL are realized by placing the dots on a wafer in a certain arrangement with the nearest neighbor exchange interaction acting as the "wire" between nearest neighbors. The advantage is that the "wire" in SSL is not a physical wire and hence does not incur any I^2R dissipation. Of course, the disadvantage is that the arrangement of the dots is fixed and cannot be changed on the fly. Therefore, a particular computer can do only one specific computation. It is not reconfigurable for a different task. This is an extreme example of an application-specific integrated circuit (ASIC).

While this is the grand vision, the more tempered vision is to fashion universal Boolean logic gates (e.g. NAND) or entire combinational/sequential logic circuits (e.g. half and full adders, or S-R flip flops) out of these interacting spins. Just three quantum dots arranged in a row can act as a NAND gate, where the two peripheral dots are the two input ports and the central dot is the output port. This is shown in Figure 8.2. References [21, 22] have established the operation of such a NAND gate using rigorous quantum mechanics and Ref. [26] has designed half and full adders, etc. based on semiempirical (heuristic) arguments. A full quantum mechanical analysis would have been computationally prohibitive in the latter case.

In any computer, there has to be a methodology for input/output (I/O). Writing (input) involves orienting the spins in selected quantum dots in one of the two orientations to write the bits 0 or 1. This requires generating extremely localized magnetic fields with spin-polarized scanning tunneling electron microscope tips. Reading (output) involves reading the spin orientations in certain other quantum dots. The reading operation can be carried out using a variety of schemes, which have been experimentally demonstrated [27–29].

The operating temperature of SSL is determined by how strong the exchange interaction between the spins in neighboring quantum dots can be and what bit error rate is tolerable. Reference [22] established the following relation between the exchange interaction energy J (between two neighboring dots) and the computational error probability p due to an array of spins being in an excited state (not the correct ground state):

$$2J = kT \ln(1/p) \tag{8.14}$$

Theoretical estimates have shown that a reasonable value of J in semiconductor quantum dots is 1 meV [30]. Therefore, if we wish to maintain an error probability p no higher than 10^{-9}, then we will have to work at an extremely low temperature of $\sim$1 K. This would be impractical. However, there has been a

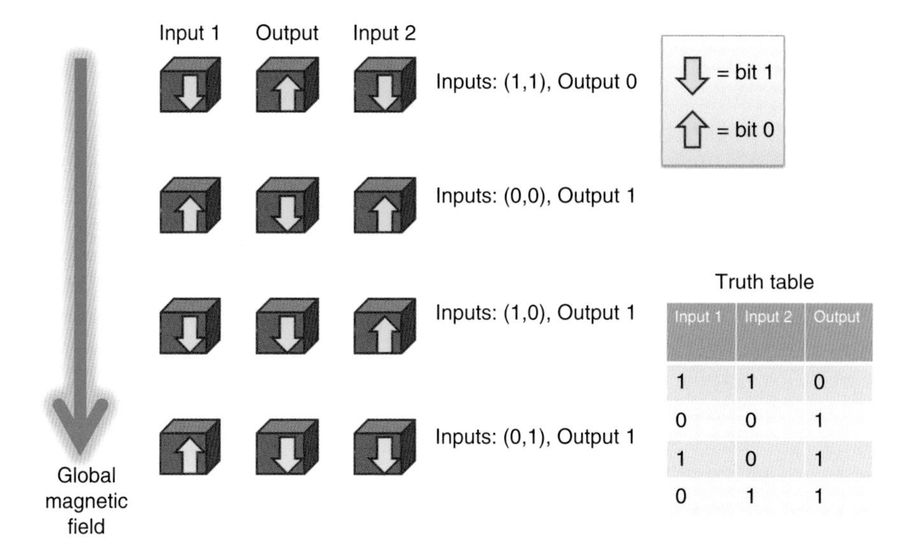

Figure 8.2 Configuring a NAND gate with exchange-coupled single-electron spins whose polarizations can point either parallel or antiparallel to a global magnetic field. The peripheral spins are the two input bits of the NAND gate and the central spin is the output bit. Down-spin encodes bit 1 and up-spin bit 0. Exchange interaction between nearest neighbors ensures that their spins are mutually antiparallel. This ensures the NAND functionality as shown. When one input is bit 0 and the other bit 1, there is seemingly a tie since the antiparallel ordering will not have any preference for the central spin to point up or down. This is resolved by the global magnetic field which, in this case, makes the down-spin preferred over the up-spin and breaks the tie.

recent claim that graphene nanoflakes can implement SSL-type logic gates with much higher exchange interaction strength ($J = 180\,\text{meV}$) [31], which would allow room-temperature operation with an error probability of 10^{-3}. This error probability is still too high for logic systems, which means that there is currently no clear pathway to take SSL to room-temperature operation.

In many ways, the SSL idea is like the SPINFET. It has little, if any, device utility, but was the first known attempt to harness single spins for performing any type of computation. It might have been the progenitor of spin-based quantum computing involving single spins in quantum dots [32–35], which is still an active field of research today despite the fact that it too has no clear pathway to room-temperature operation.

8.3 Nanomagnetic Devices: A Nanomagnet as a Giant Classical Spin

A *single-domain* nanomagnet is a nanometer scale (50–100 nm) ferromagnet consisting of many ($\sim 10^4$–10^6) electron spins. An amazing fact about them is that because of exchange interaction between the spins, all of the spins point in the *same* direction, which is the direction of the nanomagnet's magnetization. If the magnetization is forcibly rotated through an angle by some external agent like a magnetic field, then all the spins rotate *together in unison*, in perfect synchrony, acting like one giant classical spin [36, 37]. Unlike the single-electron spin, this giant classical spin is relatively robust at room temperature. That immediately raises the possibility of replicating SSL with single-domain nanomagnets and making it work at room temperature.

The SSL architectures for logic gates (e.g. the NAND gate of Figure 8.2) can be replicated with single-domain nanomagnets provided we can reproduce two features: (i) making the magnetization of the nanomagnet bistable (only two magnetization orientations are stable and allowed), and (ii) having an interaction between two nanomagnets that will tend to make their stable magnetization orientations mutually antiparallel just like exchange interaction makes two interacting spins antiparallel.

The first feature is implemented by shaping the nanomagnets as *elliptical disks* shown in Figure 8.3. Such a nanomagnet has only two stable magnetization orientations along the ellipse's major axis – pointing to the right and pointing to the left. No other orientation is stable because of the shape of the nanomagnet. The two stable orientations can encode the binary bits 0 and 1. Note that in the case of single spins, we needed a magnetic field to define the spin quantization axis and generate a stable and a metastable polarization. Here, we do not need a magnetic field and instead use the nanomagnet geometry to implement bistability of the magnetization. The two stable magnetization states are energetically degenerate, unlike in the case of single spin where the two states encoding the binary bits 0 and 1 are energetically nondegenerate.

The second feature is replicated by dipole interaction between nanomagnets. If two of them are placed in close proximity such that the line joining their centers is collinear with the minor axis of the ellipses, then dipole interaction will tend to make their magnetizations mutually antiparallel. This gives us all the ingredients necessary to implement SSL gates with nanomagnets. A nanomagnetic NAND gate is shown in Figure 8.4 and looks almost identical to the SSL NAND gate in Figure 8.2. The magnetic field in Figure 8.4 is not needed for making the magnetization of any nanomagnet bistable, but to break a "tie", i.e. for ensuring that when one input bit is 1 and the other 0, the output bit is 1 and not 0. This would ensure that the truth table is that of a NAND gate.

Table 8.2 shows a comparison between single-domain nanomagnets and single spins.

Unlike single-spin polarizations which are delicate and can be randomly flipped by thermal perturbations (single-spin lifetimes are few nanoseconds at best in room temperature), the magnetization state of a nanomagnet is very robust and

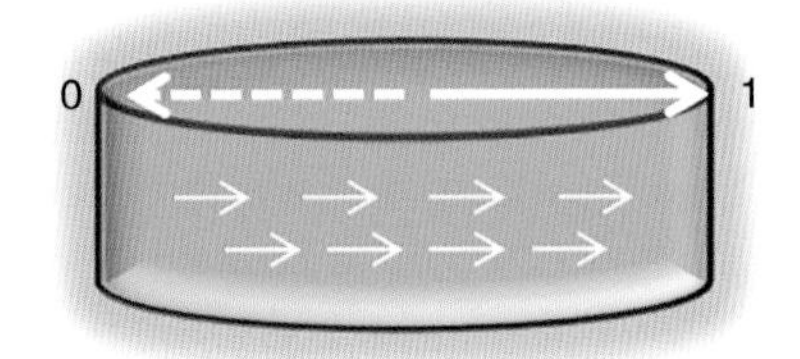

Figure 8.3 A single-domain nanomagnet shaped like an elliptical disk (dimensions not to scale). All the spins inside the nanomagnet point in the same direction. The spins have two allowed directions – pointing right or left along the major axis. These two magnetizations orientations encode the binary bits 0 and 1.

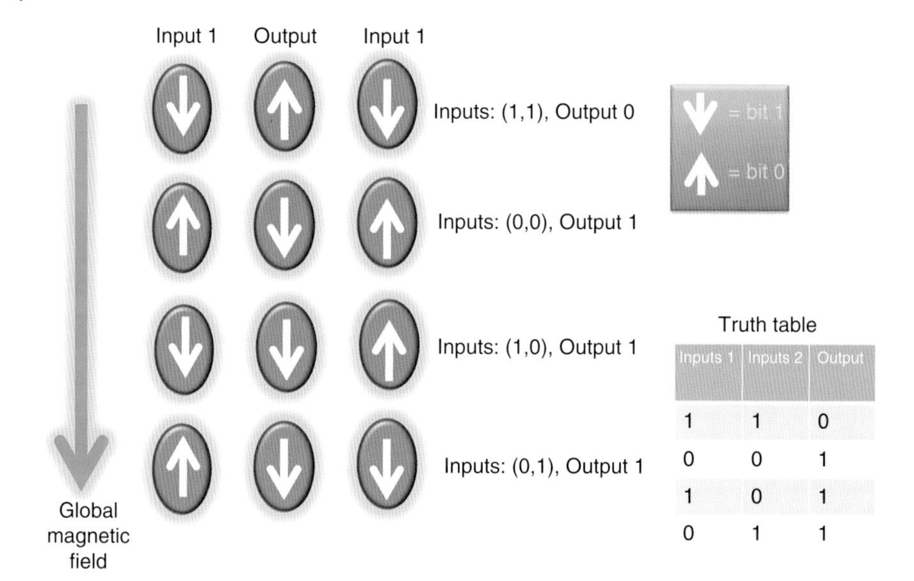

Figure 8.4 A nanomagnetic NAND gate implemented in the same way as the SSL NAND gate in Figure 8.2. Here, the global magnetic field is not needed for ensuring two stable magnetization states, but is needed for breaking the tie that occurs when the two input bits are dissimilar, meaning the magnetizations of the peripheral nanomagnets are mutually antiparallel.

Table 8.2 Comparison between single spins and single-domain nanomagnets.

	Single spin	Single-domain nanomagnet
State-encoding bits	Energetically nondegenerate	Energetically degenerate
Interaction	Exchange	Dipole
State longevity	Few nanoseconds	Centuries

can persist for centuries. This is because of a large energy barrier Δ that exists between the two stable magnetization states of an elliptical nanomagnet corresponding to the magnetization pointing right or left along the major axis. The height of this energy barrier is determined by the major and minor axis dimensions of the ellipse, as well as the thickness of the elliptical disk [38]. It is usually on the order of 1.5 eV (60 kT at room temperature). The probability of random bit flip (or magnetization flip between the two stable orientations) due to thermal perturbations is $e^{-(\Delta/kT)} = e^{-60}$, which is negligible. The mean time between two random, spontaneous, and unwanted bit flips at room temperature is $\tau_0 \exp[\Delta/kT] > 36$ centuries, where τ_0 is the inverse of the attempt frequency which is typically 1 fs to 1 ps [39]. Therefore, the binary bit state encoded in the magnetization of a single-domain nanomagnet is extremely robust. Information written into it stays intact for many centuries, and this is why the nanomagnet is considered "non-volatile."

8.3.1 Reading Magnetization States (or Stored Bit Information) in Nanomagnets

We first address how the magnetization orientation or, equivalently, the stored bit is "read" in a nanomagnet and then we discuss how to "write" a bit into a nanomagnet. Reading is accomplished with a construct known as a *magneto-tunneling junction* (MTJ) shown in Figure 8.5a. It has three layers – a hard ferromagnetic layer, a spacer layer, and a soft ferromagnetic layer (a single-domain nanomagnet) whose magnetization orientation encodes the bit to be read. The hard layer is implemented with a synthetic anti-ferromagnet consisting of exchange-coupled magnetic layers and its magnetization is fixed (hence "hard"). When a voltage is applied between the hard and soft layers, electrons tunnel between the two layers through the ultrathin spacer layer. Because the tunneling is spin-dependent, the resistance of the MTJ depends on whether the magnetizations of the hard and soft layers are mutually parallel or antiparallel. Since the hard layer's magnetization is fixed, effectively the resistance depends solely on the magnetization orientation of the soft layer, which encodes the stored bit. Therefore, we can "read" the stored bit by *measuring the MTJ resistance*. Usually, when the magnetizations of the hard and soft layers are antiparallel, the resistance is high and when they are parallel, the resistance is low. The resistance measurement thus provides an unambiguous reading of the bit stored in the magnetization orientation of the soft layer as long as the ratio of the high-to-low resistance (called the "tunneling magnetoresistance" (TMR)) is

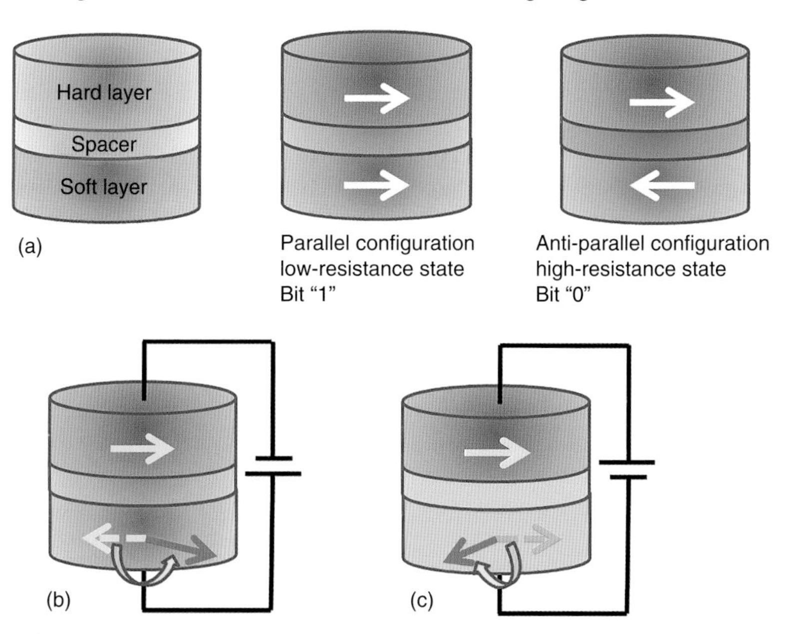

Figure 8.5 (a) A magnetotunneling junction whose two resistances encode the binary bits 0 and 1. (b) Writing bit 1 (low-resistance state). (c) Writing bit 0 (high-resistance state). Reproduced with permission from N. D'Souza, et al., Energy-efficient switching of nanomagnets for computing: Straintronics and other methodologies, to appear in *Nanotechnology* (https://doi.org/10.1088/1361-6528/aad65d). Copyright Institute of Physics.

sufficiently large to allow unambiguous determination of whether the measured resistance is "high" or "low." A TMR of 2 : 1 may be adequate and at the time of this writing, the record at room temperature is 7 : 1 [40].

8.3.2 Writing Magnetization States (or Storing Bit Information) in Nanomagnets

Writing bit information in a bistable nanomagnet (or the elliptical soft layer of an MTJ) involves orienting the magnetization in one of the two stable directions. There are many ways of achieving this and we discuss a few of them here.

8.3.2.1 Spin-Transfer Torque

One popular way to orient the magnetization of a nanomagnet in a desired direction is to pass a spin-polarized current through it. The spins of the electrons carrying the current are polarized primarily in the desired direction. These spins transfer their angular momenta to the resident electrons in the nanomagnet whose spins then orient along the polarization of the incoming spins, which is, of course, the desired direction. The process of transferring angular momenta in this manner results in the exertion of a torque on the magnetization of the nanomagnet. This torque – known as the "spin-transfer torque" (STT) – turns the magnetization to the desired orientation and writes a bit [41].

The MTJ, which reads the stored bit, can also be utilized to write a desired bit because the hard layer acts as a spin polarizer to generate a spin-polarized current. We will always want the magnetization of the soft layer to be either parallel or antiparallel to that of the hard layer's since the bit value can be either 0 or 1. Let us assume that the parallel configuration represents bit 1 and the antiparallel configuration bit 0. These two configurations are shown in Figure 8.5a.

If we wish to write the bit "1" which corresponds to the parallel configuration, then we will connect a battery across the MTJ in such a manner that the negative terminal is in contact with the hard layer and the positive terminal is in contact with the soft layer. This will inject electrons from the hard layer into the soft layer via tunneling through the intervening spacer layer. The spins of the injected electrons will be mostly polarized along the magnetization of the hard layer. They will exert an STT on the soft layer by transferring their angular momenta to the resident spins. As a result, if the soft layer's magnetization was already parallel to that of the hard layer's, then it will remain so. On the other hand, if it was antiparallel, the STT will turn the magnetization around by 180° and make it parallel to that of the hard layer's. This is shown in Figure 8.5b.

If we wish to write the bit "0," we will simply reverse the polarity of the battery as shown in Figure 8.5c. In this case, the soft layer will try to inject electrons into the hard layer by tunneling through the spacer layer. Because of spin-dependent tunneling, those electrons whose spins are aligned parallel to the hard layer's magnetization will preferentially tunnel through the spacer to enter the hard layer and leave the soft layer. This depletes the population of such spins in the soft layer, and, as a result, spins that are antiparallel to the hard layer's magnetization gradually become majority spins in the soft layer. This makes the soft layer's magnetization antiparallel to that of the hard layer's and writes the bit 0. If the soft layer's magnetization was already antiparallel to that of the hard layer's, then this process leaves it in the antiparallel state and the bit "0" is written unambiguously. Thus, we can write either bit 0 or bit 1 by simply choosing the polarity of the battery.

Note that this process of writing bits results in a *non-toggle* memory. The process of writing does not just toggle the previously stored bit. Instead, it definitively writes either bit 0 or bit 1, regardless of what the previously stored bit was. This should be contrasted with *toggle* memory, where the process of writing merely toggles the previously stored bit. If the previously stored bit was 1, it will write the bit 0, and vice versa. Therefore, in a toggle memory, the write cycle has to be always preceded by a read cycle. If the bit read is the desired bit, then do nothing. Else, toggle the previously stored bit to write the desired bit. We mention this here since many of the other writing schemes will result in toggle memory.

The advantages of STT are (i) non-toggle memory and (ii) low write error probability (WEP). Writing error occurs when the STT fails to turn the magnetization of the soft layer in the desired direction. This can happen owing to many factors, such as thermal noise that acts as a random stray magnetic field, materials defects that pin the magnetization, etc. These issues can be overcome by increasing the magnitude of the spin-polarized current. The switching speed can be also increased by increasing the magnitude of the current. Of course, all this will incur an energy penalty (reliability and speed come at the expense of energy). There is however a minimum threshold current density needed to bring about the switching. This threshold current density (also known as the critical current density J_{cr}) depends on many factors such as the energy barrier within the nanomagnet, the degree of spin polarization in the current, etc. It can be quite high ($\sim$1 MA cm^{-2}). Lowering the critical current density is the most important research objective since it will reduce the energy dissipated in writing a bit.

8.3.2.2 Spin-Transfer Torque Aided by Giant Spin Hall Effect

The critical current density can be reduced with the aid of the giant spin Hall effect (GSHE) [42–44]. The (extrinsic) spin Hall effect [45–50] is a phenomenon that is best explained with the aid of Figure 8.6, which shows a two-dimensional

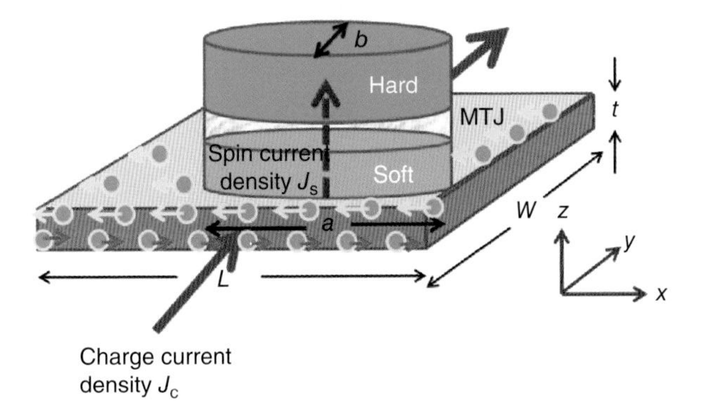

Figure 8.6 Switching the magnetization of the soft layer of a magnetotunneling junction with a spin current generated by the giant spin Hall effect in a heavy metal film. The spin current is generated by a charge current. Reproduced with permission from N. D'Souza, et al., Energy-efficient switching of nanomagnets for computing: Straintronics and other methodologies, to appear in *Nanotechnology* (https://doi.org/10.1088/1361-6528/aad65d). Copyright Institute of Physics.

slab of material that has strong spin–orbit interaction. A "charge" current of density J_c is injected into it in the y-direction with a battery and may or may not have any net spin polarization. The electrons in the injected current suffer spin-dependent scattering as they travel through the slab because of the spin–orbit interaction, as a result of which $+z$-polarized spins are deflected to one edge of the slab and $-z$-polarized spins are deflected to the opposite edge as shown in Figure 8.6. This builds up a spin imbalance in the z-direction that drives a spin current of density J_s in the z-direction, up into the soft layer. This spin current delivers an STT on the magnetization of the soft layer and rotates it in the direction of the spin polarization of the spin current.

The ratio of the spin-to-charge current density is called the "spin Hall angle" and is given by

$$\theta_{SH} = \frac{J_s}{J_c} \tag{8.15}$$

The spin Hall angle is usually quite small in most materials, but in certain heavy metals it can be large. It is reported to be 0.15 in β-Ta [42], 0.3 in β-W [43], and 0.24 in CuBi alloys [44]. These materials are therefore said to exhibit the GSHE.

Note that the spin current does not dissipate any power since the scalar product $\vec{J}_s \cdot \vec{E} = 0$, where $\vec{E}$ is the electric field driving the charge current and it is collinear with $\vec{J}_c$, which is perpendicular to $\vec{J}_s$. Any power dissipation is due to the charge current.

Let us say that the critical spin current (minimum spin current needed to switch the soft magnetic layer of the MTJ) is I_s^{cr}. The corresponding charge current is

$$I_c^{cr} = J_c^{cr} Lt = \frac{J_s^{cr}}{\theta_{SH}} Lt = \frac{I_s^{cr}}{\theta_{SH}(\pi/4)ab} Lt = \frac{I_s^{cr}}{\theta_{SH}} \frac{4Lt}{\pi ab} \tag{8.16}$$

where L is the length, W is the width, and t the thickness of the slab, and $(\pi/4)ab$ is the cross-sectional area of the MTJ through which the spin-polarized current I_s flows (a is the major axis and b is the minor axis of the elliptical cross-section of the MTJ). The power dissipation then turns out to be

$$\begin{aligned} P_d &= (I_c^{cr})^2 R = \left(\frac{I_s^{cr}}{\theta_{SH}} \frac{4Lt}{\pi ab}\right)^2 \rho \frac{W}{Lt} = (I_s^{cr})^2 \left(\frac{4}{\pi\theta_{SH}}\right)^2 \rho \frac{WLt}{a^2b^2} \\ &= (I_s^{cr})^2 \left(\frac{4}{\pi\theta_{SH}}\right)^2 \rho \frac{WLt}{a^2b^2} = (I_s^{cr})^2 \left(\frac{4}{\pi\theta_{SH}}\right)^2 \rho \frac{V_{slab}}{a^2b^2} \end{aligned} \tag{8.17}$$

where R is the resistance in the path of the charge current, V_{slab} is the slab volume, and ρ is the resistivity of the slab material (the heavy metal).

Clearly, there are two ways to make the power (and energy) dissipation small: first by using a material with large spin Hall angle, and second by using a slab with very small thickness t [3, 51]. The energy dissipation to flip the magnetization of a nanomagnet can be reduced to $\sim 10^4$ kT (1.6 fJ) or even lower by exploiting the GSHE [52].

Note also from Eq. (8.16) that the ratio of the spin current to charge current is

$$\beta = \frac{I_s}{I_c} = \theta_{SH} \frac{(\pi/4)ab}{Lt} = \theta_{SH} \frac{A_{MTJ}}{A_{slab}} \tag{8.18}$$

where A_{MTJ} is the cross-sectional area of the MTJ through which the spin current flows and A_{slab} is the cross-sectional area of the slab through which the charge current flows. The quantity β acts like a gain [3] and its value can be made much larger than unity by ensuring that $A_{slab}/A_{MTJ} \ll \theta_{SH}$.

An interesting question that arises at this point is what is the equivalent amount of charge ΔQ that has to be moved to switch the magnetization of the soft layer of an MTJ using STT? Obviously some charge has to be moved somewhere to produce the charge current I_c. Recall from Section 8.1 that the quantity ΔQ was 300–400 electrons for a modern-day transistor (the 14-nm FinFET in the Intel i7-6700K processor). In order to calculate ΔQ for STT, we proceed as [3].

We have to rotate a minimum number of spins in the soft layer of an MTJ with the spin-polarized charge current (or simply spin current in the case of GSHE-assisted STT) to make the switching occur. Let this number be N. Since the electrons passing through the soft layer impart their angular momenta to the resident electron spins to make the nanomagnet switch, it is clear that we need at least $2N$ electrons to pass through. Hence, $\Delta Q \geq 2qN$, where q is the electronic charge.

In the worst-case scenario, we have to rotate all the spins in a nanomagnet to make the magnetization flip, in which case, $N = M$, where M is the total number of spins in a nanomagnet. The latter is given by [3]

$$M = M_s\Omega/\mu_B \tag{8.19}$$

where M_s is the saturation magnetization of the nanomagnet and μ_B is the Bohr magnetron, which is the magnetic moment of a single electron. Therefore, we get that

$$\Delta Q \geq \frac{2qM_s\Omega}{\mu_B} \tag{8.20}$$

In the case of GSHE-assisted STT, the abovementioned inequality becomes

$$\Delta Q \geq \frac{2qM_s\Omega}{\beta\mu_B} = \frac{2qM_s t_m L t}{\mu_B}\frac{1}{\theta_{SH}} \tag{8.21}$$

where t_m is the thickness of the soft layer.

We cannot make M_s or Ω arbitrarily small. For one thing, making Ω too small may make the nanomagnet superparamagnetic as opposed to ferromagnetic. More importantly, these quantities determine the energy barrier within the nanomagnet and decreasing them will decrease the energy barrier height, which is damaging. The spontaneous bit flip probability increases exponentially as the energy barrier is lowered and hence we must maintain a minimum barrier height for reliability. If we use reasonable numbers in Eq. (8.21), e.g. $t = t_m = 5\,\text{nm}$, $L = 100\,\text{nm}$, $M_s = 10\,\text{A}\,\text{m}^{-1}$, and $\theta_{SH} = 0.2$, we get that $\Delta Q \sim 270\,000$ electrons! This immediately tells us that STT, even with GSHE assistance, is *not* very energy efficient. Fortunately, the energy dissipation is $\Delta Q \Delta V$ and the quantity ΔV can be quite small since the driving current passes through a heavy metal slab which has a relatively low resistance. Nonetheless, STT switching of a nanomagnet dissipates more energy than what is involved in switching a transistor, unless the soft layer's thickness or the slab's cross-sectional area is reduced significantly. The STT switching speed depends on the current (and hence the energy dissipated), but typical switching delays are on the order of 1 ns, which makes the switching much slower than that of modern transistors that typically occurs in $\sim$100 ps.

8.3.2.3 Voltage-Controlled Magnetic Anisotropy

One of the reasons why STT is relatively energy inefficient is that it is a method of switching magnetization by passing a current through an MTJ. A rather large current is needed to switch the magnetization of the soft layer. At the same time, the MTJ resistance is also quite high, which leads to a large I^2R loss. The MTJ resistance R is high because of the presence of the spacer layer which is an insulator and current has to tunnel through it. One obvious solution to this problem is to separate the read and write current paths. The read current measures the resistance of the MTJ and therefore must necessarily pass through the spacer layer, but the write current needs to pass only though the soft layer and can avoid the spacer layer as long as there is an external spin polarizer to polarize the write current. Since the read current is always much smaller than the write current (so that the read current never overwrites the stored bit), passing the write current through a low-resistance path that avoids the high-resistance spacer layer should produce an energy benefit. This is precisely what the GSHE feature does.

This idea, however, has a serious shortcoming when it comes to memory applications. In memory applications, the most important consideration is the *density* of memory cells and anything that reduces the density is an anathema. If we separate the read and write paths, we will end up with a three- or four-terminal device which has a much larger footprint than a two-terminal device. This will inevitably decrease the density of memory cells, making it an unattractive idea. Memory cells must be two-terminal to be competitive in the memory market. Hence, the idea of separating the read and write paths to reduce energy dissipation is not a viable solution. Consequently, the GSHE idea is not all that attractive for memory applications since it will result in at least a 3-terminal memory cell.

All this begs the question if there are *voltage-controlled* magnet switching schemes that could be more energy efficient than the *current-controlled* schemes. There is no fundamental reason as to why one would be more efficient than the other, but the recent spate of activity in voltage-controlled switching is inspired by experimental evidence (bolstered by theoretical simulations) that voltage-controlled switching may turn out to be less dissipative and hence more energy conserving. There are two principal voltage-controlled switching schemes – *voltage-controlled magnetic anisotropy* (VCMA) and *straintronics.* We discuss them next.

Before we discuss VCMA, we wish to make a distinction between in-plane magnetic anisotropy and perpendicular magnetic anisotropy in a nanomagnet. The total potential energy of a nanomagnet (U) depends on the orientation of its magnetization and is given by

$$U = \underbrace{\frac{\mu_0}{2} M_s^2 \Omega \left[N_{d-zz} \cos^2\theta + N_{d-yy} \sin^2\theta \sin^2\phi + N_{d-xx} \sin^2\theta \cos^2\phi\right]}_{\text{shape anisotropy energy}} \\ - \underbrace{\left[K_{s0} \frac{\Omega}{t_m}\right] \sin^2\theta \cos^2\phi}_{\text{surface anisotropy energy}} \tag{8.22}$$

where θ and ϕ are the polar and azimuthal angles of the magnetization vector shown in Figure 8.7. Here, μ_0 is the permeability of free space, N_{d-mm} is the

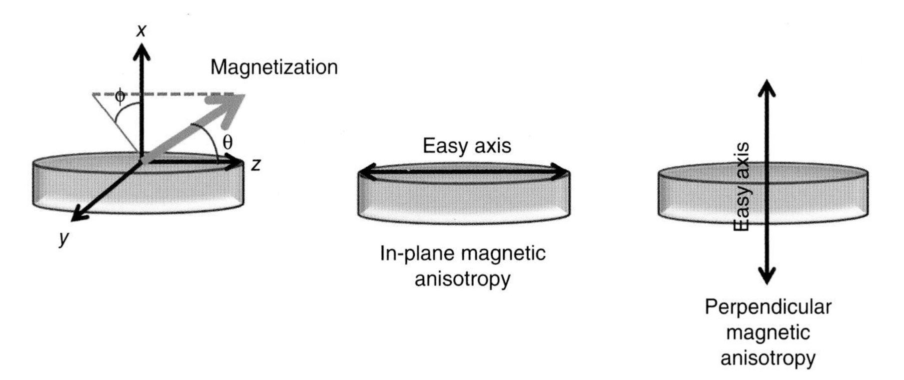

Figure 8.7 In-plane and perpendicular magnetic anisotropy. Reproduced with permission from N. D'Souza, et al., Energy-efficient switching of nanomagnets for computing: Straintronics and other methodologies, to appear in *Nanotechnology* (https://doi.org/10.1088/1361-6528/aad65d). Copyright Institute of Physics.

demagnetization factor associated with the mth coordinate axes (these factors depend on the major axis, minor axis, and thickness dimensions of an elliptical nanomagnet), t_m is the nanomagnet thickness, and the other quantities have the same meaning as before. The quantity K_{s0} is the surface anisotropy constant which depends on the interface between the nanomagnet and the material in immediate contact with it.

For an elliptical nanomagnet with major axis along the z-axis, minor axis along the y-axis, and thickness along the x-axis, as shown in Figure 8.7, $N_{d-zz} < N_{d-yy} \ll N_{d-xx}$. Therefore, as long as the shape anisotropy term in Eq. (8.22) dominates over the surface anisotropy term, i.e. K_{s0} is relatively small, it is obvious that the potential energy U is minimized when $\theta = 0°$ or $180°$, and $\phi = 90°$. That means that the magnetization will prefer to lie along the $\pm z$-axis, or the major axis, and the two directions along the major axis are the two stable orientations. The major axis is hence called the "easy axis" since it is easiest for the magnetization to align along that direction. Since the easy axis lies in the plane of the nanomagnet, the latter is said to have *in-plane anisotropy*.

Next, consider the situation when the surface anisotropy term is dominant, i.e. K_{s0} is large. In this case, the potential energy is minimized when $\theta = 90°$, and $\phi = 0°$ or $180°$, which means that the easy axis is along the x-axis and the magnetization will prefer to lie along the $\pm x$-direction. Such a nanomagnet will have *perpendicular magnetic anisotropy* (PMA). Note that a nanomagnet with PMA need not have an elliptical cross-section in order to have bistable magnetization orientations. It can very well have circular cross-section and yet be bistable since the two stable directions are perpendicular to the nanomagnet's plane and have nothing to do with the cross-section's shape. In fact, this is preferred since it is difficult to lithographically control eccentricity of the ellipse across numerous nanomagnets. Furthermore, PMA nanomagnets can have a smaller cross-section (footprint) and that is conducive to higher density. That is why PMA nanomagnets are almost universally preferred for memory applications where density is the primary consideration.

VCMA essentially consists of changing the surface anisotropy energy constant K_{s0} in Eq. (8.22) with a voltage. This is easily implemented in an MTJ whose hard

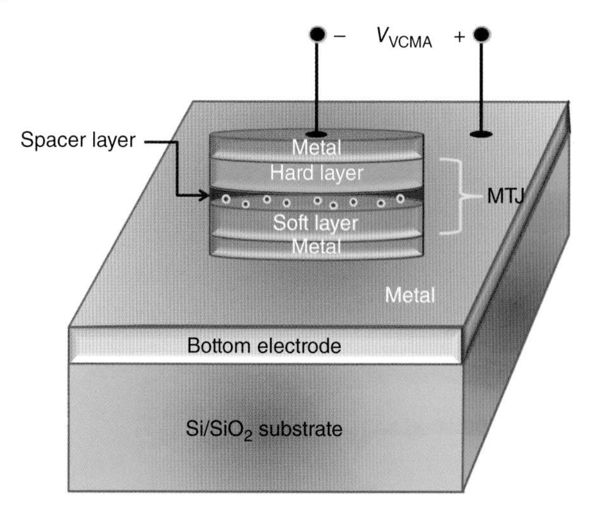

Figure 8.8 Switching the magnetization orientation of the soft layer of a p-MTJ with voltage-controlled magnetic anisotropy. Reproduced with permission from N. D'Souza, et al., Energy-efficient switching of nanomagnets for computing: Straintronics and other methodologies, to appear in *Nanotechnology* (https://doi.org/10.1088/1361-6528/aad65d). Copyright Institute of Physics.

and soft layers have perpendicular magnetic anisotropy (p-MTJ). The structure is shown in Figure 8.8. When a potential V_{VCMA} is applied between the hard and soft layers, electrons are injected into the spacer layer from either the hard or the soft layer depending on the polarity of V_{VCMA}. These electrons accumulate in the spacer layer and modify the band structure and/or occupation of "d-like" bands in the transition metal soft layer. Because of spin–orbit interaction or spin-dependent screening in the soft layer, this modifies the magnetic anisotropy perpendicular to the interface [53–56]. This effect is similar to the surface magnetoelectric effect where an electric field modifies the magnetocrystalline anisotropy and magnetization at the interface of a ferromagnetic metal and dielectric owing to spin-dependent screening [57, 58] or change in bandstructure [59]. Put succinctly, the voltage V_{VCMA} will change the surface anisotropy constant K_{s0}. This change is expressed in a linear relationship of the form [53]

$$K_{s0} \rightarrow K_{s0} + \frac{\kappa V_{VCMA}}{t_b} \tag{8.23}$$

where κ is the so-called VCMA constant and t_b is the thickness of the spacer layer.

If the applied voltage V_{VCMA} increases K_{s0} (κV_{CMA} product positive), then the surface anisotropy term in Eq. (8.22) will dominate and the easy axis of the free layer will become perpendicular to the plane of the soft layer, i.e. the material will exhibit PMA. On the other hand, if the applied voltage V_{VCMA} decreases K_{s0} (κV_{CMA} product negative), then the shape-anisotropy term in Eq. (8.22) will dominate and the easy axis will lie in the plane of the soft layer along the major axis, i.e. the material will exhibit in-plane magnetic anisotropy. Thus, by changing the applied voltage, we can switch the magnetization of the free layer in Figure 8.8 from vertical to in-plane, or vice versa. Since the resistance of the MTJ depends on the relative orientation between the hard and soft layers, this will allow switching the resistance of the MTJ deterministically between two states, thereby allowing one to write the bits 0 and 1. VCMA switching has been demonstrated and discussed by a large number of groups [53–56, 60–72].

Note that the device in Figure 8.8 has only *two terminals* which are used for both read and write operations. Since VCMA allows building two-terminal memory cells, it will be an extremely attractive paradigm for non-volatile memory applications provided the write energy dissipation is considerably less than that in STT (including GSHE-assisted STT) and the TMR (which is the resistance on/off ratio) is sufficiently large.

Normal VCMA switching rotates the magnetization of the soft layer by 90° (in-plane to out-of-plane, or vice versa), and not by full 180°. This will not result in a large TMR. The resistance of an MTJ depends on the angle φ between the magnetizations of the hard and soft layers according to [73]

$$\frac{R(\varphi) - R(0)}{R(\pi) - R(0)} \propto \frac{1 - \cos\varphi}{3 + \cos\varphi} \tag{8.24}$$

This implies that

$$\mathrm{TMR} = \frac{R(\pi/2)}{R(0)} = \frac{R(\pi)/R(0) + 2}{3} < \frac{R(\pi)}{R(0)} \tag{8.25}$$

Clearly, the TMR or resistance ratio is smaller when the angular separation is 90° than when it is full 180°. In the preceding equations, we had assumed that the product of the spin polarizations at the interfaces of the ferromagnets and spacers in the MTJ is 1/3, but the inequality above is valid for any spin polarization.

It would therefore be desirable to rotate the magnetization of the free layer by full 180° to increase the TMR. This is accomplished with the help of an in-plane magnetic field. Let us say that the soft layer has perpendicular magnetic anisotropy and, in the absence of V_{VCMA}, the easy axis (and hence the magnetization) is pointing vertically up. When V_{VCMA} is turned on, the easy axis begins to turn toward an in-plane orientation, but because of the in-plane magnetic field, the magnetization that follows the easy axis initially, begins to precess about the magnetic field. By turning off V_{VCMA} and the magnetic field at the precise moment when the magnetization has completed a half period of precession, the magnetization can be flipped or rotated by 180°. This strategy was employed in [60] to demonstrate full magnetic reversal with VCMA. The only problem with this approach is that the time taken to complete one half-period of precession is not a fixed quantity in the presence of room-temperature thermal noise, which introduces a spread in this time. However, as long as we turn off V_{VCMA} and the magnetic field when the magnetization has rotated *close* to 180°, it will ultimately settle into the stable orientation that is antiparallel to the initial orientation after withdrawal of the VCMA voltage. This is because the magnetization will always gravitate to the closest energy minimum upon removal of all excitation. Therefore, the probability of not flipping, or the write error probability (WEP) will be relatively small. By tailoring the magnetic field strength and the turn-off time, one can minimize the WEP [68, 69].

The advantage of VCMA switching is that it is much less dissipative than STT switching. While STT switching typically dissipates ~40 fJ/bit energy [70], VCMA switching can dissipate ~6 aJ/bit currently in MTJs with a relatively large resistance area product [68]. The reader will recognize from Figure 8.8 that the voltage inducing VCMA will also cause a spin-polarized current to flow between the soft and hard layers, which will exert an STT on the soft

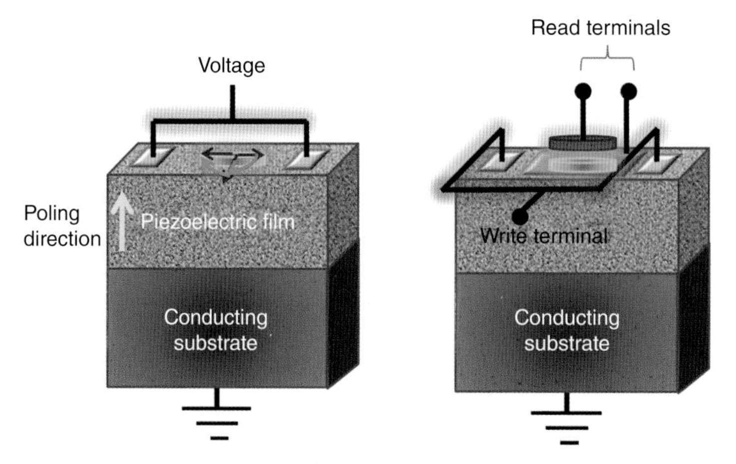

Figure 8.9 (a) Straintronic switching of an elliptical magnetostrictive nanomagnet with bistable magnetization. (b) A straintronic memory cell implemented with an MTJ whose soft layer is a magnetostrictive/piezoelectric two-phase multiferroic.

layer. Thus, VCMA and STT coexist. The total energy dissipation in switching is $I_{STT}V_{VCMA}t_s + C_{MTJ}V_{VCMA}^2 = V_{VCMA}^2 t_s/R_{MTJ} + C_{MTJ}V_{VCMA}^2$, where t_s is the switching delay, R_{MTJ} is the resistance of the MTJ, and C_{MTJ} is its capacitance. Clearly, the dissipation can be reduced by increasing R_{MTJ} and decreasing C_{MTJ}, both of which can be accomplished by reducing the cross-sectional area of the MTJ. This is the approach adopted in [68] and it has the additional benefit of decreasing the memory cell footprint.

8.3.2.4 Straintronics

Perhaps the most energy-efficient way to rotate the magnetization of a *magnetostrictive* nanomagnet (and thus write a bit into a magnetic memory cell) is to use electrically generated mechanical strain. Strain can rotate the magnetization of a magnetostrictive nanomagnet via the inverse magnetostriction effect, also known as the Villari effect, and the strain can be generated electrically as shown in Figure 8.9a.

The structure consists of an elliptical magnetostrictive disk (with bistable magnetization directions along the major axis due to in-plane anisotropy) in elastic contact with an underlying poled piezoelectric film deposited on a conducting substrate. They form a two-phase (piezoelectric/magnetostrictive) multiferroic. Two electrodes are delineated on the surface of the piezoelectric such that the line joining their centers is collinear with the major axis. These two electrodes are shorted together and a voltage is applied between them and ground. Depending on the polarity of the voltage, an electric field is generated in the piezoelectric, which is mostly either parallel or antiparallel to the direction of poling.

The electric field generates biaxial strain in the piezoelectric via d_{33} and d_{31} coupling. This strain is tensile along the major axis and compressive along the minor axis, or vice versa, depending on the polarity of the applied voltage [74–76]. It is partially or fully transferred to the elliptical nanomagnet and rotates its magnetization via the Villari effect. The term "straintronics" was coined to describe this method of rotating a magnetostrictive nanomagnet's magnetization.

The maximum rotation angle is 90°, which will place the magnetization along the minor axis of the ellipse. Normally, if the stress is withdrawn long after the 90° rotation is complete, the magnetization will either return to the original orientation along the major axis or to the opposite direction along the major axis with equal probability. Thus, the probability of bit flip will be only 50%, which is, of course, unacceptable. However, when the magnetization rotates, it also lifts out of the nanomagnet's plane. This introduces a demagnetizing field and a resulting torque that will continue to rotate the magnetization past 90° if the stress is withdrawn *at the precise moment* that the projection of the magnetization vector on the nanomagnet's plane aligns with the minor axis [77]. This will complete 180° rotation and flip the magnetization with a very high probability (~99.9999% at room temperature in the presence of thermal noise [77]).

Such a construct can be used to write bits in a *toggle* memory cell composed of an MTJ built atop the multiferroic nanomagnet. The multiferroic will serve as the soft layer. One would first read the initial magnetization orientation (and hence the stored bit) by measuring the MTJ resistance with a small read current. If the stored bit is already the desired bit, do nothing. Otherwise, pulse the voltage to generate stress and turn off the pulse at the precise moment (as indicated earlier) to flip the magnetization and write the desired bit.

Non-toggle Memory It is also possible to implement non-toggle memory with straintronics. The basic principle behind this is that if the magnetostrictive material has positive magnetostriction, then uniaxial compressive stress along any direction will drive the magnetization to a state that is approximately perpendicular to the stress axis and tensile stress will align it along the stress axis. The opposite will be true if the magnetostriction coefficient is negative. This principle is adopted to devise a non-toggle memory cell [78].

In an elliptical nanomagnet, we can bring the two stable directions out of the major axis and make them subtend an angle of ~90° with each other if we apply a fixed magnetic field of appropriate strength along the minor axis as shown in Figure 8.10.

Let us suppose that the magnetostriction coefficient of the nanomagnet is positive. We can apply uniaxial stress along one of the stable directions, say, direction "0." If the stress is compressive (applied with, say, a positive voltage), then the

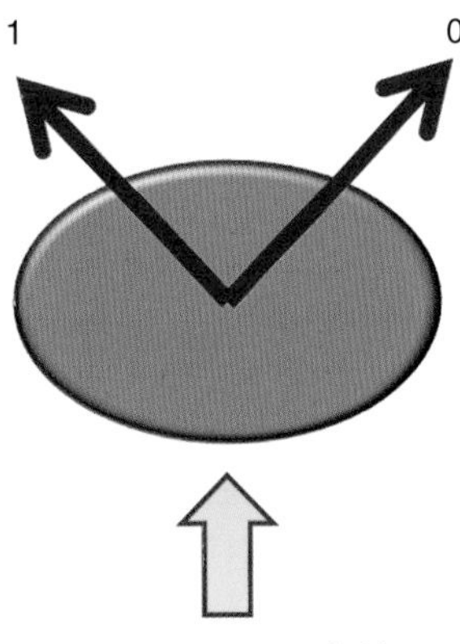

Figure 8.10 Applying a magnetic field along the minor axis of an elliptical nanomagnet to create two stable magnetization directions that are at 90° with each other.

magnetization will go to the state “1,” while if the stress is tensile (applied with a negative voltage), then the magnetization will align along the direction “0” [78]. These two directions encode the bits 0 and 1. Thus, by choosing the polarity of the voltage, we can deterministically write the bits 0 or 1, without having to know the previously stored bit. This constitutes a non-toggle memory [78]. It is also possible to increase the angular separation between the two stable orientations to obtain a higher resistance ratio (TMR) in an MTJ [79].

The reader will understand that a straintronic memory cell cannot be two-terminal. Figure 8.9b shows a four-terminal realization. Since the piezoelectric layer will be much more resistive than the MTJ, it cannot be placed in series with the MTJ for the read current to pass through. If we do, the read current will not be able to discriminate between the high- and low-resistance states of the MTJ (and read the stored bit) since the resistance in the path of the read current will be dominated by the piezoelectric and not by the MTJ. Thus, we will have to separate the read and write paths as shown in Figure 8.9b, so that the read current does not have to pass through the piezoelectric. An ultrathin metallic layer is placed between the magnetostrictive layer and the piezoelectric for the purpose of selectively contacting the MTJ for the read operation. This layer will undoubtedly partially impede strain transfer from the piezoelectric film to the magnetostrictive soft layer; but if the metallic layer is sufficiently thin, this will be a small effect. The “more than two” terminal implementation is certainly not conducive to high-density memory, but there is still some interest in this methodology of writing bits because the write energy can be extremely low.

Straintronic writing can result in extremely low write energy dissipation. There are three primary sources of dissipation during the write operation: the dissipation associated with magnetization damping (or Gilbert damping) in the magnetostrictive nanomagnet, the charging dissipation CV^2 associated with applying a voltage across the piezoelectric layer (where C is the capacitance of the electrodes to which voltage V is applied to generate the strain in the piezoelectric film in Figure 8.9a), and the mechanical energy (stress × strain × nanomagnet volume). These quantities are plotted in Figure 8.11 for a 6-nm-thick Terfenol-D nanomagnet (Terfenol-D has one of the highest magnetostriction among all materials) of elliptical shape (major axis = 100 nm, minor axis = 90 nm) delineated on a 24-nm-thick PZT (lead zirconate titanate) thin film. These plots were generated by solving the stochastic Landau–Lifshitz–Gilbert simulation in the presence of room-temperature thermal noise [80].

Figure 8.11 shows that the energy dissipation in straintronic switching is about 500 kT at room temperature, which is roughly 2 aJ of energy. Other theoretical estimates, carried out for specific cases, support this ultralow estimate [81–83]. Recent experiments to demonstrate straintronic switching in Co and FeGa nanomagnets fabricated on piezoelectric PMN-PT substrate also bear out this estimate for properly scaled (~100 nm thick) piezoelectric thin films [84–88].

Straintronic Logic So far, we have discussed only straintronic non-volatile “memory,” which involves writing bits in a nanomagnetic cell with strain. They have been experimentally demonstrated in Refs [84, 85]. Even a full straintronic MTJ has been experimentally demonstrated and comprises a complete unit for writing and reading a “bit” [86, 87]. In this subsection, we address straintronic “logic.”

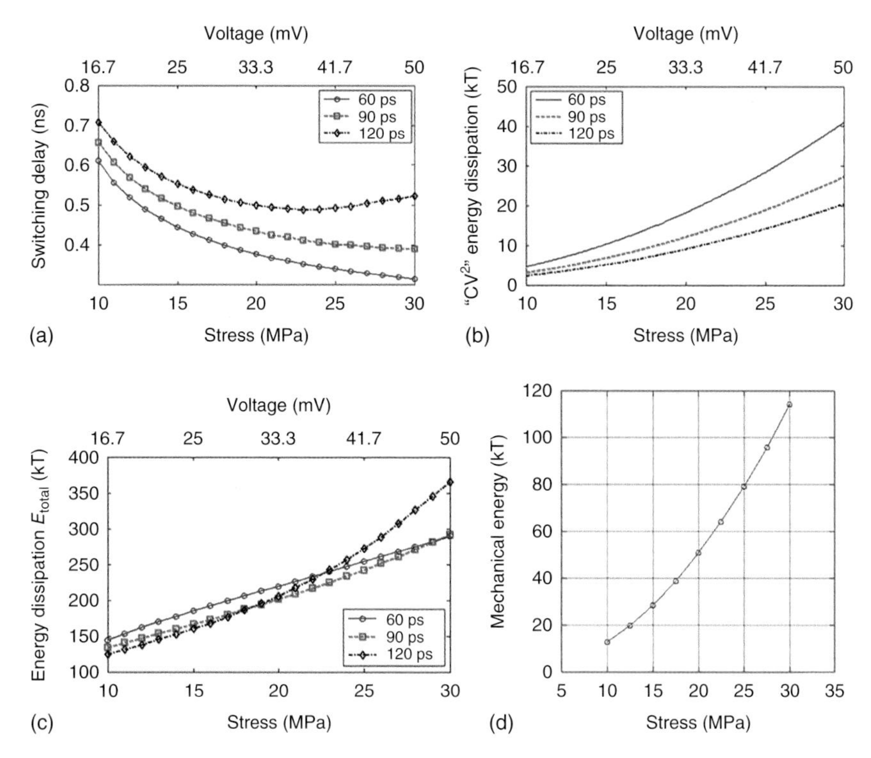

Figure 8.11 (a) The mean switching delay associated with rotating the magnetization of a magnetostrictive nanomagnet as a function of stress (in the presence of room-temperature thermal noise which introduces a spread in the switching delay). The nanomagnet is made of Terfenol-D and is shaped like an elliptical disk of major axis 100 nm, minor axis 90 nm, and thickness 6 nm. The plots are for three different rise times of the stress pulse: 60, 90, and 120 ps. (b) The charging (CV^2) energy dissipation (in units of room-temperature kT) as a function of stress. (c) The mean nonmechanical energy (Gilbert damping + CV^2) as a function of stress. (d) The mechanical energy dissipation as a function of stress. In panels (a), (b), and (c), the voltage required to generate the corresponding stress is plotted along the top horizontal axis. Source: Reproduced from Biswas et al. 2014 [79] with permission of the American Institute of Physics.

There are two types of straintronic logic that have been considered in the literature: dipole coupled logic and MTJ-based logic. An example of a dipole-coupled NAND gate is shown in Figure 8.4. Such gates and strain-based Bennett clocking schemes to steer logic bits from one logic stage to the next have been examined in Refs [82, 83, 88–90] and experimentally demonstrated in [91–93]. However, dipole-coupled schemes are rather error prone in the presence of room-temperature thermal noise, which can easily scuttle the conditional switching dynamics required in logic implementations [94–97]. In contrast, logic schemes based on MTJs are less error prone. An MTJ-based straintronic NAND gate has been proposed in [98]. It fulfills all the basic requirements of "logic" and calculations indicate that the energy-delay product associated with carrying out a NAND operation is $\sim 10^{-26}$ J s. This is comparable to that of a single transistor's energy-delay product. The bit error probability per logic

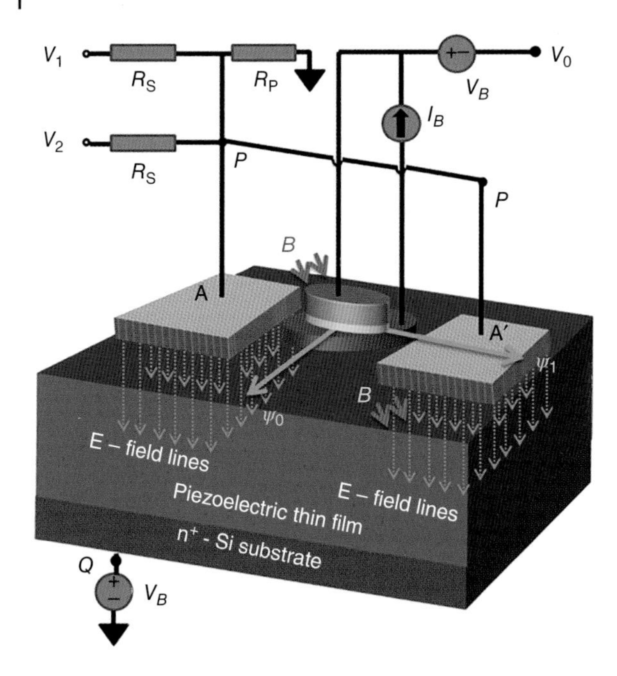

Figure 8.12 A straintronic NAND gate implemented with a "skewed" MTJ, where the major axes of the hard and soft layers are at an acute angle greater than zero. Source: Adapted from Biswas et al. 2014 [98].

operation is however still very high: 10^{-9}–10^{-8} at room temperature in the presence of thermal noise [98].

Figure 8.12 shows the schematic of the NAND gate proposed in [98]. The two input bits are encoded in voltage levels V_1 and V_2, while the output voltage level V_0 is the NAND function of V_1 and V_2. This logic gate has the special property that it is non-volatile, meaning that the output bit is stored locally in the magnetization states of the hard and soft layers of the MTJ and is not lost when the device is powered off. Therefore, such a construct can double as both logic and memory. These dual-function elements are very desirable since they can implement non-von-Neumann architectures where there is no physical partition between processor and memory. That reduces energy consumption, delay, and error rate. They can also lead to such curiosities as "instant-on" computers with insignificant boot delay. The Achilles' heel however is that the error probability is rather high, about 10^{-8} at room temperature while the switching speed (and hence clock rate) is rather slow, about 0.75 GHz [98]. These shortcomings may offset the advantage of non-volatility in many traditional Boolean computing applications.

Non-Boolean Straintronic Architectures Boolean "logic" has stringent requirements for reliability and is much less forgiving of errors than "memory." In memory, if a particular bit is in error, it does not corrupt any other bit since all bits are independent and isolated from each other. Errors are not contagious. However, in logic gates, if the output bit of a gate is corrupted, it feeds into the next stage and corrupts all subsequent bits. Errors are therefore infectious. That is why it is possible to detect and correct errors in memory chips with algorithmic remedies (e.g. parity check), but logic chips do not have such protection. Therefore, logic devices have to be extremely reliable with error probabilities less than 10^{-18} in any switching cycle. With that low error probability and with a 4 GHz clock, the frequency of errors will be about one in every eight years, which is tolerable.

Note that the error probability that matters is *not* the "static" error probability, but the "dynamic" error probability. In a nanomagnet with bistable magnetization states, the static error probability is determined by the energy barrier Δ separating the two stable states. It is $e^{-\Delta/kT}$ at a temperature T. This can be very small, e.g. 5.7×10^{-19} if $\Delta = 42\,kT$ at the operating temperature. However, the static error probability is immaterial. What matters is the dynamic error that occurs when the switching takes place between two states and the energy barrier is lowered or eliminated (e.g. with strain) to allow the switching to occur. The dynamic error probability in nanomagnets switched with strain or any other mechanism is several orders of magnitude larger than the required 10^{-18}, because magnetization dynamics is very vulnerable to thermal noise, which acts as a stray magnetic field. In strain-based switching, the dynamic error probability is typically $>10^{-9}$ at room temperature [94–98]. Because of this reason alone, nanomagnetic devices are often not preferred for Boolean *logic*, despite their superior energy efficiency and the wonderful property of non-volatility.

Recently, attention has been focused on non-Boolean architectures such as artificial neurons and synapses implemented with nanomagnetic devices such as MTJs [99–107]. The advantage of these constructions is supposedly that they are more energy efficient than CMOS-based implementations of neurons, although this may not be true of magnetic devices switched with STTs [99–101]. They are more likely to be true of magnetic devices switched with strain [103]. In these applications, the high error rate is not a showstopper since neuromorphic networks do not demand extreme precision and are fault tolerant.

Figure 8.13 shows a threshold firing neuron implemented with an MTJ whose soft layer is a two-phase multiferroic (magnetostrictive/piezoelectric) switched with electrically generated strain.

The threshold neuron fires (i.e. its output changes) when the weighted sum of the inputs exceeds a threshold value. This is expressed as

$$O = f\left(\sum_{i=1}^{N} w_i x_i + b\right) \tag{8.26}$$

where O is the output, x_i-s are the inputs, w_i-s are the weights, and b is a fixed bias. The function f is a nonlinear function, usually a step function. Figure 8.13b shows the straintronic MTJ-based implementation of a threshold firing neuron. Note that the MTJ resistance switches abruptly when the voltage applied at the electrodes A and A′ exceeds a threshold value that makes the magnetization of the soft layer rotate. Therefore, we can write

$$R_{\mathrm{MTJ}} = f(V_P) \tag{8.27}$$

where R_{MTJ} is the MTJ resistance measured between the bottom soft layer and the top hard layer, while V_P is the voltage at node P which is also the voltage applied at the electrodes A and A′.

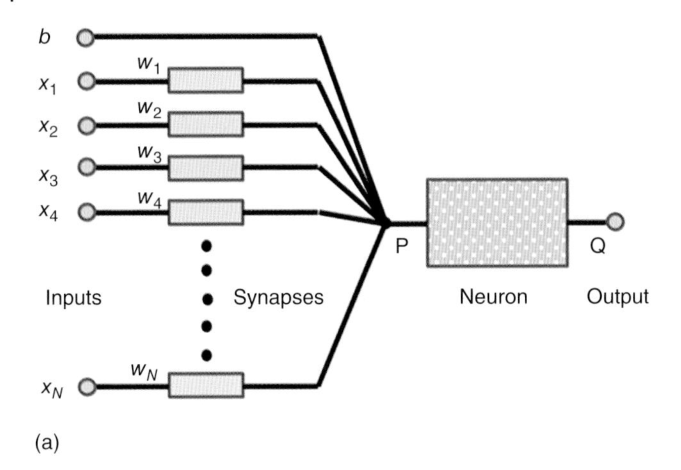

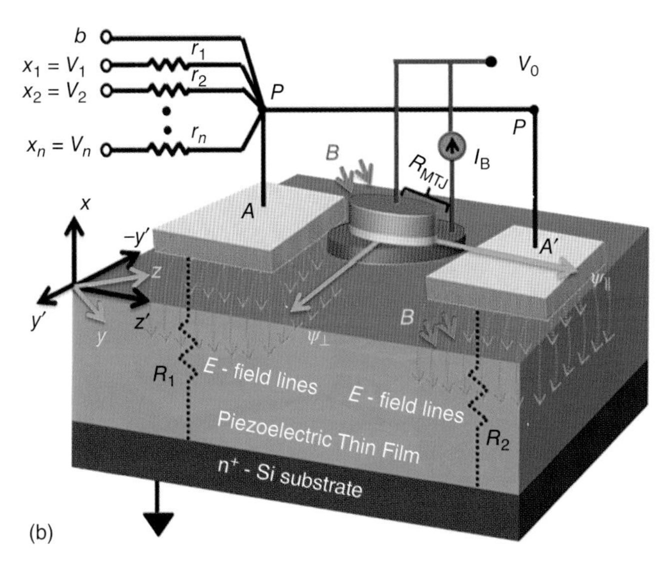

Figure 8.13 (a) A threshold firing neuron. The neuron's output changes when the weighted sum of the inputs exceeds a threshold value. (b) An implementation of a threshold neuron with a "skewed" MTJ, whose hard and soft layers are noncollinear. Source: Biswas et al. 2015 [103]. Reproduced with permission of Institute of Physics.

Using superposition, it is easy to show that

$$V_P = \sum_{i=1}^{N} w_i V_i + b \tag{8.28}$$

where

$$w_i = \frac{R_1 \| R_2 \| r_1 \| r_2 \| \cdots \| r_{i-1} \| r_{i+1} \| \cdots \| r_N}{R_1 \| R_2 \| r_1 \| r_2 \| \cdots \| r_{i-1} \| r_{i+1} \| \cdots \| r_N + r_i} \tag{8.29}$$

Since $V_0 = I_B R_{MTJ}$, we get from Eqs. (8.27) and (8.28) that

$$V_0 = f\left(\sum_{i=1}^{N} w_i V_i + b\right) \tag{8.30}$$

which replicates Eq. (8.26). This system therefore mimics a threshold firing neuron. Analysis in Ref. [103] shows that it is much more energy efficient than threshold neurons implemented with CMOS or even those implemented with STT-switched MTJs [99–101].

Other Non-Boolean Applications An array of nanomagnets, which interact with each other via dipole interactions, can implement a variety of image processing [108, 109] and computer vision functions [110]. The image processing functions can be of different types: de-noising, image restoration, pattern recognition, edge-enhancement detection, etc. [108, 109]. More recently, straintronic MTJs have been utilized to design Bayesian inference engines, which are essential elements of belief networks and are ideal for computing in the presence of uncertainty (e.g. disease progression, stock market behavior). By virtue of their non-volatility and peculiar transfer characteristics, skewed straintronic MTJs can implement Bayesian network elements with orders of magnitude savings in energy-delay product and circuit footprint [111, 112]. Recently, there has also been an effort to implement highly efficient ternary content addressable memory (TCAM) with skewed straintronic MTJs [113, 114]. In TCAM, a memory cell is searched based on its content, rather than its row and column address. TCAMs are useful for high-speed and parallel data processing and have been used in network routers, IP filters, virus-detection processors, lookup tables, etc. The challenges are to increase the search speed, reduce the cell footprint, and lower the energy dissipation. These objectives can be met with skewed straintronic MTJs [113, 114]. For example, 16 CMOS transistors will be needed to implement a static TCAM cell in CMOS-based implementation, but it can be implemented with a single skewed straintronic MTJ by virtue of the latter's peculiar non-monotonic transfer characteristic, resulting in significant improvement in energy dissipation and cell density [113, 114].

8.4 Conclusion

In this chapter, I have briefly discussed devices and architectures predicated on using the spin degree of freedom of an electron (as opposed to the charge degree of freedom) to store, process and communicate information. By its nature, such a review cannot be comprehensive and I have deliberately omitted discussions of devices that are also spin based (and equally important) but somewhat removed from the central theme of this review. The spin-based devices that have been discussed here are not always more energy efficient than their charge-based counterparts and they are certainly not faster or as reliable, but they possess the very important property of non-volatility and often have unusual transfer characteristics that make them eminently suitable for certain types of special purpose circuits and architectures. While the jury is still out on whether they have an important role to play in electronics, they certainly deserve a close look because they may bring about significant energy saving and area reduction in certain types of applications.

Acknowledgments

Some of the SPINFET work was carried out in collaboration with Prof. Marc Cahay of the University of Cincinnati. Almost all of the straintronic work was carried out in collaboration with Prof. Jayasimha Atulasimha of Virginia Commonwealth University. Many of my students have been involved in the research reported here. They are Prof. Sandipan Pramanik of the University of Alberta, Dr. Kuntal Roy of Purdue, Dr. Ayan Kumar Biswas of Carnegie Mellon University, and Dr. Hasnain Ahmad of IMEC, Belgium. The work has been mostly funded by the US National Science Foundation under grants ECCS-1124614, ECCS-0196554, CMMI-1301013, CCF-0726373, CCF-1216614, ECCS-0608854 and ECCS-0089893.

References

1 Moore, G.E. (1965). *Electronics Magazine*, 4. New York: McGraw Hill.
2 Bandyopadhyay, S. (2007). *J. Nanosci. Nanotechnol.* 7: 168.
3 Datta, S., Diep, V.Q., and Behin-Aein, B. (2015). Emerging Nanoelectronic Devices. In: *IEEE design automation conference* (ed. G. Bourianoff, V. Zhirnov, J. Hutchby and A. Chen). New York: John Wiley & Sons Chapter 2.
4 Yogendra, K., Fan, D., Shim, Y. et al. (2016). IEEE Design Automation Conference (ASP-DAC), 21st Asia and South Pacific, 2016.
5 Locatelli, N., Cros, V., and Grollier, J. (2014). *Nat. Mater.* 13: 11.
6 Datta, S. and Das, B. (1990). *Appl. Phys. Lett.* 56: 665.
7 Bandyopadhyay, S. and Cahay, M. (2015). *Introduction to Spintronics*, 2e. Boca Raton: CRC Press.
8 Bychkov, Y.A. and Rashba, E.I. (1984). *J. Phys. C* 17: 6039.
9 Pierret, R.F. (1996). *Semiconductor Device Fundamentals*. Reading, MA: Addison-Wesley.
10 Bournel, A., Dollfus, P., Galdin, S. et al. (1997). *Solid State Commun.* 104: 85.
11 Kazakova, L.A., Kostsova, V.V., Karymshakov, R.K. et al. (1972). *Sov. Phys. Semicond.* 5: 1495.
12 Salis, G., Wang, R., Jiang, X. et al. (2005). *Appl. Phys. Lett.* 87: 262503.
13 Trivedi, A., Bandyopadhyay, S., and Cahay, M. (2007). *IET Circuits Devices Syst.* 1: 395.
14 Schliemann, J., Egues, J.C., and Loss, D. (2003). *Phys. Rev. Lett.* 90: 146801.

15 Hall, K.C. and Flatté, M.E. (2006). *Appl. Phys. Lett.* 88: 162503.
16 Cartoixá, X., Tang, D.Z.Y., and Chang, Y.-C. (2003). *Appl. Phys. Lett.* 83: 1462.
17 Bandyopadhyay, S. and Cahay, M. (2004). *Appl. Phys. Lett.* 85: 1814.
18 Bandyopadhyay, S. and Cahay, M. (2005). *Physica E: Low Dimensional Systems and Nanostructures* 25: 399.
19 Bandyopadhyay, S. and Cahay, M. (2004). *Appl. Phys. Lett.* 85: 1433.
20 Bandyopadhyay, S., Das, B., and Miller, A.E. (1994). *Nanotechnology* 5: 113.
21 Molotkov, S.N. and Nazin, S.S. (1995). *JETP Lett.* 62: 256.
22 Agarwal, H., Pramanik, S., and Bandyopadhyay, S. (2008). *New J. Phys.* 10: 015001.
23 Bandyopadhyay, S. and Roychowdhury, V.P. (1996). *Jpn. J. Appl. Phys.* 35: 3350.
24 Molotkov, S.N. and Nazin, S.S. (1996). *Zh. Eksp. Teor. Fiz.* 110: 1439.
25 Bandyopadhyay, S. (2005). *Superlat. Microstruct.* 37: 77.
26 Basu, T., Sarkar, S.K., and Bandyopadhyay, S. (2007). *IET Circuits Devices Syst.* 1: 194.
27 Rugar, D., Budakian, R., Mamin, H.J., and Chui, B.H. (2004). *Nature* 430: 329.
28 Xioa, M., Martin, I., Yablonovitch, E., and Jiang, H.W. (2004). *Nature* 430: 435.
29 Elzerman, J.M., Hanson, R., Williams van Beveren, L.H. et al. (2004). *Nature* 430: 431.
30 Melnikov, D.V. and Leburton, J.-P. (2006). *Phys. Rev. B* 73: 155301.
31 Wang, W.L., Yazyev, O.V., Meng, S., and Kaxiras, E. (2009). *Phys. Rev. Lett.* 102: 157201.
32 Loss, D. and DiVincenzo, D.P. (1998). *Phys. Rev. A* 57: 120.
33 Bandyopadhyay, S., Balandin, A., Roychowdhury, V.P., and Vatan, F. (1998). *Superlatt. Microstruct.* 23: 445.
34 Burkard, G., Loss, D., and DiVincenzo, D.P. (1999). *Phys. Rev. B* 59: 2070.
35 Loss, D., Burkard, G., and DiVincenzo, D.P. (2000). *J. Nanopart. Res.* 2: 401.
36 Brown, W.F. Jr., (1968). *J. Appl. Phys.* 39: 993.
37 Cowburn, R.P., Koltsov, D.K., Adeyeye, A.O. et al. (1999). *Phys. Rev. Lett.* 83: 1042.
38 Chikazumi, S. (1964). *Physics of Magnetism*. New York: Wiley.
39 Gaunt, P. (1977). *J. Appl. Phys.* 48: 3470.
40 Ikeda, S., Hayakawa, J., Ashizawa, Y. et al. (2008). *Appl. Phys. Lett.* 93: 082508.
41 Ralph, D.C. and Stiles, M.D. (2008). *J. Magn. Magn. Mater.* 320: 1190.
42 Liu, L., Pai, C.-F., Li, Y. et al. (2012). *Science* 336: 555.
43 Pai, C.-F., Liu, L., Li, Y. et al. (2012). *Appl. Phys. Lett.* 101: 122404.
44 Niimi, Y., Kawanishi, Y., Wei, D.H. et al. (2012). *Phys. Rev. Lett.* 109: 156602.
45 D'yakonov, M. and Perel', V.I. (1971). *JETP Lett.* 13: 467.
46 D'yakonov, M. and Perel', V.I. (1971). *Phys. Lett. A* 35: 459.
47 Hirsh, J.E. (1999). *Phys. Rev. Lett.* 83: 1834.
48 Zhang, S. (2000). *Phys. Rev. Lett.* 85: 393.

49 Kato, Y.K., Myers, R.C., Gossard, A.C., and Awschalom, D.D. (2004). *Science* 306: 1910.
50 Wunderlich, J., Kaestner, B., Sinova, J., and Jungwirth, T. (2005). *Phys. Rev. Lett.* 94: 047204.
51 Wang, K.L., Alzate, J.G., and Khalili-Amiri, P. (2013). *J. Phys. D: Appl. Phys.* 46: 074003.
52 Bhowmik, D., You, L., and Salahuddin, S. (2014). *Nat. Nanotech.* 9: 59.
53 Kyuno, K., Ha, J.-G., Yamamoto, R., and Asano, S. (1996). *J. Phys. Soc. Jpn.* 65: 1334.
54 Maruyama, T., Shiota, Y., Nozaki, T. et al. (2009). *Nat. Nanotechnol.* 4: 158.
55 Shiota, Y., Maruyama, T., Nozaki, T. et al. (2009). *Appl. Phys. Express* 2: 063001.
56 Pertsev, N.A. (2013). *Sci. Rep.* 3: 2757.
57 Duan, C.-G., Velev, J.P., Sabirianov, R.F. et al. (2008). *Phys. Rev. Lett.* 101: 137201.
58 Niranjan, M.K., Duan, C.-G., Jaswal, S., and Tsymbal, E.Y. (2010). *Appl. Phys. Lett.* 96: 222504.
59 Nakamura, K., Shimabukuro, R., Fujiwara, Y. et al. (2009). *Phys. Rev. Lett.* 102: 187201.
60 Kanai, S., Yamanouchi, M., Ikeda, S. et al. (2012). *Appl. Phys. Lett.* 101: 122403.
61 Wang, W.-G., Li, M., Hageman, S., and Chien, C.L. (2012). *Nat. Mater.* 11: 64.
62 Rajanikanth, A., Hauet, T., Montaigne, F. et al. (2013). *Appl. Phys. Lett.* 103: 062402.
63 Nozaki, T., Shiota, Y., Shiraishi, M. et al. (2010). *Appl. Phys. Lett.* 96: 022506.
64 Miwa, S., Matsuda, K., Tanaka, K. et al. (2016). *Appl. Phys. Lett.* 107: 162402.
65 Shiota, Y., Nozaki, T., Bonell, F. et al. (2012). *Nat. Mater.* 11: 39.
66 Stöhr, J., Siegmann, H.C., Kashuba, A., and Gamble, S.J. (2009). *Appl. Phys. Lett.* 94: 072504.
67 Kanai, S., Nakatani, Y., Yamanouchi, M. et al. (2013). *Appl. Phys. Lett.* 103: 072408.
68 Grezes, C., Ebrahimi, F., Alzate, J.G. et al. (2016). *Appl. Phys. Lett.* 108: 012403.
69 Grezes, C., Rijas Rozas, A., Ebrahimi, F. et al. (2016). *AIP Adv.* 6: 075014.
70 Endo, M., Kanai, S., Ikeda, S. et al. (2010). *Appl. Phys. Lett.* 96: 212503.
71 Amiri, P.K. and Wang, K.L. (2012). *Spin* 2: 1240002.
72 Amiri, P., K., Alzate, J.G., Cai, X.Q. et al. (2015). *IEEE Trans. Magn.* 51: 3401507.
73 Camsari, K.Y., Ganguly, S., and Datta, S. (2015). *Sci. Rep.* 5: 10571.
74 Cui, J., Hockel, J.L., Nordeen, P.K. et al. (2013). *Appl. Phys. Lett.* 103: 232905.
75 Liang, C.Y., Keller, S.M., Sepulveda, A.E. et al. (2014). *J. Appl. Phys.* 116: 123909.
76 Cui, J.Z., Liang, C.Y., Paisley, E.A. et al. (2015). *Appl. Phys. Lett.* 107: 092903.
77 Roy, K., Bandyopadhyay, S., and Atulasimha, J. (2013). *Sci. Rep.* 3: 3038.
78 Tiercelin, N., Dusch, Y., Klimov, A. et al. (2011). *Appl. Phys. Lett.* 99: 192507.

79 Biswas, A.K., Bandyopadhyay, S., and Atulasimha, J. (2014). *Appl. Phys. Lett.* 105: 072408.
80 Roy, K., Bandyopadhyay, S., and Atulasimha, J. (2012). *J. Appl. Phys.* 112: 023914.
81 Roy, K., Bandyopadhyay, S., and Atulasimha, J. (2011). *Appl. Phys. Lett.* 99: 063108.
82 Salehi Fashami, M., Roy, K., Atulasimha, J., and Bandyopadhyay, S. (2011). *Nanotechnology* 22: 155201.
83 Salehi Fashami, M., Atulasimha, J., and Bandyopadhyay, S. (2012). *Nanotechnology* 23: 105201.
84 Ahmad, H., Atulasimha, J., and Bandyopadhyay, S. (2015). *Nanotechnology* 26: 401001.
85 Ahmad, H., Atulasimha, J., and Bandyopadhyay, S. (2015). *Sci. Rep.* 5: 18264.
86 Zhao, Z.Y., Jamali, M., D'Souza, N. et al. (2016). *Appl. Phys. Lett.* 109: 092403.
87 Li, P., Chen, A., Li, D. et al. (2014). *Adv. Mater.* 26: 4320.
88 Atulasimha, J. and Bandyopadhyay, S. (2010). *Appl. Phys. Lett.* 97: 173105.
89 D'Souza, N., Atulasimha, J., and Bandyopadhyay, S. (2011). *J. Phys. D: Appl. Phys.* 44: 265001.
90 D'Souza, N., Atulasimha, J., and Bandyopadhyay, S. (2012). *IEEE Trans. Nanotechnol.* 11: 418.
91 D'Souza, N., Salehi Fashami, M., Bandyopadhyay, S., and Atulasimha, J. (2016). *Nano Lett.* 16: 1069.
92 Sampath, V., D'Souza, N., Bhattacharyya, D. et al. (2016). *Nano Lett.* 16: 5681.
93 Sampath, V., D'Souza, N., Bhattacharyya, D. et al. (2016). *Appl. Phys. Lett.* 109: 123.
94 Salehi Fashami, M., Munira, K., Bandyopadhyay, S. et al. (2013). *IEEE Trans. Nanotechnol.* 12: 1206.
95 Salehi Fashami, M., Atulasimha, J., and Bandyopadhyay, S. (2013). *Sci. Rep.* 3: 3204.
96 Munira, K., Xie, Y.K., Nadri, S. et al. (2015). *Nanotechnology* 26: 245202.
97 Alrashid, M.M., Bhattacharyya, D., Bandyopadhyay, S., and Atulasimha, J. (2015). *IEEE Trans. Elec. Dev.* 62: 2978.
98 Biswas, A.K., Bandyopadhyay, S., and Atulasimha, J. (2014). *Sci. Rep.* 4: 7553.
99 Sharad, M., Fan, D., Aitken, K., and Roy, K. (2014). *IEEE Trans. Nanotechnol.* 13: 23.
100 Sharad, M., Fan, D., and Roy, K. (2014). *Appl. Phys. Lett.* 114: 234906.
101 Sengupta, A., Choday, S.H., Kim, Y., and Roy, K. (2015). *Appl. Phys. Lett.* 106: 143701.
102 Sengupta, A. and Roy, K. (2016). *Phys. Rev. Appl.* 5: 024012.
103 Biswas, A.K., Atulasimha, J., and Bandyopadhyay, S. (2015). *Nanotechnology* 26: 281001.
104 Locatelli, N., Vincent, A.F., Mizrahi, A. et al. (2015). Proceedings of Design Automation & Test in Europe Conference (DATE), p. 994. IEEE.
105 Vincent, A.F., Larroque, J., Locatelli, N. et al. (2015). *IEEE Trans. Biomed. Circ. Syst.* 9: 166.

106 Burr, G.W., Shelby, R.M., Sebastian, A. et al. (2017). *Adv. Phys. X* 2: 89.

107 Huang, Y., Kang, W., Zhang, X. et al. (2017). *Nanotechnology* 28: 08LT02.

108 D'Souza, N., Atulasimha, J., and Bandyopadhyay, S. (2012). *IEEE Trans. Nanotechnol.* 11: 896.

109 Abeed, M.A., Biswas, A.K., Al-Rashid, M.M. et al. (2017). *IEEE Trans. Electron Dev.* 64: 2417.

110 Bhanja, S., Karunaratne, D.K., Panchumarthy, R. et al. (2016). *Nat. Nanotechnol.* 11: 177.

111 Khasanvis, S., Li, M.Y., Rahman, M. et al. (2015). *IEEE Trans. Nanotechnol.* 14: 980.

112 Khasanvis, S., Li, M.Y., Rahman, M. et al. (2015). *Computer* 48: 54.

113 Dey Manasi, S., Al-Rashid, M.M., Atulasimha, J., et al., (2017). *IEEE Trans. Elec. Dev.* 64: 2835.

114 Dey Manasi, S., Al-Rashid, M.M., Atulasimha, J., et al., (2017). *IEEE Trans. Elec. Dev.* 64: 2842.

9

Terahertz Properties and Applications of GaN

Berardi Sensale-Rodriguez

The University of Utah, Department of Electrical and Computer Engineering, 50 S. Central Campus Drive, Salt Lake City, UT 84112, USA

9.1 Introduction

The terahertz frequency range is the region of the electromagnetic spectrum lying between the microwaves and the infrared. As a spectral range, terahertz is usually defined between the frequencies of 300 GHz and 3–30 THz [1]. Owing to the fact that at terahertz wavelengths both electrical and optical phenomena are significant, terahertz is widely considered as the melting pot between electronics and photonics. Promising applications in many diverse areas including astronomy, security, wide-bandwidth communications, chemical and biomedical sensing, etc., terahertz technology has turned into a very important field of scientific research during the past three decades [2–5]. However, progress in "practical" terahertz technologies has been constrained due to the lack of materials and devices efficiently responding at these frequencies. This chapter reviews the state of the art of terahertz technology and discusses progress and perspectives of gallium nitride (GaN) and GaN-based devices toward terahertz applications. This chapter is organized in three sections, which are outlined as follows: Section 9.1 introduces the topic of terahertz technology and provides an overview of applications, devices, and challenges. Section 9.2 discusses the materials properties of GaN and gives an overview of the charge-transport mechanisms relevant to terahertz applications in this material. Finally, Section 9.3 provides an overview of the state of the art in terms of GaN-based terahertz devices, future perspectives, and limitations.

9.1.1 Applications of Terahertz Technology

Terahertz waves strongly interact with matter. For instance, terahertz radiation penetrates through dielectric materials as well as nonpolar liquids, whereas it attenuates through/reflects from conductive materials. Furthermore, many materials exhibit strong absorption and dispersion signatures at terahertz wavelengths. These features, like our fingerprints, are unique and characteristic. From this perspective, terahertz technology has found important applications in the

Advanced Nanoelectronics: Post-Silicon Materials and Devices,
First Edition. Edited by Muhammad Mustafa Hussain.

fields of spectroscopy, sensing, and imaging. In this context, terahertz imaging has been proposed as a suitable alternative to X-rays and ultrasound for security, quality control, and nondestructive testing applications [1–3]. In contrast to other imaging techniques, such as X-rays, terahertz radiation has the property of being nonionizing. When compared to X-rays, with photon energies on the order of kiloelectronvolt, the energy of the terahertz photon is very small, i.e. on the order of just a few millielectronvolt. Therefore, terahertz photons are associated with energies much smaller than those required for ionization of chemical bonds [6]; this is critical for in vivo applications such as security inspection of travelers, medical analysis, and so on. Furthermore, the diffraction-limited spatial resolution allowed by terahertz waves (ten to hundreds of micrometers, as dictated by the terahertz wavelength) can be similar to that of the human eye. Owing to these properties, terahertz imaging has gained significant appeal for in vivo medical and security applications, e.g. dentistry [7, 8], cancer detection [9–12], and personal scanners [13, 14]. However, the applications of terahertz imaging are restricted since (i) terahertz radiation is not able to penetrate through metals and (ii) water has a strong absorption at terahertz frequencies.

Very important spectroscopic information lies at terahertz frequencies; for instance, the rotational and vibrational modes of biological molecules as well as DNAs are located in the terahertz band [15–18]. From this point of view, terahertz spectroscopy can provide important physical and chemical information, for instance, for the purpose of disease diagnosis. In a broader sense, terahertz spectroscopy has become a popular technique for thin-film characterization [19], biological and chemical sensing [20, 21], and as a method for detecting illegal substances (e.g. explosives and drugs) [22]. Furthermore, using broadband pulses, terahertz imaging is not only able to provide information about the shape (or presence) of different objects but can also enable identifying the composition of these objects [23].

Compared with radiofrequency and microwaves, terahertz waves have higher frequency and thus can enable larger communications bandwidth. From this perspective, terahertz technology has potential applications in high-capacity wireless communications [24–29]. In addition, terahertz waves are less directional and less susceptible to scintillation effects than are infrared waves. As a result, certain low-attenuation frequency bands lying in the terahertz range have been proposed for short-range wireless links. Examples of these bands include [0.38, 0.44], [0.45, 0.52], [0.52, 0.72], and [0.77, 092] THz. In this context, the application of terahertz for long-range communications (or for hazardous materials detection from far away) is limited by the large atmospheric attenuation at terahertz frequencies, especially at frequencies above 1 THz. Free space loss, given by the Friis formula: $Loss(\mathrm{dB}) = 20\log_{10}(c/4\pi fr)$, where c is the speed of light, f is the frequency of the electromagnetic wave, and r is the distance between the transmitter and the receiver, is also significant at terahertz frequencies. For instance, as discussed by Kleine-Ostmann and Nagatsuma [29], the power attenuation for a 1-THz carrier propagating over a 100-m distance is $\sim$180 dB.

Finally, terahertz technology is also very important in the field of radio astronomy [30]. In fact, radio astronomy has been historically the field in which terahertz technology has been most widely applied. A large number of atmospheric molecules such as water, carbon monoxide, nitrogen, oxygen, and so on can be detected in the terahertz band due to their strong spectral signatures. The spectral features associated with these molecules can be used for atmospheric monitoring of our planet as well as for ultrahigh spatial resolution mapping of the universe; in this regard, terahertz technology has truly enabled space research and the study of interstellar formation.

9.1.2 Terahertz Devices and Challenges

In order to realize the multiple application possibilities that terahertz technology has to offer, there is a need for developing devices capable of interacting with terahertz waves as well as responding to terahertz frequencies in a controllable manner. However, fundamental physical limitations severely constrain the development of practical terahertz technologies.

First, because of the small energy of terahertz photons (on the order of a few millielectronvolt), terahertz absorption and emission is usually largely owed to either intraband transitions or intersubband transitions rather than to interband transitions. However, to take advantage of intersubband transitions, devices should be cooled down to cryogenic temperatures ($kT <$ terahertz photon energy) [31]. For instance, although devices such as intersubband detectors and modulators can efficiently operate at smaller wavelengths, e.g. at the infrared, in the terahertz frequency range their operation principles have been solely demonstrated at cryogenic temperatures [32]. Since terahertz systems ideally require room-temperature operation, the dominating processes in most types of terahertz devices are typically intraband transitions, which are in turn closely related to the carrier transport properties.

Second, the optical phonon energies of semiconductor materials typically lie in the terahertz range. Shown in Table 9.1 are the Reststrahlen bands for several semiconductors [33]; for all the listed materials, it is observed that the longitudinal-optical (LO)-phonon frequencies lie in the 5–40 THz range; thus, LO phonons can severely degrade the performance of terahertz devices. For instance, non-radiative relaxation via LO-phonon scattering is widely reported as one of the main reasons for performance degradation in terahertz quantum cascade lasers (QCLs) [34].

Table 9.1 Reststrahlen band for different semiconductor materials.

Material	Reststrahlen band (TO–LO phonon energy) (meV)	Associated frequency (THz) and wavelength (μm)
Si	57–63	13.9–15.5/20–22
GaN	66–92	16–22/19–22
GaAs	31–36	7.5–8.5/34–40
InAs	27–29	6.4–7.0/43–46
Graphene	160	38.6–8

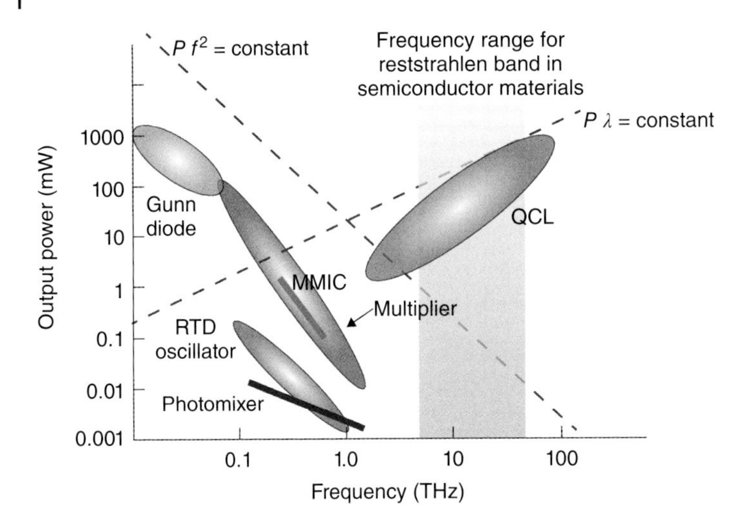

Figure 9.1 Output power for several terahertz emitters as a function of frequency. In the center of the terahertz range, the attainable power levels sharply drop. Source: Adapted by permission from MacMillan Publishers Ltd. Ref. [3].

There are two classes of solid-state terahertz sources; those based on optical-like phenomena (mainly QCLs), and those based on purely electrical phenomena. Figure 9.1 shows the typical output powers (versus frequency) for several different types of terahertz emitters. It is observed that the device performance heavily drops when reaching the center of the terahertz band (frequencies around 2–5 THz). Terahertz QCLs, which rely on electron relaxation between subbands of quantum wells via terahertz photon emission, have been recently demonstrated to provide significant terahertz output powers. However, the maximum temperatures at which they can directly operate are only as high as ~200 K [35]. Several popular electronic-based emitters include Gunn diodes [36, 37], resonant tunneling diode (RTD) oscillators [38], multiplier sources [39, 40], etc. Recent experimental demonstrations of compact room-temperature terahertz emitters based on RTD oscillators in the InGaAs materials system have provided output power levels of 7 μW at 1 THz [41]. RTD-based devices operating at higher frequencies (1.92 THz) were also reported, but achieving much smaller output powers (0.4 μW) [42]. Room-temperature terahertz emission was claimed as well in high-electron-mobility transistors (HEMTs) by means of electron-plasma wave instability [43]. The principles of operation of these devices are discussed in later sections. It is worth mentioning that terahertz emission can be also obtained from monolithic integrated microwave integrated circuit (MMIC) oscillators made of traditional transistors (heterojunction bipolar transistors – HBTs, HEMTs, or FETs (field-effect transistors)). However, devices with f_{max} well above 1 THz are required to achieve acceptable output powers with good efficiency – via fundamental harmonics. In practice, the highest oscillation frequency (first harmonic) so far reported for an MMIC oscillator is ~600 GHz with an output power of 12 μW [44]. In general, the output power of electronic-based emitters decreases inversely with frequency squared, which is a result of an RC time constant limitation of these devices. To overcome this RC time constant limitation, active traveling-wave devices should be designed. In traveling-wave approaches the RC time constant can be circumvented since the operation speed is set by the local distributed device parameters rather than by the lumped total device capacitance and resistance. This is a reason why electron-plasma-wave-based devices could potentially overcome this RC time constant limitation and thus operate at terahertz frequencies.

From the perspective of terahertz detectors, there is a need for developing devices with high responsivity, low noise equivalent power (NEP), and high-speed operation at room temperature (RT). In this regard, it is worth noting that state-of-the art terahertz detectors are very sensitive, but they operate at milli-Kelvin temperature. Several popular direct detectors include Schottky diode detectors, hot-electron bolometers (HEBs), pyroelectric detectors, Golay cells, and so on [45]. Even, single-photon detectors have been recently reported using single-electron transistors (SETs) [46]. Electron-plasma-wave-enabled high-responsivity room-temperature terahertz detection was also recently reported in HEMTs, owing to the fact that electron plasma waves can achieve velocities much higher than the electron drift velocity [47]. In addition to their record high room-temperature responsivity $>2\,\mathrm{kV\,W^{-1}}$, these devices can also allow for very low room-temperature NEP $\sim 15\,\mathrm{pW/Hz^{0.5}}$ at 1 THz [48], which is one order of magnitude lower than what is achievable by other types of RT terahertz detectors such as Golay cells ($\sim 200\,\mathrm{pW/Hz^{0.5}}$) and similar to/or better than that of Schottky diode detectors (~ 5 to $40\,\mathrm{pW/Hz^{0.5}}$).

By taking advantage of the interaction between terahertz waves and the surrounding environment, terahertz devices can be employed as sensors for diverse applications. A simple example of a terahertz biomedical sensor is a glycoprotein detector [49]. The operation of this device is based on monitoring the terahertz transmittance. A film of biotin (vitamin H) is deposited on a substrate. This substrate is mounted on a galvanometric shaker, and half of the film is exposed to the solution of interest. Sensing is performed by measuring transmittance through the two halves of the film. Another sensor application is real-time monitoring of microfluids/gases, which can be achieved in parallel-plate [50] and dielectric pipe [51] waveguide resonant cavities by exploiting the change in the resonant frequency of the structure due to the presence of different liquids/vapors. Very sensitive biochemical sensing can be accomplished using other resonant structures such as metal-hole arrays and metal meshes; owing to the very sensitive dependence of the terahertz transmittance on the properties of the surrounding dielectric media. In general, metamaterials can provide a field enhancement and thus improve the terahertz sensitivity to changes in the surrounding refractive index. Furthermore, devices like these can be fabricated even on paper substrates, showing excellent sensitivity to glucose and urea solution concentration [52].

In this context of recent progress on terahertz devices, GaN is an emerging material that has some key advantages for many of these applications. In the next sections, we discuss the physical properties of GaN relevant to terahertz applications as well as how we can harness these properties in order to develop efficient terahertz devices.

9.2 GaN: Properties and Transport Mechanisms Relevant to THz Applications

With the fast growing need for photonic and power electronics technologies, there is an enormous need for high-power high-frequency semiconductor devices. In this regard, GaN provides among the highest electron saturation velocity, breakdown-voltage and operation temperature, and thus one of the best combined frequency-power performance among semiconductor materials. Promising many practical applications, as described in Section 9.1, terahertz technology is one of the most emerging areas in electronics and photonics. The need for compact, low-cost, high-resolution, and high-power terahertz systems has fueled the utilization of GaN as the active material in terahertz devices. The abovementioned characteristics of GaN, together with the possibility of attaining high two-dimensional electron densities in AlGaN/GaN heterostructures as well as its large LO phonon energy (~90 meV), make it one of the most promising semiconductor materials for terahertz applications. In this section, we discuss the electronic properties of GaN and provide an overview of physical phenomena occurring in this material that is relevant to terahertz applications.

9.2.1 Mobility and Injection Velocity

Wide bandgap semiconductors, such as III-nitrides, offer an interesting alternative to traditional III–V materials due to the possibility of operation with a higher output power resulting from the increased critical field, higher saturation velocity, and better thermal conductivity in III-nitride materials. Thus, not only high-power but also high-frequency terahertz operation is expected from GaN-based semiconductors. GaN showcases excellent transport properties such as large peak electron velocity ($\sim 2.5 \times 10^7$ cm s^{-1}), large saturation velocity ($\sim 2.0 \times 10^7$ cm s^{-1}), and excellent carrier transport properties as evidenced by a high electron mobility ($\mu > 2000$ cm^2 V^{-1} s^{-1}) in GaN two-dimensional electron gases (2DEGs); these properties make GaN suitable for use as channel material in high-frequency power devices [53–56].

In AlGaN/GaN 2DEGs, the electron mobility at high 2DEG densities is limited by alloy scattering, whereas for AlN/GaN 2DEGs, interface roughness scattering is what limits mobility at high electron densities [57]. From this perspective, there is a large improvement by the removal of the alloy barrier. In contrast, at low 2DEG densities, dislocation scattering from charged cores and strain fields are the dominant scattering mechanisms. As discussed by Jena and coworker [58], it is possible to calculate the effect of all relevant scattering processes affecting the electron mobility in GaN 2DEGs. Considering that scattering by background unintentional donors, scattering by the alloy disorder due to finite penetration of the electron wavefunction into the alloy AlGaN barrier (this is absent for AlN barriers), interface roughness at the Al(Ga)N/GaN heterojunction, scattering from remote surface donors, acoustic phonon scattering, and dislocation scattering, the mobility curves depicted in Figure 9.2 are obtained. The temperature dependence of mobility reveals important facts about impurities in these heterostructures and gives valuable clues for the design of

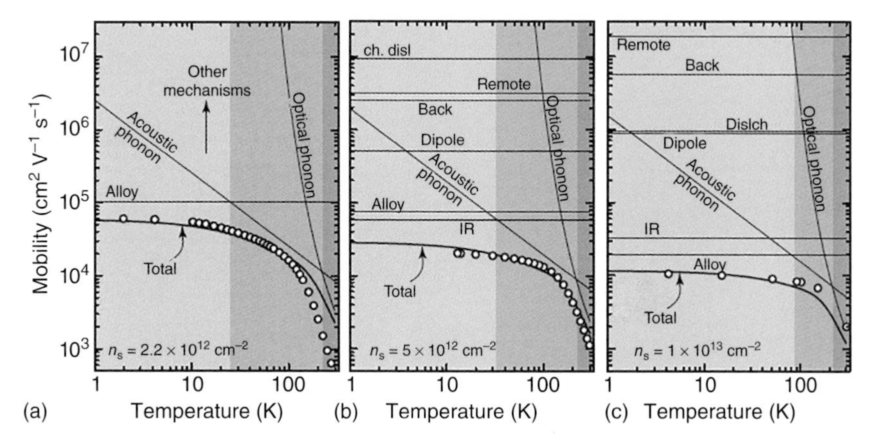

Figure 9.2 Mobility versus temperature for three AlGaN/GaN samples with different electron densities. Contributions of several scattering mechanisms are depicted. Source: Wood and Jena 2007 [58]. Reproduced with permission of Springer.

better devices. Experimentally extracted electron mobilities are also plotted in the same figure. From the plot, it is clear that alloy scattering is severe, and is the main low-temperature mobility limiting scattering mechanism. At high carrier densities ($n_s \sim 10^{13}$ cm^{-2}), it starts affecting even the room-temperature electron transport. Furthermore, the electron mobility limited by polar optical phonon scattering at RT is ~2000 cm^2 V^{-1} s^{-1}. Removal of the alloy barrier by growing thin AlN layers can lead to a larger mobility at high electron densities, especially at low temperatures. Experimentally, high-conductivity 2DEGs at AlN/GaN heterojunctions have been reported by many authors. The sheet densities in these heterostructures can be tuned from 5×10^{12} to 5×10^{13} cm^{-2} by varying the AlN thickness from ~1 to ~7 nm [59]. A critical thickness is observed beyond which biaxial strain relaxation and cracking of AlN occurs, and a degradation of carrier mobility is observed at extremely high sheet densities. A high-mobility window, corresponding to AlN thicknesses ~3 to 5 nm, was identified, within which room-temperature mobility exceeding 1000 cm^2 V^{-1} s^{-1} and sheet densities ~1–3 $\times 10^{13}$ cm^{-2} can be obtained, thereby yielding low sheet resistances ($R_s < 200\ \Omega/\square$). Interface roughness scattering and strain relaxation are, in this case, the main factors preventing even lower sheet resistance in these structures.

Furthermore, besides mobility, in nanoscale devices (as dictated, e.g. by the gate length of high-frequency HEMTs), the effective velocity at which electrons are injected into the device channel has a major impact on the overall device speed and thereby on high-frequency performance. This is particularly important in materials systems where electron–optical-phonon scattering is strong, which is the case in GaN. In this regard, unlike Si MOSFETs and most other III–V-based HEMTs, electron–optical-phonon interaction is exceptionally strong in III-nitrides. This effect has been studied in GaN HEMTs by Fang et al. [60, 61]. As a result of this interaction, the mean free path of hot electrons in GaN is ~3.5 nm, thus much smaller than typical gate length dimensions for high-frequency HEMTs. Therefore, whereas Si-MOSFETs and other III–V HEMTs can approach near-ballistic behavior by reduction of

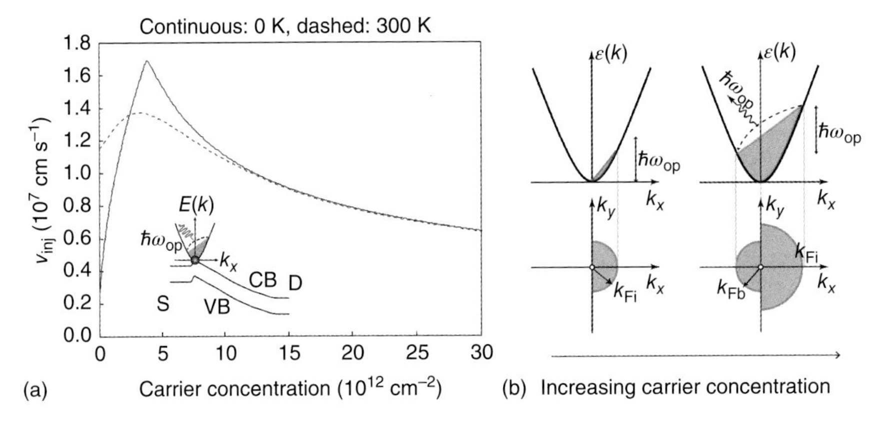

Figure 9.3 (a) Calculated injection velocity into GaN HEMT channels. (b) Evolution of k-space occupation of electrons as charge concentration in the channel is increased. Source: Fang et al. 2012 [60]. Adapted with permission of IEEE.

parasitic delays and gate length, the situation is starkly different in GaN-based HEMTs. By incorporating the effect of polar optical-phonon backscattering into a quasi-ballistic model, it is found that the injection velocity into GaN channels increases with channel charge density until optical-phonon emission occurs, at which point injection velocity reaches $\sim 1.6 \times 10^7$ cm s^{-1} at 0 K, but degrades to $\sim 1.3 \times 10^7$ cm s^{-1} at RT [61]. For reference, these maximum injection velocity levels are on the order of 2× smaller than the peak drift velocity in GaN. Depicted in Figure 9.3a is the calculated injection velocity versus charge density. This peak injection velocity leads to a peak transconductance (g_m) and limits the device cutoff frequency f_T in HEMTs. As a result, not only the electron mobility but also the effective injection velocity is important for predicting the fundamental speed limits in GaN channel devices. The physical origin of this effect is straightforward; electrons are injected from the source (source injection point) as depicted in the inset in Figure 9.3a. At high 2DEG densities, electrons from the highest right-going energy state emit an optical phonon and scatter into the highest empty left-going state. Scattering into the bottom of the subband is possible when the condition $2k_F < k_{op}$ is satisfied, where k_F is the radius of the Fermi circle at equilibrium and k_{op} is such that $\hbar k_{op}/2m^* = \hbar\omega_{op}$, $\hbar\omega_{op}$ being the optical-phonon energy. This sets a charge density level $n_s = n_0$, above which backscattering occurs. For $n_s < n_0$, the highest energy right-going state has an energy $\hbar\omega_{op}$, and the lowest occupied state also moves to the right, as shown in the left panel in Figure 9.3b. However, if $n_s > n_0$, backscattering takes place as evidenced by the left-going occupation depicted in the right panel in Figure 9.3b.

Once the carrier distribution in k-space at the source injection point is known, the current flowing into the channel can be evaluated as well as the effective speed at which electrons are injected. The current is found by summing over the group velocities of all occupied states in k-space; for 2D carriers in a parabolic band, the current density is given by [60]:

$$J = \frac{J_{th}}{2\sqrt{\pi}} \left[\mathcal{F}_{1/2}(\eta) - \mathcal{F}_{1/2}\left(\eta - \frac{\hbar\omega_{op}}{k_B T} \right) \right] \quad (9.1)$$

where $J_{th} = qv_{th}k_BTm^*/\pi\hbar^2$ is an effective thermal current, with $v_{th} = \sqrt{2k_BT/m^*}$ being the thermal velocity, $\mathcal{F}_{1/2}(\eta)$ is the Fermi–Dirac integral, and $\eta = E_{in}/k_BT$, where E_{in} is the injection electron quasi-Fermi level. The two Fermi–Dirac integrals are for the right- and left-going carriers, respectively. The carrier concentration is the sum of the right- and left-going electrons, given by [60]:

$$n_s = \frac{m^* k_B T}{2\pi\hbar^2}\left[\log(1+e^{\eta}) + \log\left(1+e^{\eta-\frac{\hbar\omega_{op}}{k_BT}}\right)\right] \tag{9.2}$$

Finally, the injection velocity can be computed from Eqs. (9.3) and (9.2) as [60]:

$$v_{inj} = \frac{\partial J}{q\partial n_s} \tag{9.3}$$

Note that the injection velocity is strongly dependent on the carrier density at the source injection point. Conceptually, injection velocity is different from saturation velocity, which in turn is a materials constant arising from the velocity field characteristics in materials.

9.2.2 Drift Velocity and Negative Differential Resistance

When analyzing the velocity versus electric field characteristics in GaN, it is observed that GaN exhibits negative differential mobility, i.e. drift velocity overshoots at a certain field and afterwards decreases [62, 63]. Various explanations have been proposed in order to explain this negative differential resistance (NDR) effect, including electron intervalley transfer and inflection of the central valley. Furthermore, when comparing the transit times for electrons in GaN and GaAs, based on Monte Carlo simulations, GaN shows substantially shorter transit times, owing to the higher velocity in this material [64]. Depicted in Figure 9.4a are the velocity field characteristics for various semiconductors, including GaN [65]. As a result of these short transit times, since operation

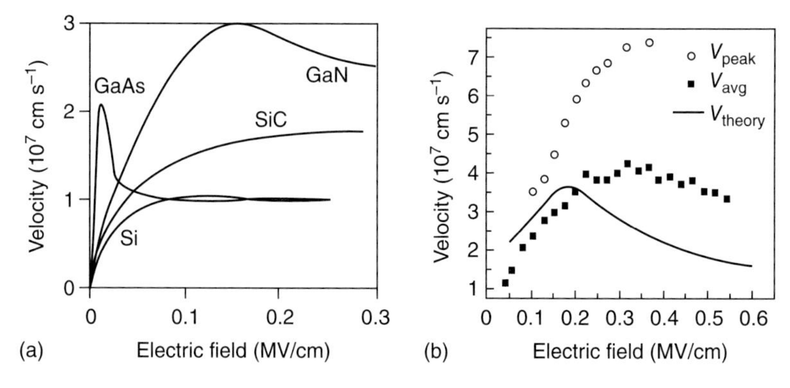

Figure 9.4 (a) Calculated velocity field characteristics for various semiconductors. Source: Jain et al. 2000 [65]. Reproduced with permission of AIP Publishing. (b) Calculated and measured velocity field characteristics for GaN. Source: Figure extracted from Ref. [66] with permission from Wiley.

frequency is directly proportional to the transit time, GaN-based NDR devices are expected to attain higher frequencies of operation than GaAs-based devices. Experimentally, the electron transit time and velocity-field characteristic in GaN have been determined by several authors employing various techniques, for instance, using an optically detected time-of-flight technique with femtosecond resolution, which monitors the change in the electro absorption due to charge transport in an AlGaN/GaN heterojunction p–i–n diode [66]. In this measurement, it was found that electron velocity overshoot can occur at electric fields as low as $105\,\text{kV cm}^{-1}$, with the peak transient velocity increasing with E up to $\sim 320\,\text{kV cm}^{-1}$, at which field a peak velocity of $7.25 \times 10^7\,\text{cm s}^{-1}$ is attained. At higher fields, the increase in transit time with increasing field suggests the onset of NDR due to intervalley transfer. Furthermore, full-zone Monte Carlo calculations suggested that at fields smaller than $325\,\text{kV cm}^{-1}$, the overshoot is a result of band nonparabolicity in the Γ valley rather than due to intervalley transfer. Depicted in Figure 9.4b are the experimental and theoretical electron velocities in GaN as a function of electric field, which show a very good agreement.

9.2.3 Transport due to Electron Plasma Waves

Two-dimensional electron gases in semiconductor heterostructures can allow for collective motion of electrons, i.e. electron plasma waves, whose group velocity is >10× larger than typical electron drift velocities (i.e. $v_g > 10^8\,\text{cm s}^{-1}$) [47]. These electron-plasma-wave group velocities are associated with terahertz frequencies of operation. Devices based on electron plasma waves can overcome the fundamental limits of operation of classical electron devices and thus have attracted significant attention during recent years for terahertz generation, detection, and amplification [67]. In this context, the properties of electron plasma waves in gated semiconductor heterostructures can be intuitively understood from a transmission line (TL) model. For this purpose, let us consider the 2DEG in an HEMT structure. The transistor will be assumed to be operating in the linear regime with a "small" drain-to-source voltage. At terahertz frequencies, the 2DEG conductivity should be modeled as frequency dependent, for instance, employing a Drude model:

$$\sigma(\omega) = \frac{\sigma_0}{1 + j\omega\tau} \tag{9.4}$$

where σ_0 is the 2DEG DC electrical conductivity and τ is the electron momentum relaxation time. Therefore, it is observed that the 2DEG resistivity is of the form:

$$\sigma(\omega) = \frac{1}{\sigma\,\text{DC}} + j\omega\frac{\tau}{\sigma\,\text{DC}} = r + j\omega l \tag{9.5}$$

here r and l represent the 2DEG "distributed" resistance and inductance (kinetic inductance). But the gate stack in the HEMT also provides for a "distributed" (gate) capacitance and conductance (leakage). Therefore, an HEMT at terahertz frequencies can be considered and modeled as a TL. The equivalent TL parameters are given by

$$r = \frac{1}{e\mu n_s W},\quad l = \frac{m^*}{e^2 n_s W}, c = c_0 W, g = g_0 W \tag{9.6}$$

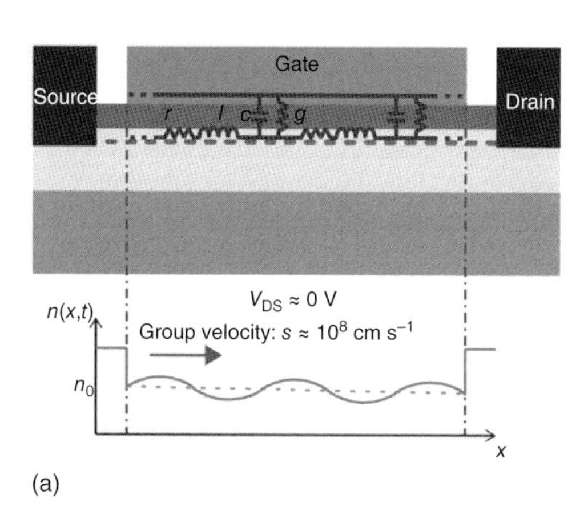

$n_{s,max}$ (cm^{-2})	Si GaN GaAs InAs	2×10^{13} 3×10^{13} 5×10^{12} 2×10^{12}
Max. operating freq. (THz)	Si GaN GaAs InAs	14 16 7.5 6.4
s_{max} (cm s^{-1})	Si GaN GaAs InAs	1.8×10^8 2.5×10^8 1.7×10^8 1.9×10^8

(b)

Figure 9.5 (a) High-frequency TL for an HEMT; waves of electron density, i.e. electron plasma waves can take place in the 2DEG. (b) Plasmonic properties for several semiconductors.

where c and g represent the distributed gate capacitance and conductance, respectively, e and m^* are the electron charge and effective mass, μ is the 2DEG electron mobility, W the transistor width, n_s the 2DEG sheet charge density, c_0 the gate capacitance per unit area, and g_0 the gate conductance per unit area. The fact that the transistor channel can be modeled as a TL means that voltages from gate to channel, currents along the channel, electron velocities, and electron densities depend on time and space: they are waves. Therefore, waves of electron density, or electron plasma waves can take place in the 2DEG, as depicted in Figure 9.5a. In this regard, Burke et al. have shown experimental demonstrations of this TL model in the GaAs materials system in 2000 [68]. Later works have also experimentally demonstrated the validity of this model in GaN-based HEMT structures [69]. In general, the TL propagation constant and characteristic impedance can be defined as

$$\gamma = \sqrt{(r + j\omega l)(g + j\omega c)}, \quad Z = \sqrt{\frac{r + j\omega l}{g + j\omega c}} \tag{9.7}$$

This high-frequency TL model for the HEMT can alternatively be derived from the hydrodynamic equations that govern the electron transport (Euler and continuity equations). The nonlinearity of these partial differential equations (PDEs) leads to the explanation of rectification phenomena and thereby terahertz detection in HEMTs. However, at RT, electron plasma wave oscillations are typically overdamped, even in high-mobility materials as is the case in GaN. There are two operation regimes possible: resonant regime and nonresonant regime depending on the damping of the electron plasma waves, resonant operation requires: $\omega\tau \gg 1$ and $s\tau/L \gg 1$, where s is the electron plasma wave group velocity and L is the gate length [70]. By assuming that the electron band diagram in the region under the gate consists of quantized energy levels within the channel, the charge in the 2DEG is given by

$$n_s(E_f) = \sum_i \int_{E_i}^{\infty} \mathrm{DOS} \cdot f(E)\mathrm{d}E \tag{9.8}$$

where f is the Fermi–Dirac distribution, $i = 1, \ldots, M$ with M being the number of quantized energy levels, and DOS the 2D density of states, which is given by

$$\mathrm{DOS} = m^*_{\mathrm{DOS}}/\pi\hbar^2 \tag{9.9}$$

where m^*_{DOS} is the 2D DOS electron effective mass. By assuming a single band occupation and taking into account the Fermi–Dirac distribution, the quantum capacitance (as a function of 2DEG concentration n_s) can be derived as

$$C_q = q^2(g_v m^*_{\mathrm{DOS}}/\pi\hbar^2)(1 - \exp(-n_s\pi\hbar^2/kTm^*_{\mathrm{DOS}})) \tag{9.10}$$

The total capacitance can be therefore written from the contributions from gate and quantum capacitances as

$$C_{\mathrm{total}} = (C_{\mathrm{gate}}^{-1} + C_q^{-1})^{-1} \tag{9.11}$$

It is observed from Eqs. (9.10) and (9.11) that C_{total} is a function of n_s, i.e. $C_{\mathrm{total}} = C_{\mathrm{total}}(n_s)$, and thus the electron plasma wave group velocity (s) is given by [33]

$$s^2 = q^2 n_s/m^* \cdot C_{\mathrm{total}}\,(n_s) \tag{9.12}$$

Therefore, the attainable velocity for plasma waves in a semiconductor is determined by the charge density and its effective mass. At high frequencies, HEMT-based terahertz devices can operate in a resonant mode in nanoscale devices with high carrier mobilities. It is then expected that narrow bandgap semiconductor FETs with high carrier mobility 2DEGs (e.g. InAs) should outperform wide bandgap FETs, such as GaN. However, GaN-based devices have unique advantages with respect to other materials for operation frequencies higher than 7 THz. There are two distinct effects that affect the high-frequency behavior of these detectors: (i) plasmon and optical phonon coupling and (ii) maximum achievable 2D carrier concentration $n_{s,max}$. While the effect of phonons is common to both resonant and nonresonant modes of operation, $n_{s,max}$ is particularly important when considering resonant detection, where detection frequency depends on channel charge density. Depicted in Figure 9.5b are the typical values of $n_{s,max}$ in 2DEGs and the corresponding maximum plasmon group velocity achievable; owing to its large attainable charge density, GaN offers superior velocities than other common semiconductors, including high-mobility III–Vs [33]. In this regard, listed in Table 9.1 (Section 9.1) are the Reststrahlen band energies (between the transverse optical (TO) and LO phonon energies) of Si, GaN, GaAs, InAs, and graphene, and their correspondent associated frequencies and wavelengths. Electromagnetic waves in the Reststrahlen band cannot propagate in a material due to the strong interaction between photons and phonons. Similarly, a strong plasmon-phonon coupling is expected in this energy range; thus, the corresponding plasmon modes are forbidden. Plasmons with energy above Reststrahlen bands most likely suffer from strong scattering with LO phonons, similar to the case for electrons. As a result, the maximum allowed plasmon frequency in these HEMTs is the lower of the two aforementioned effects, thus setting the fundamental upper operation frequency of the HEMT-based terahertz detectors. It can be noticed that the maximum plasmon frequency allowed by $n_{s,max}$ is always higher than that due to plasmon-phonon coupling. For instance, the upper frequency limit for GaN is 16 THz, while for GaAs and InAs is much lower, ~7 THz, due to their low TO/LO phonon energies. Graphene is also a very interesting material for terahertz applications since its LO phonon energy is the highest. However, the biggest challenge at present is how to harvest the intrinsic high-mobility nature of graphene in a realistic device layout. Out of today's semiconductors, GaN is most attractive for HEMT-based terahertz detectors in the frequency range from 7 to 16 THz from this point of view.

9.3 GaN-based Terahertz Devices: State of the Art

GaN-based terahertz devices have been demonstrated to be capable of operating at multiple terahertz wavelengths, from the lower end of the terahertz range, i.e. W-band, with high-power MMICs, to the upper end, i.e. 5–12 THz with relatively high photon densities and at RT. As a result, GaN-based devices are promising for communications application as well as for imaging and spectroscopy systems with high resolutions and depth of penetration. This section provides a review on the state of the art of GaN-based terahertz devices, including HEMTs, NDR devices, QCLs, and electron-plasma-wave-based devices.

9.3.1 High-Electron Mobility Transistors

As discussed in previous sections, GaN 2DEGs can provide a very large electron density, on the order of 10^{13} cm^{-2} due to strong polarization effects. Furthermore, GaN also provides a high electron mobility up to 2200 cm^{2} V^{-1} s^{-1} [71]. The combination of high carrier concentration and high electron mobility results in high current density, e.g. >4 A mm^{-1} output current density levels, and low channel resistance, which are especially important for high-frequency operation and power switching applications [71]. GaN-based HEMTs also benefit from large breakdown voltages, large thermal conductivities, and therefore can handle large power densities at high voltages. These fundamental properties of GaN enable small-footprint power amplification with high-power added efficiency in cellular devices, base stations, wireless networks, and defense systems. From an application perspective, GaN-based MMICs have been reported attaining output power densities of 30 W mm^{-1} in the X-band [72] and 3 W mm^{-1} in the W-band [73]. In this regard, higher impedance (owing to the reduced footprint in GaN devices) allows for easier matching with lower loss in amplifiers. Depicted in Figure 9.6a is a typical cross-section of a GaN HEMT. The device operation relies on the modulation of the 2DEG charge (thus current) by the gate electrode. From a historical perspective, first efforts toward HEMTs date back to the first observation of a 2DEG in an AlGaN/GaN heterostructure (with carrier density ~10^{11} cm^{-2} and mobility ~800 cm^{2} V^{-1} s^{-1} at RT), which was reported in 1992 [75]. Later, in 1993, the first DC operation of an AlGaN/GaN HEMT was demonstrated, with output current densities as high as 40 mA mm^{-1} [75]. Over the recent years, the performance of GaN-based HEMTs, as quantified by current gain cutoff frequency (f_T) and power gain cutoff frequency (f_{max}), have significantly increased. In this

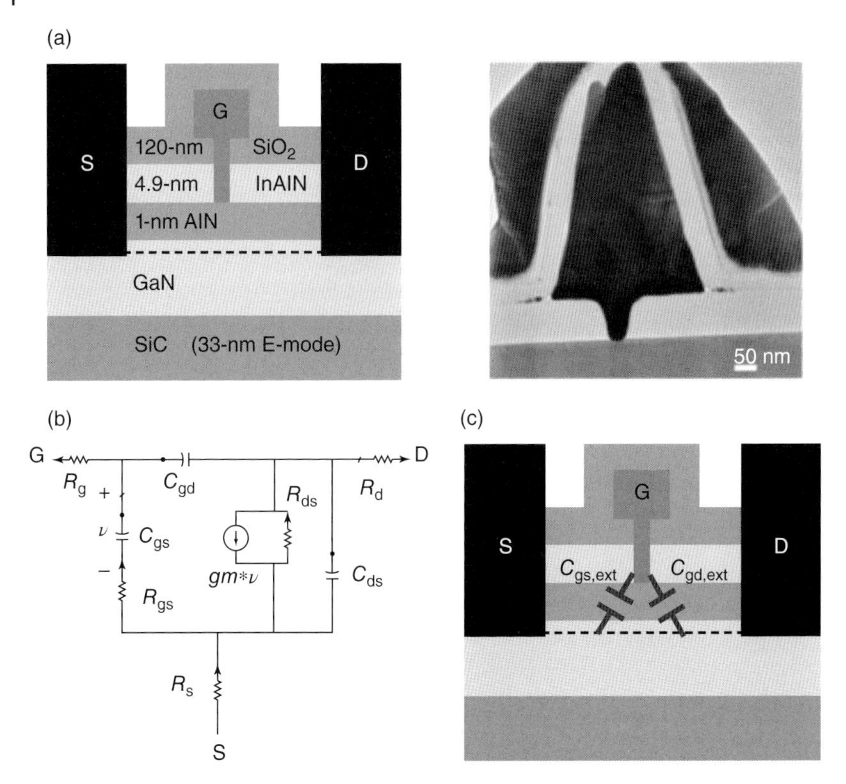

Figure 9.6 (a) Schematic cross-section of an E-mode GaN HEMT (left) and cross-sectional TEM image (right). Source: Figure extracted from Ref. [74] with permission from Elsevier. (b) Equivalent circuit model for an HEMT. (c) Schematic depicting the origin of the extrinsic capacitances in GaN HEMTs. Sensale-Rodriguez et al. 2013 [74]. Reproduced with permission of Elsevier.

regard, from the small signal equivalent circuit model for an HEMT shown in Figure 9.6b, the following expressions could be derived for f_T and f_{max}:

$$f_{max} \approx \frac{f_T}{2\sqrt{\frac{(R_g+R_s+R_{gs})}{R_{ds}} + 2\pi f_T R_g C_{gd}}} \tag{9.13}$$

$$f_T \approx \frac{1}{2\pi} \frac{g_{m,i}}{C_{gs} + C_{gd} + g_{m,i} C_{gd}(R_s + R_d)\left[1 + \left(1 + \frac{C_{gs}}{C_{gd}}\right)\frac{1}{g_{m,i}R_{ds}}\right]} \tag{9.14}$$

In general, increasing the transconductance (g_m) together by decreasing the effect of parasitic elements has been followed as a way to improve these cutoff frequencies, and thus the high-frequency operation of GaN-based HEMTs. In this regard, high current-gain/power-gain cutoff frequencies have been achieved primarily by an aggressive gate length scaling, assisted by a channel depth scaling through design of back barriers. Following this approach, I-gate InAlN HEMTs with a gate length of 30 nm showing a maximum current gain cutoff frequency (f_T) of 370–400 GHz were demonstrated in 2012 and 2013 [76, 77]. More recently, in 2015, T-gate AlN/GaN HEMTs with a gate length of 20 nm and f_T as well as f_{max} of ~450 GHz were demonstrated [78]. The large speed attained in these devices is the result of innovative device scaling technologies such as self-aligned gate processes, development of n^+-GaN regrown ohmic contacts, use of thin AlN and InAlN top barriers, use of back barriers in AlGaN or InGaN, etc. The device speed, given by f_T, is linked to the total delay (τ) by: $f_T = 1/(2\pi\tau)$. Rearranging terms in Eq. (9.14), the following expression is derived for the delay:

$$\tau = \frac{C_{gs} + C_{gd}}{g_{m,i}} + \frac{C_{gs} + C_{gd}}{g_{m,i}} \left(\frac{R_s + R_d}{R_{ds}} \right) + C_{gd}(R_s + R_d) \tag{9.15}$$

As discussed in Ref. [79] the capacitances C_{gs} and C_{gd} can be decomposed in intrinsic and extrinsic components (see Figure 9.6c), that is

$$C_{gs} + C_{gd} = (C_{gs,i} + C_{gd,i}) + (C_{gs,ext} + C_{gd,ext}) \tag{9.16}$$

By substituting Eq. (9.16) in Eq. (9.15), the following expression is derived for the delay:

$$\tau = \tau_i + \tau_{par} \tag{9.17}$$

where

$$\tau_i = \frac{C_{gs,i} + C_{gd,i}}{g_{m,i}} \tag{9.18}$$

is defined as the intrinsic delay, and

$$\tau_{par} = \frac{C_{gs,ext} + C_{gd,ext}}{g_{m,i}} + \frac{C_{gs} + C_{gd}}{g_{m,i}} \left(\frac{R_s + R_d}{R_{ds}} \right) + C_{gd}(R_s + R_d) \tag{9.19}$$

is defined as the parasitic delay. The substantial parasitic delay (τ_{par}) limits the improvement in f_T that can be achieved by gate length scaling [74]; further improvements in terms of device speed require a reduction in these parasitic capacitances. Reduction in the extrinsic capacitance can be achieved by many ways, for instance by raising the T-gate stem height, etch-back of the dielectric, by means of employing ultrathin passivation schemes, and even by eliminating the T-gate cap [79]. If neglecting short channel effects, by knowing C_{ext}, $C_{gs,i}$, $g_{m,i}$, R_s, and R_d, it is possible to determine f_T. The intrinsic transconductance ($g_{m,i}$) can be expressed as

$$g_{m,i} = v_{inj} C_{gs,i} \tag{9.20}$$

where v_{inj} is the effective injection velocity defined in Eq. (9.3) and $C_{gs,i}$ is the intrinsic gate capacitance. $C_{gs,i}$ can be obtained from 1D self-consistent Schrodinger–Poisson simulations for different remaining barrier thicknesses after gate recess. Depicted in Figure 9.7a are the results for $C_{gs,i}$ as a function of gate recess (assuming an initial 6.1 nm InAlN barrier). Furthermore, assuming a v_{inj} of 1.5×10^7 cm s^{-1}, as per the discussion in Section 9.2.1, $g_{m,i}$ can be calculated as a function of gate recess (see Figure 9.7b). Since both $C_{gs,i}$ and $g_{m,i}$ scale inversely with the remaining recessed barrier thickness, the intrinsic delay is independent of recess depth. On the other hand, the parasitic delay due to the extrinsic capacitances decreases as the remaining barrier thickness is reduced. As a result, the device speed increases with decreasing the barrier thickness as

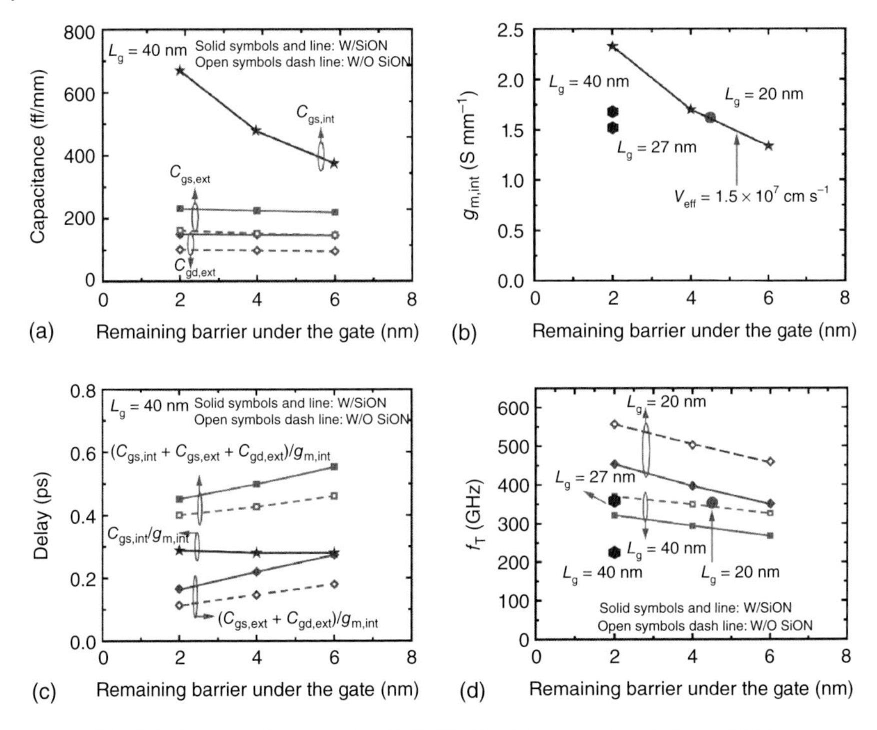

Figure 9.7 Calculated performance for GaN HEMTs with a gate stem height of 200 nm, recessed gate, and a thin barrier. For a gate length of 40 nm (a) extrinsic and intrinsic capacitances, (b) estimated intrinsic g_m, and (c) delays. (d) Projected f_T. Source: Song et al. 2014 [79]. Reproduced with permission of IEEE.

shown in Figure 9.7c, i.e. an aggressive gate recess can lead to a reduced delay, provided that v_{inj} does not degrade with gate recess. Using Eqs. (9.17)–(9.20) it is possible to predict the degree of improvement in f_T expected from these HEMTs by means of gate length scaling, gate recess, ohmic contact improvement, as well as dielectric etch-back. It was found that for a 40-nm device, $f_T \sim 300$–400 GHz can be achieved, which is close to the best reported GaN HEMT device with T-gates [77]. Since the intrinsic $C_{gs,i}$ scales with the gate length, i.e. the intrinsic capacitance of a 20-nm gate length device is half of that for a 40-nm device, the projected speed performance for a 20-nm device can reach 500 GHz by employing low-k dielectric passivation. Depicted in Figure 9.7d are the calculated f_T as a function of gate recess. To further improve the GaN HEMT speed, it is very important to seek approaches that could enhance the injection velocity and thus g_m, such as the use of InGaN or isotope-disordered channels [79].

9.3.2 NDR and Resonant Tunneling Devices

As discussed in previous sections, GaN has several advantages over higher mobility semiconductors such GaAs and InP for NDR applications. Higher power and higher operation frequency are possible in GaN, which is directly related to its larger intervalley energy gap and higher peak and saturation velocities. Work to date on NDR- based devices has been mostly simulation based. Monte Carlo simulations predict a high NDR relaxation frequency approaching 700 GHz in GaN, which is much larger than what is possible in GaAs (~100 GHz) [64]. Based on this mechanism, experimental demonstrations of Gunn diodes have been reported with bias oscillations observed using a series inductance [80]. However, multiple challenges remain in terms of device fabrication and materials growth quality that prevent terahertz operation.

Besides the NDR effect in bulk GaN, resonant tunneling in GaN heterostructures is another alternative mechanism that could lead to terahertz operation. Resonant tunneling of electrons in compound semiconductors has been extensively studied since the foundational work by Tsu and Esaki [81]. In quantum wells in semiconductor heterostructures, quantum confinement introduced by the barriers results in a discrete electronic energy spectrum. By means of applying bias across the structure, it is possible to bring the bound states in resonance with electrons at the injector side. Resonant tunneling has been exploited to design highly efficient charge injectors into the upper lasing level in terahertz QCLs [82]. However, no room-temperature operation has been yet achieved in terahertz QCLs. Furthermore, the operation frequencies of QCLs demonstrated to date are limited to <5 THz as a result of phonon absorption in the usually employed materials systems such as AlGaAs/GaAs and InGaAs/InAlAs. NDR in QW heterostructures has been exploited to realize high-frequency RTD oscillators [38]. However, demonstration of RTD oscillators providing milliwatt output powers at frequencies higher than 1 THz, which is required for most practical applications, is still lacking. GaN has emerged as an attractive materials system due to the large conduction band energy offset possible between GaN and AlGaN compounds. In this regard, the conduction band offset between GaN and AlN is on the order of 1.75 eV, which is much larger than what is possible in the arsenides. In addition, the high LO phonon energy in GaN is expected to prevent the depopulation of the upper lasing level, thus raising the hopes for RT operation of nitride terahertz-QCLs [83] as discussed in Section 9.3.3. Double-barrier Al(Ga)N/GaN/Al(Ga)N RTDs have been explored during the past decade by many authors. Experiments on AlN-barrier RTDs grown on sapphire templates have shown a region of NDR under forward bias. However, the current-voltage measurements are characterized by a hysteresis behavior and lack of repeatability [84–86]. This has been attributed to a high density of defects in GaN films grown on sapphire. These defects can act as electron traps, leading to self-charging and preventing coherent transport of carriers. Several methods have been investigated to reduce the defect density and enhance resonant transport, such as low-temperature growth of superlattices and lateral epitaxial overgrowth; however, the NDR experimentally observed in these devices degraded after consecutive voltage sweeps [87]. Later works focused on GaN-based RTD growth on free-standing GaN substrates with low dislocation densities [88]. These devices contained low Al composition AlGaN barriers and exhibited NDR features; however, operation has been shown only at cryogenic temperatures (below 110 K for a RTD with 18% AlGaN barriers and below 130 K for a diode with 35% AlGaN barriers). More recently, repeatable room-temperature NDR was reported in AlN/GaN/AlN RTDs [89, 90]. However, current levels are still relatively low and no high-frequency operation has been yet demonstrated.

9.3.3 Quantum Cascade Lasers

The operation of QCLs relies on intersubband transitions in multiple quantum well structures. At terahertz wavelengths, GaAs-based QCLs have been demonstrated with output powers as high as 250 mW at 10 K and operation at temperatures approaching 200 K. However, the relatively small LO phonon energy in GaAs is an obstacle for further temperature improvement and RT operation has remained a challenge. In this regard, GaN is a promising candidate for higher temperature (even RT) terahertz QCLs as a result of its much larger LO phonon energy as well as intersubband scattering times that are more than an order of magnitude shorter than those in GaAs [83]. However, the development of GaN-based terahertz QCLs has been hindered owing to the difficulty in growing high-quality AlGaN/GaN heterostructures. In addition, the large built-in polarization field makes precise bandstructure design extremely challenging in GaN. Experimentally, early demonstrations of emission in GaN-based structures were reported in InGaN/GaN multiple quantum wells [91]. Later, spontaneous emission at 1.4–2.8 THz originating from intersubband transitions was reported on GaN-based terahertz QCL structures with active regions consisting of four QW structures for one period [92]. However, the observed peak frequencies in the emission spectra were different from the target ones and no stimulated emission was observed. More recently, in order to avoid unexpected emission from undegenerate levels, GaN-based terahertz QCLs were reported on a three-level laser system quantum cascade structure [93]. Lasing action was observed from the designed levels in these structures. Furthermore, the emission frequencies (5.5 and 7.0 THz at ~5 K) are the highest reported in terahertz QCLs to date for any kind of terahertz QCL. It is expected that by improving the crystal quality through optimized epitaxial growth processes and by means of optimizing the quantum cascade device structure, lasing at higher frequencies and operation at higher temperatures will be possible in the future.

9.3.4 Electron-Plasma-Wave-based Devices

FETs based on the Dyakonov–Shur electron plasma wave theory have been touted in the past decade as an attractive candidate for terahertz detection; very high responsivities have been predicted ($>10^7$ V W^{-1} has been predicted assuming unity coupling efficiency of incident signal to the device [47]; ~1000 V W^{-1} has been experimentally reported at 1.4 THz at RT working in the resonant regime [94]). This detection mechanism has been reported by various research groups using Si- [95], GaAs- [96], and GaN-based devices [94]. As discussed in Section 9.2.3, operation of electron-plasma-wave-based devices could take place either in a resonant or in a nonresonant mode. For FET detectors, as is the case in GaN HEMTs, the responsivity is predicted to peak near the threshold voltage. As a result, this region of operation is very important, and the effects of quantum capacitance and gate leakage should be accounted for when calculating the detector responsivity [33]. Furthermore, the higher 2DEG density in AlGaN/GaN heterostructures as well as the large LO-phonon energy in GaN lead to higher plasma wave frequencies and faster electron plasma wave group velocities than in other traditional semiconductors, thus promising better performance in resonant terahertz detectors. Experimentally, El Fatimy et al. reported on both resonant and nonresonant plasmonic detection in GaN HEMTs with 150-nm gates [97]. Resonant response was observed at cryogenic temperatures, with clear resonant peaks at 4 K. Furthermore, at RT, where the detector response was nonresonant, a minimum NEP of 5×10^{-9} W/Hz$^{0.5}$ was reported. Later work by Tanigawa et al. reported a responsivity of 1100 V W^{-1} using an 80-nm gate length GaN HEMT [94]. The gate electrode was used as a matched dipole antenna to couple parallel polarized terahertz radiation into electron plasma waves. In this context, efficient coupling of external terahertz radiation into and out of plasmons is essential for the operation of electron-plasma-wave-based devices. An alternative approach to excite plasmons in the 2DEG is via a grating gate coupler as illustrated in Figure 9.8a. Using a grating gate topology, Muravjov et al. observed strong plasmon resonances in the terahertz transmission spectra through GaN 2DEGs [101]. Multiple resonances in the frequency range between 1 and 5 THz were observed at temperatures from 10 to 170 K. The observed resonance frequencies corresponded to the excitation of electron plasma waves with wave vectors equal to the reciprocal lattice vectors of the metal grating. These

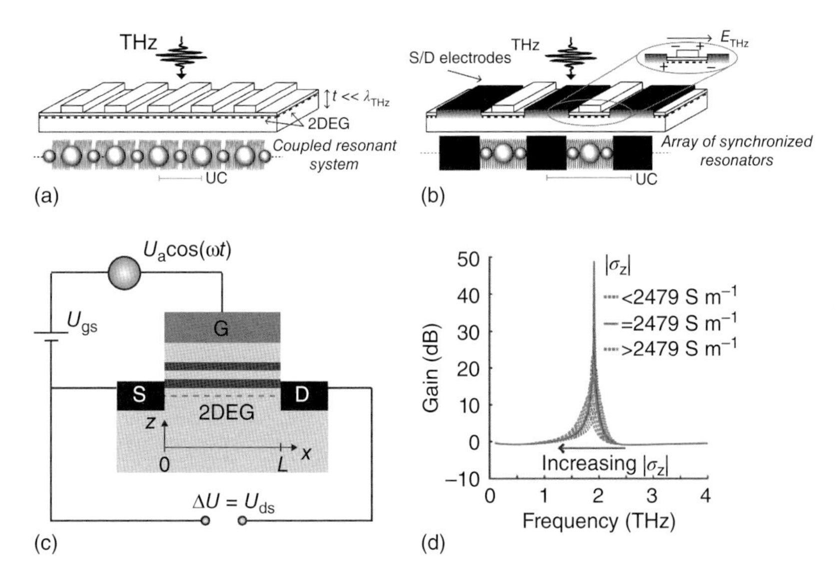

Figure 9.8 (a) Sketch of a grating gate HEMT structure. This configuration operates as a coupled resonant system where the unit cell (UC) period defines the plasmon resonance. (b) Sketch of an HEMT array configuration; the addition of S/D electrodes enhances the terahertz to plasmon coupling by properly synchronizing electron plasma waves in each unit cell. This configuration operates as an array of independent resonators; the plasmon resonance is defined by both the gated 2DEG and the access region 2DEG rather than by the unit cell period. Source: (Panels a and b) Condori Quispe et al. 2017 [98]. Reproduced with permission of IEEE. (c) Cross-section sketch of an RTD-gated HEMT. Source: (Panel c) Sensale-Rodriguez et al. 2012 [99]. Reproduced with permission of Electrochemical Society. (d) Simulated gain versus frequency for an RTD-gated HEMT array (σ_z represents negative differential conductance from gate to 2DEG). Source: (Panel d) Condori Quispe et al. 2016 [100]. Reproduced with permission of AIP Publishing.

resonances were tunable by gate voltage. In a grating gate configuration, adjacent unit cells (UCs) interact with each other, making this a coupled resonant system. However, in contrast, via addition of source (S) and drain (D) electrodes, in an HEMT array configuration as depicted in Figure 9.8b, every unit cell becomes effectively independent. In this configuration, the terahertz to plasmon coupling is enhanced due to a cooperative effect by synchronizing the electron plasma waves in each UC of the array, as theoretically discussed by Popov et al. [102]. Experimental demonstration of terahertz coupling to electron plasma waves in ultrathin membrane HEMT arrays via plasmon synchronization was demonstrated by Condori Quispe et al. [98]. A thin-membrane configuration enabled to remove substrate effects and further enhance the coupling. This approach allows (i) more efficient excitation of high-order plasmonic modes than what is otherwise possible in grating gate approaches and (ii) superior overall coupling, even in configurations having less number of devices per unit area.

In practice, the detector responsivities experimentally reported to date have been modest in comparison with the very high responsivities theoretically predicted. As proposed by Ryzhii and Shur [103] and Sensale-Rodriguez et al. [99], by providing a gain medium, such as an element providing negative differential conductance (NDC) from gate to channel in an HEMT, plasmonic losses due to electron scattering in the semiconductor 2DEG can be counteracted, which in turn leads to stronger responses in terahertz detector configurations (i.e. Responsivity $> 10\,\mathrm{kV\,W^{-1}}$ and NEP $< 1\ \mathrm{pW/Hz^{0.5}}$) [99]. This can be understood by simply observing that a condition for the electron plasma wave damping to be counteracted is that $\mathrm{Re}(\gamma) = \sqrt{(r + j\omega l)(g + j\omega c)} = 0$; thus: $rc = -lg$; which can be alternatively expressed as $c/g = -\tau$. Therefore, the design of the gate stack plays an important role in these devices, and by providing NDC from gate to channel, enhanced terahertz performance can be achieved. In this context, Sensale-Rodriguez et al. [104] showed the possibility of stable terahertz power amplification in GaN-based HEMT-like structures exhibiting NDC from the transistor gate to its 2DEG, as depicted in Figure 9.8c; these devices were called RTD-gated plasma-wave HEMTs. The fundamental mechanism enabling gain at terahertz frequencies in these devices is thus the interplay between the gate NDC and the electron plasma waves in the 2DEG, i.e. the NDC provides a gain medium for the plasma waves excited in the semiconductor 2DEG. Simulations predicted that power gains on the order of 10 dB are possible in these devices via optimized antenna coupling [104]. Furthermore, larger power gain is possible via efficient coupling of terahertz waves into and out of plasmons in the RTD-gated HEMT 2DEG via the grating gate itself, part of the active device, rather than by an external antenna structure. Numerical simulations by Condori Quispe et al. predicted the possibility of attaining gain >40 dB in these devices at frequencies >1 THz (see Figure 9.8d) [100].

References

1 Siegel, P.H. (2002). Terahertz technology. *IEEE Trans. Microw. Theory Tech.* 50 (3): 910–928.

2 Siegel, P.H. (2004). Terahertz technology in biology and medicine. *IEEE Trans. Microw. Theory Tech.* 52 (10): 2438–2447.
3 Tonouchi, M. (2007). Cutting-edge terahertz technology. *Nat. Photonics* 1 (2): 97–105.
4 Koch, M. (2007). Terahertz technology: a land to be discovered. *Opt. Photonics News* 18 (3): 20–25.
5 Ferguson, B. and Zhang, X.-C. (2002). Materials for terahertz science and technology. *Nat. Mater.* 1 (1): 26–33.
6 Fitzgerald, A.J., Berry, E., Zinovev, N.N. et al. (2002). An introduction to medical imaging with coherent terahertz frequency radiation. *Phys. Med. Biol.* 47 (7): R67–R84.
7 Crawley, D.A., Longbottom, C., Cole, B.E. et al. (2003). Terahertz pulse imaging: a pilot study of potential applications in dentistry. *Caries Res.* 37 (5): 352–359.
8 Pickwell, E., Wallace, V.P., Cole, B.E. et al. (2007). A comparison of terahertz pulsed imaging with transmission microradiography for depth measurement of enamel demineralisation in vitro. *Caries Res.* 41 (1): 49–55.
9 Nakajima, S., Hoshina, H., Yamashita, M. et al. (2007). Terahertz imaging diagnostics of cancer tissues with a chemometrics technique. *Appl. Phys. Lett.* 90 (4): 41102.
10 Zaytsev, K.I., Kudrin, K.G., Karasik, V.E. et al. (2015). In vivo terahertz spectroscopy of pigmentary skin nevi: pilot study of non-invasive early diagnosis of dysplasia. *Appl. Phys. Lett.* 106 (5): 53702.
11 Brun, M.-A., Formanek, F., Yasuda, A. et al. (2010). Terahertz imaging applied to cancer diagnosis. *Phys. Med. Biol.* 55 (16): 4615.
12 Fitzgerald, A.J., Wallace, V.P., Jimenez-Linan, M. et al. (2006). Terahertz pulsed imaging of human breast tumors. *Radiology* 239 (2): 533–540.
13 Appleby, R. and Wallace, H.B. (2007). Standoff detection of weapons and contraband in the 100 GHz to 1 THz region. *IEEE Trans. Antennas Propag.* 55 (11): 2944–2956.
14 Cooper, K.B., Dengler, R.J., Llombart, N. et al. (2011). THz imaging radar for standoff personnel screening. *IEEE Trans. Terahertz Sci. Technol.* 1 (1): 169–182.
15 Globus, T.R., Woolard, D.L., Khromova, T. et al. (2003). THz-spectroscopy of biological molecules. *J. Biol. Phys.* 29 (2–3): 89–100.
16 Markelz, A.G., Roitberg, A., and Heilweil, E.J. (2000). Pulsed terahertz spectroscopy of DNA, bovine serum albumin and collagen between 0.1 and 2.0 THz. *Chem. Phys. Lett.* 320 (1): 42–48.
17 Nagel, M., Haring Bolivar, P., Brucherseifer, M. et al. (2002). Integrated THz technology for label-free genetic diagnostics. *Appl. Phys. Lett.* 80 (1): 154–156.
18 Nagel, M., Richter, F., Haring-Bolívar, P., and Kurz, H. (2003). A functionalized THz sensor for marker-free DNA analysis. *Phys. Med. Biol.* 48 (22): 3625.
19 O'Hara, J.F., Withayachumnankul, W., and Al-Naib, I. (2012). A review on thin-film sensing with terahertz waves. *J. Infrared Millim. Terahertz Waves* 33 (3): 245–291.

20 Walther, M., Fischer, B.M., Ortner, A. et al. (2010). Chemical sensing and imaging with pulsed terahertz radiation. *Anal. Bioanal. Chem.* 397 (3): 1009–1017.

21 Fischer, B.M., Walther, M., and Jepsen, P.U. (2002). Far-infrared vibrational modes of DNA components studied by terahertz time-domain spectroscopy. *Phys. Med. Biol.* 47 (21): 3807.

22 Dobroiu, A., Sasaki, Y., Shibuya, T. et al. (2007). THz-wave spectroscopy applied to the detection of illicit drugs in mail. *Proc. IEEE* 95 (8): 1566–1575.

23 Fischer, B., Hoffmann, M., Helm, H. et al. (2005). Chemical recognition in terahertz time-domain spectroscopy and imaging. *Semicond. Sci. Technol.* 20 (7): S246.

24 Hasan, M., Arezoomandan, S., Condori, H., and Sensale-Rodriguez, B. (2016). Graphene terahertz devices for communications applications. *Nano Commun Netw.* 10: 68–78.

25 Song, H.J. and Nagatsuma, T. (2011). Present and future of terahertz communications. *IEEE Trans. Terahertz Sci. Technol.* 1 (1): 256–263.

26 Federici, J. and Moeller, L. (2010). Review of terahertz and subterahertz wireless communications. *J. Appl. Phys.* 107 (11): 111101.

27 Akyildiz, I.F., Jornet, J.M., and Han, C. (2014). Terahertz band: next frontier for wireless communications. *Phys. Commun.* 12: 16–32.

28 Fitch, M.J. and Osiander, R. (2004). Terahertz waves for communications and sensing. *J. Hopkins APL Tech. Dig.* 25 (4): 348–355.

29 Kleine-Ostmann, T. and Nagatsuma, T. (2011). A review on terahertz communications research. *J. Infrared Millim. Terahertz Waves* 32 (2): 143–171.

30 Kulesa, C. (2011). Terahertz spectroscopy for astronomy: from comets to cosmology. *IEEE Trans. Terahertz Sci. Technol.* 1 (1): 232–240.

31 Soref, R.A., Friedman, L.R., Sun, G. et al. (1999). Quantum well intersubband THz lasers and detectors. *Proc. SPIE* 3795: 516–527.

32 Machhadani, H., Kotsar, Y., Sakr, S. et al. (2010). Terahertz intersubband absorption in GaN/AlGaN step quantum wells. *Appl. Phys. Lett.* 97 (19): 191101.

33 Sensale-Rodriguez, B., Liu, L., Wang, R. et al. (2011). FET THz detectors operating in the quantum capacitance limited region. *Int. J. High Speed Electron. Syst.* 20 (3): 597–609.

34 Carosella, F., Ndebeka-Bandou, C., Ferreira, R. et al. (2012). Free-carrier absorption in quantum cascade structures. *Phys. Rev. B* 85 (8): 85310.

35 Kumar, S., Chan, C.W.I., Hu, Q., and Reno, J.L. (2011). A 1.8-THz quantum cascade laser operating significantly above the temperature of $\hbar\omega/kB$. *Nat. Phys.* 7 (2): 166–171.

36 Li, C., Khalid, A., Pilgrim, N. et al. (2009). Novel planar Gunn diode operating in fundamental mode up to 158 GHz. *J. Phys. Conf. Ser.* 193 (1): 12029.

37 Khalid, A., Li, C., Pilgrim, N.J. et al. (2011). Novel composite contact design and fabrication for planar Gunn devices for millimeter-wave and terahertz frequencies. *Phys. Status Solidi C* 8 (2): 316–318.

38 Asada, M., Suzuki, S., and Kishimoto, N. (2008). Resonant tunneling diodes for sub-terahertz and terahertz oscillators. *Jpn. J. Appl. Phys.* 47: 4375.

39 Schiller, S., Roth, B., Lewen, F. et al. (2009). Ultra-narrow-linewidth continuous-wave THz sources based on multiplier chains. *Appl. Phys. B Lasers Opt.* 95 (1): 55–61.

40 Rodriguez-Morales, F., Yngvesson, K.S., Zannoni, R. et al. (2006). Development of integrated HEB/MMIC receivers for near-range terahertz imaging. *IEEE Trans. Microw. Theory Tech.* 54 (6): 2301–2311.

41 Suzuki, S., Asada, M., Teranishi, A. et al. (2010). Fundamental oscillation of resonant tunneling diodes above 1 THz at room temperature. *Appl. Phys. Lett.* 97 (24): 242102.

42 Maekawa, T., Kanaya, H., Suzuki, S., and Asada, M. (2016). Oscillation up to 1.92 THz in resonant tunneling diode by reduced conduction loss. *Appl. Phys. Express* 9: 24101.

43 El Fatimy, A., Dyakonova, N., Meziani, Y. et al. (2010). AlGaN/GaN high electron mobility transistors as a voltage-tunable room temperature terahertz sources. *J. Appl. Phys.* 107 (2): 24504.

44 Kim, M., Rieh, J.S., and Jeon, S. (2012). Recent progress in terahertz monolithic integrated circuits. 2012 IEEE International Symposium on Circuits and Systems, 746–749.

45 Sizov, F. and Rogalski, A. (2010). THz detectors. *Prog. Quantum Electron.* 34 (5): 278–347.

46 Komiyama, S., Astafiev, O., Antonov, V. et al. (2000). A single-photon detector in the far-infrared range. *Nature* 403 (6768): 405–407.

47 Dyakonov, M. and Shur, M. (1996). Detection, mixing, and frequency multiplication of terahertz radiation by two-dimensional electronic fluid. *IEEE Trans. Electron Devices* 43 (3): 380–387.

48 Watanabe, T., Tombet, S.B., Tanimoto, Y. et al. (2012). Ultrahigh sensitive plasmonic terahertz detector based on an asymmetric dual-grating gate HEMT structure. *Solid State Electron.* 78: 109–114.

49 Mickan, S.P., Menikh, A., Liu, H. et al. (2002). Label-free bioaffinity detection using terahertz technology. *Phys. Med. Biol.* 47 (21): 3789.

50 Mendis, R., Astley, V., Liu, J., and Mittleman, D.M. (2009). Terahertz microfluidic sensor based on a parallel-plate waveguide resonant cavity. *Appl. Phys. Lett.* 95 (17): 171113.

51 You, B., Lu, J.-Y., Yu, C.-P. et al. (2012). Terahertz refractive index sensors using dielectric pipe waveguides. *Opt. Express* 20 (6): 5858–5866.

52 Tao, H., Chieffo, L.R., Brenckle, M.A. et al. (2011). Metamaterials on paper as a sensing platform. *Adv. Mater.* 23 (28): 3197–3201.

53 Mishra, U.K., Likun, S., Kazior, T.E., and Yi-Feng, W. (2008). GaN-based RF power devices and amplifiers. *Proc. IEEE* 96 (2): 287–305.

54 Mishra, U.K., Parikh, P., and Wu, Y.-F. (2002). AlGaN/GaN HEMTs – an overview of device operation and applications. *Proc. IEEE* 90 (6): 1022–1031.

55 Zimmermann, T., Deen, D., Cao, Y. et al. (2008). AlN/GaN insulated-gate HEMTs with 2.3 A/mm output current and 480 mS/mm transconductance. *IEEE Electron Device Lett.* 29 (7): 661–664.

56 Ambacher, O., Smart, J., Shealy, J.R. et al. (1999). Two-dimensional electron gases induced by spontaneous and piezoelectric polarization charges in N- and Ga-face AlGaN/GaN heterostructures. *J. Appl. Phys.* 85 (6): 3222–3233.
57 Jena, D., Smorchkova, I., Gossard, A.C., and Mishra, U.K. (2001). Electron transport in III–V nitride two-dimensional electron gases. *Phys. Status Solidi B* 228 (2): 617–619.
58 Wood, C. and Jena, D. (eds.) (2007). *Polarization Effects in Semiconductors: From Ab Initio Theory to Device Applications.* Springer Science & Business Media.
59 Cao, Y. and Jena, D. (2007). High-mobility window for two-dimensional electron gases at ultrathin AlN/GaN heterojunctions. *Appl. Phys. Lett.* 90 (18): 182112.
60 Fang, T., Wang, R., Xing, H. et al. (2012). Effect of optical phonon scattering on the performance of GaN transistors. *IEEE Electron Device Lett.* 33 (5): 709–711.
61 Fang, T., Wang, R., Li, G. et al. (2011). Effect of optical phonon scattering on the performance limits of ultrafast GaN transistors. Device Research Conference (DRC), 2011 69th Annual, 273–274.
62 Bhapkar, U.V. and Shur, M.S. (1997). Monte Carlo calculation of velocity-field characteristics of wurtzite GaN. *J. Appl. Phys.* 82 (4): 1649–1655.
63 Krishnamurthy, S., van Schilfgaarde, M., Sher, A., and Chen, A.-B. (1997). Bandstructure effect on high-field transport in GaN and GaAlN. *Appl. Phys. Lett.* 71 (14): 1999–2001.
64 Foutz, B.E., Eastman, L.F., Bhapkar, U.V., and Shur, M.S. (1997). Comparison of high field electron transport in GaN and GaAs. *Appl. Phys. Lett.* 70 (21): 2849–2851.
65 Jain, S.C., Willander, M., Narayan, J., and Overstraeten, R.V. (2000). III–nitrides: growth, characterization, and properties. *J. Appl. Phys.* 87 (3): 965–1006.
66 Wraback, M., Shen, H., Rudin, S., and Bellotti, E. (2002). Experimental and theoretical studies of transient electron velocity overshoot in GaN. *Phys. Status Solidi B* 234 (3): 810–816.
67 Otsuji, T. and Shur, M. (2014). Terahertz plasmonics: good results and great expectations. *IEEE Microw. Mag.* 15 (7): 43–50.
68 Burke, P.J., Spielman, I.B., Eisenstein, J.P. et al. (2000). High frequency conductivity of the high-mobility two-dimensional electron gas. *Appl. Phys. Lett.* 76 (6): 745–747.
69 Zhao, Y., Chen, W., Li, W. et al. (2014). Direct electrical observation of plasma wave-related effects in GaN-based two-dimensional electron gases. *Appl. Phys. Lett.* 105 (17): 173508.
70 Knap, W., Dyakonov, M., Coquillat, D. et al. (2009). Field effect transistors for terahertz detection: physics and first imaging applications. *J. Infrared Millim. Terahertz Waves* 30 (12): 1319–1337.
71 Shinohara, K., Regan, D., Corrion, A. et al. (2012). Self-aligned-gate GaN-HEMTs with heavily-doped n^+-GaN ohmic contacts to 2DEG. 2012 International Electron Devices Meeting, 27.2.1–27.2.4.

72 Wu, Y.F., Saxler, A., Moore, M. et al. (2004). 30-W/mm GaN HEMTs by field plate optimization. *IEEE Electron Device Lett.* 25 (3): 117–119.
73 Makiyama, K., Ozaki, S., Ohki, T. et al. (2015). Collapse-free high power InAlGaN/GaN-HEMT with 3 W/mm at 96 GHz. 2015 IEEE International Electron Devices Meeting (IEDM), 9.1.1–9.1.4.
74 Sensale-Rodriguez, B., Guo, J., Wang, R. et al. (2013). Time delay analysis in high speed gate-recessed E-mode InAlN HEMTs. *Solid State Electron.* 80: 67–71.
75 Khan, M.A., Kuznia, J.N., Van Hove, J.M. et al. (1992). Observation of a two-dimensional electron gas in low pressure metalorganic chemical vapor deposited GaN–$Al_xGa_{1-x}N$ heterojunctions. *Appl. Phys. Lett.* 60 (24): 3027–3029.
76 Yue, Y., Hu, Z., Guo, J. et al. (2013). Ultrascaled InAlN/GaN high electron mobility transistors with cutoff frequency of 400 GHz. *Jpn. J. Appl. Phys.* 52 (8S): 08JN14.
77 Yue, Y., Hu, Z., Guo, J. et al. (2012). InAlN/AlN/GaN HEMTs with regrown ohmic contacts and f_T of 370 GHz. *IEEE Electron Device Lett.* 33 (7): 988–990.
78 Tang, Y., Shinohara, K., Regan, D. et al. (2015). Ultrahigh-speed GaN high-electron-mobility transistors with f_T/f_{max} of 454/444 GHz. *IEEE Electron Device Lett.* 36 (6): 549–551.
79 Song, B., Sensale-Rodriguez, B., Wang, R. et al. (2014). Effect of fringing capacitances on the RF performance of GaN HEMTs with T-gates. *IEEE Trans. Electron Devices* 61 (3): 747–754.
80 Yilmazoglu, O., Mutamba, K., Pavlidis, D., and Karaduman, T. (2008). First observation of bias oscillations in GaN Gunn diodes on GaN substrate. *IEEE Trans. Electron Devices* 55 (6): 1563–1567.
81 Tsu, R. and Esaki, L. (1973). Tunneling in a finite superlattice. *Appl. Phys. Lett.* 22 (11): 562–564.
82 Köhler, R., Tredicucci, A., Beltram, F. et al. (2002). Terahertz semiconductor-heterostructure laser. *Nature* 417 (6885): 156–159.
83 Belkin, M.A., Wang, Q.J., Pflugl, C. et al. (2009). High-temperature operation of terahertz quantum cascade laser sources. *IEEE J. Sel. Top. Quantum Electron.* 15 (3): 952–967.
84 Foxon, C.T., Novikov, S.V., Belyaev, A.E. et al. (2003). Current–voltage instabilities in GaN/AlGaN resonant tunnelling structures. *Phys. Status Solidi C* 0 (7): 2389–2392.
85 Boucherit, M., Soltani, A., Monroy, E. et al. (2011). Investigation of the negative differential resistance reproducibility in AlN/GaN double-barrier resonant tunnelling diodes. *Appl. Phys. Lett.* 99 (18): 182109.
86 Nagase, M. and Tokizaki, T. (2014). Bistability characteristics of GaN/AlN resonant tunneling diodes caused by intersubband transition and electron accumulation in quantum well. *IEEE Trans. Electron Devices* 61 (5): 1321–1326.
87 Bayram, C., Vashaei, Z., and Razeghi, M. (2010). Reliability in room-temperature negative differential resistance characteristics of low-aluminum

content AlGaN/GaN double-barrier resonant tunneling diodes. *Appl. Phys. Lett.* 97 (18): 181109.

88 Li, D., Shao, J., Tang, L. et al. (2013). Temperature-dependence of negative differential resistance in GaN/AlGaN resonant tunneling structures. *Semicond. Sci. Technol.* 28 (7): 74024.

89 Growden, T.A., Storm, D.F., Zhang, W. et al. (2016). Highly repeatable room temperature negative differential resistance in AlN/GaN resonant tunneling diodes grown by molecular beam epitaxy. *Appl. Phys. Lett.* 109 (8): 83504.

90 Encomendero, J., Faria, F.A., Islam, S.M. et al. (2017). New tunneling features in polar III-nitride resonant tunneling diodes. *Phys. Rev. X* 7 (4): 041017.

91 Turchinovich, D., Monozon, B.S., and Jepsen, P.U. (2006). Role of dynamical screening in excitation kinetics of biased quantum wells: nonlinear absorption and ultrabroadband terahertz emission. *J. Appl. Phys.* 99 (1): 13510.

92 Terashima, W. and Hirayama, H. (2011). Spontaneous emission from GaN/AlGaN terahertz quantum cascade laser grown on GaN substrate. *Phys. Status Solidi C* 8 (7–8): 2302–2304.

93 Terashima, W. and Hirayama, H. (2015). GaN-based terahertz quantum cascade lasers. *Proc. SPIE* 9483: 948304–948308.

94 Tanigawa, T., Onishi, T., Takigawa, S., and Otsuji, T. (2010). Enhanced responsivity in a novel AlGaN/GaN plasmon-resonant terahertz detector using gate-dipole antenna with parasitic elements. 68th Device Research Conference, 167–168.

95 Tauk, R., Teppe, F., Boubanga, S. et al. (2006). Plasma wave detection of terahertz radiation by silicon field effects transistors: responsivity and noise equivalent power. *Appl. Phys. Lett.* 89 (25): 253511.

96 Antonov, A.V., Gavrilenko, V.I., Demidov, E.V. et al. (2004). Electron transport and terahertz radiation detection in submicrometer-sized GaAs/AlGaAs field-effect transistors with two-dimensional electron gas. *Phys. Solid State* 46 (1): 146–149.

97 El Fatimy, A., Tombet, S.B., Teppe, F. et al. (2006). Terahertz detection by GaN/AlGaN transistors. *Electron. Lett.* 42 (23): 1342–1343.

98 Condori Quispe, H.O., Chanana, A., Encomendero, J. et al. (2017). Experimental demonstration of enhanced terahertz coupling to plasmon in ultra-thin membrane AlGaN/GaN HEMT arrays. 75th Annual Device Research Conference (DRC).

99 Sensale-Rodriguez, B., Fay, P., Liu, L. et al. (2012). Enhanced terahertz detection in resonant tunnel diode-gated HEMTs. *ECS Trans.* 49 (1): 93–102.

100 Condori Quispe, H.O., Encomendero-Risco, J.J., Xing, H.G., and Sensale-Rodriguez, B. (2016). Terahertz amplification in RTD-gated HEMTs with a grating-gate wave coupling topology. *Appl. Phys. Lett.* 109 (6): 63111.

101 Muravjov, A.V., Veksler, D.B., Popov, V.V. et al. (2010). Temperature dependence of plasmonic terahertz absorption in grating-gate gallium-nitride transistor structures. *Appl. Phys. Lett.* 96 (4): 42105.

102 Popov, V.V., Tsymbalov, G.M., Fateev, D.V., and Shur, M.S. (2006). Cooperative absorption of terahertz radiation by plasmon modes in an array of

field-effect transistors with two-dimensional electron channel. *Appl. Phys. Lett.* 89 (12): 123504.

103 Ryzhii, V. and Shur, M. (2001). Plasma instability and nonlinear terahertz oscillations in resonant-tunneling structures. *Jpn. J. Appl. Phys.* 40: 546.

104 Sensale-Rodriguez, B., Liu, L., Fay, P. et al. (2013). Power amplification at THz via plasma wave excitation in RTD-gated HEMTs. *IEEE Trans. Terahertz Sci. Technol.* 3 (2): 200–206.

Index

Advanced Nanoelectronics: Post-Silicon Materials and Devices,
First Edition. Edited by Muhammad Mustafa Hussain.

g